ENVIRONMENT, POWER, AND SOCIETY

FOR THE TWENTY-FIRST CENTURY

ENVIRONMENT, POWER, AND SOCIETY FOR THE TWENTY-FIRST CENTURY

The Hierarchy of Energy

Howard T. Odum

COLUMBIA UNIVERSITY PRESS

NEW YORK

Columbia University Press
Publishers Since 1893
New York   Chichester, West Sussex

Copyright © 2007 Columbia University Press
All rights reserved
Library of Congress Cataloging-in-Publication Data

Odum, Howard T., 1924–
Environment, power, and society for the twenty-first century ; the hierarchy of energy /
Howard T. Odum.
p. cm.
Includes bibliographical references and index.
ISBN-13: 978-0-231-12886-5 (cloth : alk. paper)
ISBN-13: 978-0-231-12887-2 (pbk. : alk. paper)
1. Power resources. 2. Energy consumption—Environmental aspects. I. Title.
TJ163.2.03835   2007
333.79—dc22   2006038538

Columbia University Press books are printed on permanent and durable acid-free paper.

Printed in the United States of America

To Howard Washington Odum

CONTENTS

WITH THE publication of *Environment, Power, and Society* in 1971, H. T. Odum changed the lives of countless individuals; altering their worldview by starting them along a quantitative, systems-oriented path toward holistic thinking. He introduced them to the Energy Systems Language, a visual mathematics capable of representing the details and bringing into focus the complexities of any system, and to the "macroscope," his tool for eliminating details and gaining an overview of the entire system. For many of us *Environment, Power, and Society* was profound, a book that cleared away much of the mystery about how the world works and that provided extraordinary insights into why things are the way they are. It was in *Environment, Power, and Society* that we learned "industrial man . . . eats potatoes partly made of oil," and "money . . . is fed back in reward for work done" and therefore acts as a pathway selection mechanism for society. In this book, we first read the "ten commandments of the energy ethic for survival of man and nature" and took to heart "thou shall not waste potential energy." It was in *Environment, Power, and Society* that H. T. Odum developed his concept of "ecological engineering," which later became the basis for academic programs at several major universities; a journal by that name; and an international society. His statement in 1971 that, "The science of economics may profit by restating more of its theorems to include power principles," and subsequent years of doing just that, eventually led to the new field of ecological economics, which now has its own international society and journal.

Most important, however, *Environment, Power, and Society* helped many of us to understand the interrelationships of energy and environment and their importance to the well-being of humanity and the planet. His goal was always to gain understanding through unifying rather than dissecting, through aggregating rather than disaggregating. In his life he was constantly engaged in a zealous search for truth and understanding regardless of where that search carried him. H. T. often wrote of his desire to simplify to increase understanding . . .

"If the bewildering complexity of human knowledge developed in the twentieth century is to be retained and well used, unifying concepts are needed to consolidate the understanding of systems of many kinds and to simplify the teaching of general principles" (Odum 1994).

H. T. Odum was a scientist and teacher who left an incredible legacy of books, ideas, and teachings. His love of teaching, his creative and imaginative way of viewing the biosphere, his grasp of many different fields of science, and his drive and unbounded energy have left many students, colleagues, and associates awestruck. His unique way of understanding the biosphere and humanity's place within it was his gift to us all. This gift will, without question, endure and expand as it is more fully understood by this and succeeding generations. Although he left us a legacy of books, research publications, and scientific papers (see Brown and Hall 2004), his devotion to the welfare of his students, close associates, family, and humankind stands far beyond these tangible products of his scientific inquiry. In his own words in *A Prosperous Way Down* (Odum and Odum 2001), his most recent book with his wife Betty, he said, "As sometimes attributed to past cultures, people may again find glory in being an agent of the earth." H. T. Odum was an agent of the earth, striving always to teach partnership with nature that encompasses and transcends good stewardship and a profound respect for the cycles and hierarchies of the biosphere.

In the summer of 2002, H. T. Odum was diagnosed with brain cancer. Much earlier, he had arranged with Columbia University Press for the publication of this second edition and was in the process of revising the manuscript when this devastating news was revealed. He worked diligently on the final revision until his death in the fall of that year. Mark Brown; Elisabeth Odum, his wife; and Dan Campbell made a promise to finalize the manuscript and see it published. Our deep gratitude is due to Dr. Robin Smith, former Senior Executive Editor for the Sciences, at Columbia University Press, for his patience and continued encouragement toward finalizing the manuscript, and to Patrick Fitzgerald, science publisher, who worked tirelessly with us to make this book a reality.

More than thirty years have passed since publication of *Environment, Power, and Society*. With the publication of *Environment, Power, and Society for the Twenty-first Century*, we see how the ideas and themes introduced thirty-five years ago have played out and been advanced by H. T. Odum, his colleagues, and his students. In our estimation, this book is even more profound than his first book, as it contains the fruits of thirty more years of systems thinking by one of the world's most revolutionary systems scientists and ecologists. We hope that *Environment, Power, and Society for the Twenty-first Century* will have an even greater impact throughout the world and that once again there will be students who find their lives changed after reading this book, as happened to so many of us lucky enough to find Odum's first book in the early 1970s.

Mark T. Brown
Gainesville, FL

Elisabeth C. Odum
Gainesville, FL

Daniel E. Campbell
Narragansett, RI

Bibliography

Brown, M. T. and C. A. S. Hall, eds. 2004. A tribute to H. T. Odum. Special issue of *Ecological Modeling*, 78(1–2).

Odum, H. T. 1994. *Ecological and General Systems: An Introduction to Systems Ecology*. Niwot: University Press of Colorado.

Odum, H. T. and E. C. Odum. 2001. *A Prosperous Way Down: Principles and Policies*. Boulder: University Press of Colorado.

THIRTY YEARS have passed since the original edition of *Environment, Power, and Society* was published. Since that time the world has had a taste of living with global fuel shortage, high prices, and the ensuing inflation of 1973–1983. Accelerated economic growth resumed aided by new discoveries of natural gas. The Persian Gulf wars have been fought, in part, to keep global fuel reserves on the free market. With the spread of computers and the internet, many authors wrote of the "unlimited potentials of information," just as they wrote of the potentials of nuclear energy four decades ago. But a look at nature shows limits to information. Belief spread in the paradigm that all systems pulse, and many writers warned of the downturns ahead in the global pulse of affluence in developed countries, based on converging resources from the rest of the world.

People with knowledge of the details of our planet have mixed feelings when much of the detail is aggregated in order to develop simple views from a larger scale. In the quest for knowledge and in the practicality of earth management, all scales are of importance, each in its place. What has been missing in much of the past century in science and public education is teaching an imperative to view each scale in aggregated simplicity from the next one larger. Science has become conservatively rigid by overemphasizing the half right cliché that to be basic is to look smaller to the fundamental parts.

The new title, like the original book of some thirty years ago, is based on energy systems synthesis of the parts and processes of systems, a diagrammatic way of showing important relationships constrained by the limitations of materials, energy, money, and information. Whereas many inferences in 1970 were made by qualitative study of systems diagrams, simulation of models has since become a general practice. Models like those previously diagrammed, and thousands of others like it, have been published widely by a generation of systems thinkers. The energy systems language has been widely used to show main parts, pathways, and relationships of systems.

In the original volume, published when simulation was new, there was a chapter on concepts for analog and digital simulation of energy models over time. In the intervening years we have simulated most of our overview minimodels, and these are included where relevant in this book rather than in a separate chapter. Details on simulation methods are given in our recent book: *Modeling for All Scales* (Academic Press, 2000).

For the progress in developing and applying concepts over these intervening years since the publication of the first volume, I gratefully acknowledge the shared mission with students and faculty associates at the University of Florida. Elisabeth Chase Odum was a partner in our efforts to develop energy systems concepts for global education. Joan Breeze was editorial assistant.

Howard T. Odum,
Gainesville, Florida
November 2000

CHAPTER 1

THIS WORLD SYSTEM

THIS BOOK is about nature and humanity. Nature consists of animals, plants, microorganisms, earth processes, and human societies working together. These parts are joined by invisible pathways over which pass chemical materials that cycle around and around, being used and reused, and through which flow potential energies that cannot be reused. The network of these pathways forms an operating system from the parts. Behavioral cues pass between animals; human discourse and money organize society. A study of humanity and nature is thus a study of systems of energy, materials, money, and information. Therefore, we approach nature and people by studying energy systems networks. The idea is to use general systems principles to understand and predict what is possible for society and environment.

Figure 1.1 shows the essence of our system of environment, power, and society. Energy from the sun and from the earth is running the landscape and its links to humanity. The quantity of useful energy determines the amount of structure that can exist and the speed at which processes can function. The small areas of nature, the large panoramas that include civilization, and the whole biosphere of Earth and the miniature worlds of ecological microcosms are similar.[1] All use energy resources to produce, consume, recycle, and sustain.

FIGURE 1.1 System of environment, power, and society in the geobiosphere, which has inflowing solar energy, earth energy from below, cycling of materials, circulation of money, and feedback of human services to nature. This diagram, drawn by the author on a computer, is the logo for the Center for Environmental Policy at the University of Florida.

THE MACROSCOPE

In the 1600s, when Leeuwenhoek ushered in the Enlightenment through the study of the invisible world with the microscope, and when some of the atomistic theories of the Greeks received step-by-step observable verification in chemical studies, concepts of the structure and function of the natural world emerged as parts within parts within parts. Many of the advances of human civilization have come from these microscopic dissections. Yet in the 21st century the ever-accelerating knowledge of the microscopic view has not provided us with the solutions to problems with the human environment, social systems, economics, and survival, for the missing information is not wholly in the microscopic components or in identification of the parts. On the familiar scale of human life, we see the parts very well (people, economic assets, environmental components), but rarely do we think of it as a single-system operation. Pioneering thinkers such as V. I. Vernadsky (1926) recognized the intricate interdependence of humans in the processes of the earth, but many regard the scale of human life as free of controlling principles.

Astronomical systems, although infinitely larger, are seen through such distances that only the main features show (chapter 4). But on Earth, progress in understanding is slow because we are too close to see. As in the old adage about the forest and the trees, we cannot see the pattern for the parts. Figure 1.2 is a cartoon view of the steps we must take in going from detailed data to system viewing and prediction, a process we call using the macroscope. Whereas people often search

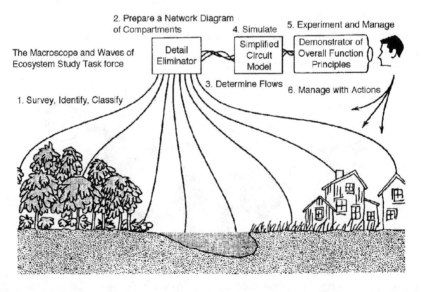

FIGURE 1.2 Cartoon of the macroscope and the steps in its use. The detail eliminator simplifies by grouping parts into compartments of similar function.

among the parts to find mechanistic explanations, the macroscopic view is the reverse. Humans, already having a clear view of the parts in their fantastically complex detail, must somehow get away, rise above, step back, group parts, simplify concepts, and interpose frosted glass to somehow see the big picture.

Since *Environment, Power, and Society* was published, many sciences have found better lenses for the macroscope, better ways to see how the parts form larger wholes and patterns. Networks are better understood with new mathematics, models, and computer simulations that join the parts to show how the larger systems perform. The world is finally using the macroscope by viewing global problems on television, traveling worldwide, exchanging information on the Internet, and discussing global policies, politics, and international trade. The daily maps of worldwide weather, the information received from the high-flying satellites, macroeconomic statistical summaries, the combined efforts of international geophysical collaborations, and the studies of cycling chemicals in the great oceans all stimulate the new view.

In this book, therefore, the reader is invited to view the world and society through the macroscope. Many explanations use the principles of energy hierarchy (chapter 4). We hope to show the great wheels of the machinery in which we are small but important cogs and perhaps in the end predict how the earth will program human services in the drama of the earth.

THE BIOSPHERE

We can begin to gain a systems view of the earth by looking through the macroscope of the astronaut high above the earth. From an orbiting satellite, the earth's living zone appears to be very simple (fig. 1.3c). The thin water- and air-bathed shell covering the earth—the biosphere—is bounded on the inside by dense solids and on the outside by the near vacuum of outer space. Past the orbiting capsule radiant energy from the sun enters the biosphere, and soon equal amounts pass outward as flows of heat radiation. Through the haze of height only a few features of energy flows can be observed. There is a great sheet of green chlorophyll and cyclones of spiraling clouds in weather belts, but the miraculously cascading machinery of parts within parts within parts is not even visible. From the heavens it is easy to talk of gaseous balances, energy budgets per million years, and the magnificent simplicity of the overall metabolism of the earth's thin outer shell. With the exception of energy flow, the geobiosphere for the most part is a closed system of the type whose materials are cycled and reused.

The biosphere is the largest ecosystem, but the forests, the seas, and the great cities are systems also. Large and small parts operate on their budgets of energy, and what can and cannot be done is determined by energy laws (chapter 3). Any phenomenon is controlled both by the working of its smaller parts and by its role in the larger system of which it is a part.

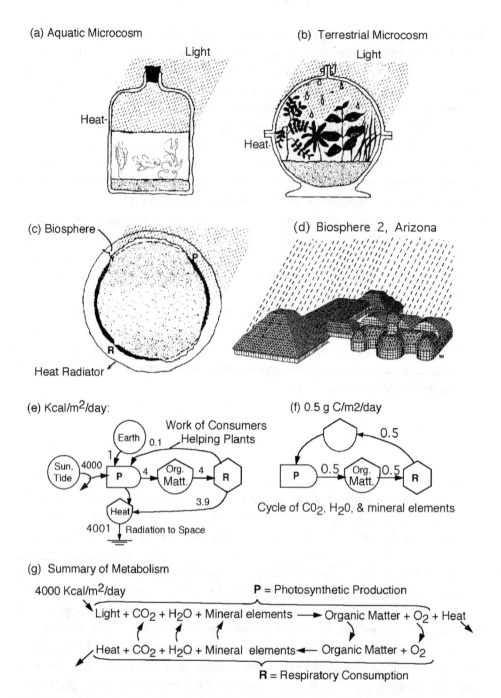

(a) Aquatic Microcosm

Light

Heat-

(b) Terrestrial Microcosm

Light

Heat

(c) Biosphere

P

R

Heat Radiator

(d) Biosphere 2, Arizona

(e) Kcal/m^2/day:

Earth 0.1

Work of Consumers
Helping Plants

Sun, Tide 4000

1

P

4

Org. Matt.

4

R

Heat

3.9

4001 Radiation to Space

(f) 0.5 g C/m2/day

P

0.5

Org. Matt.

0.5

R

0.5

Cycle of C0$_2$, H$_2$0, & mineral elements

(g) Summary of Metabolism

4000 Kcal/m^2/day

P = Photosynthetic Production

Light + CO$_2$ + H$_2$O + Mineral elements ⟶ Organic Matter + O$_2$ + Heat

Heat + CO$_2$ + H$_2$O + Mineral elements ⟵ Organic Matter + O$_2$

R = Respiratory Consumption

FIGURE 1.3 Closed-to-matter systems supported entirely by sunlight and their metabolic reactions. **(a)** Aquatic closed system; **(b)** terrestrial microcosm; **(c)** biosphere of Earth; **(d)** Biosphere 2 in Arizona; **(e)** diagram showing energy flows for the systems **(a–d)**; **(f)** cycle of materials between photosynthetic production (*P*) and respiratory consumption (*R*); **(g)** summary equations for photosynthetic production and the reverse process of respiratory consumption.

The work that results from energy flow is inherently hierarchical, with many calories of one kind required to produce a few calories of another. Much of the organization of the geobiosphere and the human economy is understandable from the energy hierarchy concepts (chapter 4).

In earlier times, energy available to human management was insufficient to control the biosphere, and people were protected in their ignorance by the great stabilizing storages of the oceans and atmosphere. In recent years, however, the accelerating growth of fossil fuel use has allowed civilization to interfere with life support, outdistancing our knowledge of the consequences. The gaseous emissions of civilization are changing the climate (chapter 5).

THE LIVING METABOLISM OF THE EARTH

With the turning of the earth, the sun comes up on fields, forests, and fjords of the biosphere, and everywhere within the light there is a great breath as tons upon tons of oxygen are released from the living photochemical surfaces of green plants, which are becoming charged with food storages by the onrush of solar photons. Then, when the sun passes in shadows before the night, there is a great exhalation of carbon dioxide (CO_2) that pours out as the oxygen (O_2) is burned, the net result of the maintenance activity of the living machinery. During the day, while oxygen is generated, a great sheet of new chemical potential energy in the form of organic matter lies newborn about the earth, but in the darkness, the new organic matter and oxygen disappear in hot and cold consumption processes that release heat through the night. Figure 1.3e and 1.3g summarizes earth metabolism.

The living process in green chlorophyll of forests, lakes, oceans, and deserts during the day is called photosynthesis by those who study a small segment of it and primary production by those who consider great masses of it. In photosynthetic production (abbreviated P in fig. 1.3), carbon dioxide, water, and nutrients are combined to make organic matter and oxygen.

Consumption of organic matter and oxygen by living organisms and by cities goes on day and night but is masked by primary production of these substances during the day. It can be measured at night. Although *respiration* usually is used for the cool fires of biological consumption, let us also allow the word to include consumption by the hot fires of autos and industry. The letter R is the abbreviation used here for all such consumption. Respiration transforms the energy of food and fuel consumption into useful work when the organic molecules are combined with oxygen to make carbon dioxide and water (fig. 1.3g). In the overall equation for consumption, materials are returned to the inorganic state, ready for primary production again (fig. 1.3f). The materials generated by consumption are the ones used by production and vice versa.[2]

The average P and R of the biosphere has been about 1 g of organic matter per square meter per day. This is about 4 kcal of organic potential energy stored daily as organic matter and burned again later (Behrenfeld et al. 2001). Together the P and R processes generate a cycle of materials. Systems running on sunlight that are closed to matter tend to develop a balance between production and consumption. Like the biosphere, three other solar-based systems that are mostly closed to matter are shown in fig. 1.3: an aquatic microcosm, a terrestrial microcosm, and the 3-acre Biosphere 2 in Arizona (Marino et al. 1999).

Resources from Above and Below

About half of the energy that comes to the earth from the sun is in the visible wavelengths, which are used in photosynthesis. The other part of solar energy is in infrared and ultraviolet light wavelengths that are absorbed as heat in the ocean, soils, and vegetation. This directly absorbed heat, plus the heat released from photosynthetic machinery using the visible wavelengths, plus heat released when organic matter is consumed, all add together to cause higher temperatures near the equator and lower temperatures in shaded regions. Between high- and low-temperature areas a giant heat engine operates, creating the great wind and water current systems of the earth. The energies processed by the earth's heat engines contribute to photosynthetic and respiratory processes by causing winds and water currents to bring raw materials such as rain and fertilizer minerals more rapidly to the plants for production.

Some of the sun's heat drives the plant's uptake of water from the soil and the transpiration of water from its leaves. Other winds and currents aid food chains by moving organic matter to the sites of consumption. Hence, the P and R processes of the biosphere are closely linked energetically with the earth's physical processes, its heat, winds, and water currents (see chapter 5).

The biosphere also receives potential energy from the pull of the sun and moon that drives the tidal currents. Available energy with the potential to do work also enters the biosphere from the earth below. The land is renewed by geologic cycles that push land to the surface to keep up with the land that is eroded away. Some of this work is driven by the circulation of hot, fluid rock deep in the earth. The geobiosphere depends on the support of the earth bringing resources up from the earth below and the tide and solar inputs from above (figs. 1.1 and 1.3e).

Solar Society

During agrarian regimes, with people and domestic animals living off the land, there was often a balance of primary production and total respiration (consumption) in the course of the year. The net annual effect on the gases of the atmosphere, on the concentration of minerals, or on the future was small, for the system was balanced as an aquarium is balanced. Some organic matter was stored as fuel and oil, but the amount per year was tiny. As agrarian systems multiplied the world over, the biosphere was little disturbed; production and consumption were similar on average.

When human societies first evolved as a significant part of the systems of nature, people had to adapt to the food and fuel energy flows available to them, developing the familiar agrarian patterns of human culture. Ethics, folkways, mores, religious teachings, and social psychology guided the individual's participation in the group and provided means for using energy sources effectively. Sunlight is spread out over the earth's surface so evenly that it is not directly available to people until after some of it has been concentrated. Much of the sun's potential is necessarily used up by the concentrating processes through plants and animals. Societies that were able to survive had to gather food and distribute energies within the social system for their successful continuance, and they developed the group organization neces- sary for these purposes. The social systems adapted to meet changing conditions such as overcrowding, fluctuations of yield, crises from competitors, and threats from internal disorder. The pattern of solar society is shown in fig. 1.4a. Its main parts and processes run on outside resources, the sunlight from above, the tides driving the ocean, and the geologic processes bringing energy and materials up from the earth below. The economic system was simple, and economic reward often reflected the energy control gained.

Humanity Takes Over Nature with Fossil Fuels

Over the last two centuries, society's basis has changed, for now much more fuel energy is coming from concentrated sources within the earth. The economy's in- dustrialized system now gets its energies from fossil fuels (coal, natural gas, oil) and nuclear fuels. Much of this energy flow goes back into our environmental system to increase the yield of food and critical materials. Figure 1.4 contrasts the new industrial system and the old agrarian society. The earlier society was dis- persed over the landscape because the energy sources were spread over the earth's surface.

Few understand that cheap food, clothing, and housing depend on cheap energy and that potatoes are really made from fossil fuel. High agricultural yields are fea- sible only because fossil fuels are put back into the farms through the use of farm equipment, manufactured chemicals, and plant varieties kept adapted by armies of agricultural specialists supported by the fossil fuel–based economy.

Adding industrial society to the biosphere suddenly is like adding large animals to a balanced aquarium (Beyers 1963b). Consumption temporarily exceeds produc- tion, the balance is upset, food and fuels for consumption become scarce as the products of respiration accumulate. These stimulate production, and the balance of respiration is restored. In some experimental systems, balance is achieved only

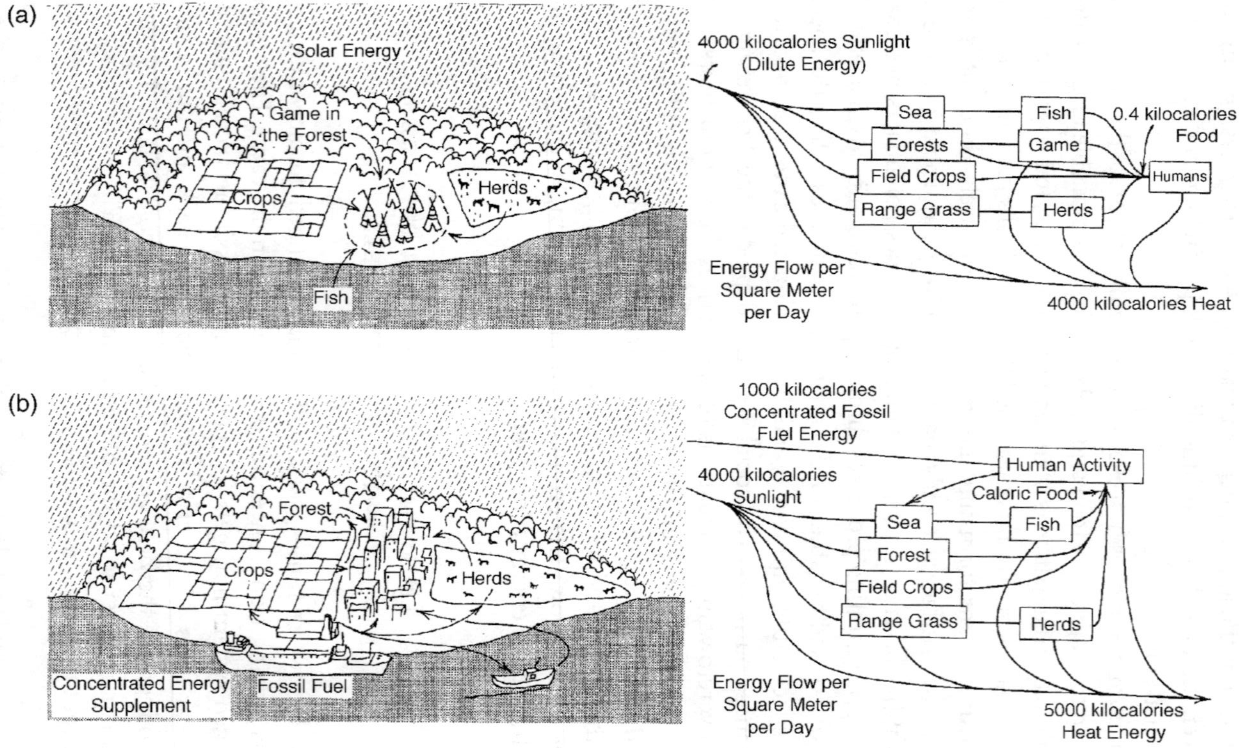

FIGURE 1.4 Comparison of **(a)** an agrarian system with **(b)** an industrialized system. (See chapter 8.)

after the large consumers that originally started the imbalance are dead. Will this happen to the present human culture?

URBAN DEVELOPMENT AND ANIMAL CITIES

Now, the highly concentrated fossil fuels cause developments to be concentrated in urban centers. Sometimes we can comprehend the essence of complex human phenomena by looking at the similar but simpler systems of nature (fig. 1.5). The energetic processes of an industrial city are like those in a dense reef of oysters, an animal city (fig. 1.5). Both are operating at high intensity based on an inflow of resources. Both are centers of energy hierarchy. The fossil record of animal cities is beautiful, with the remnant structures of ancient reef ecosystems. But many of the species of these animal cities are extinct. What does the future hold for the industrial economic reefs, our present cities?

During industrial regimes with the system running mostly on fuels, people manage affairs with technology. Even agriculture is dominated by machinery and industries supplying equipment, poisons, genetic varieties, and high-tech services. The industrial society has respiratory consumption greater than photosynthetic production. The products of respiration—carbon dioxide, metabolic water, and mineralized inorganic wastes—are discharged at rates greater than their incorporation into organic matter by photosynthesis. If the industrialized urban system were enclosed in a chamber with only the air above it at the time, it would quickly exhaust its oxygen, be stifled with waste, and destroy itself because it does not have the balanced recycling pattern of the agrarian system.

The problems with life support in the 1970s on *Apollo* space flights and the later experiments with Biosphere 2 in Arizona dramatized this principle to the world (chapter 12). At present, the biological cycles of the environment are barely able to absorb and regenerate agricultural and urban wastes. New kinds of landscapes and interface ecosystems need to evolve with the help of ecological engineering (chapter 13).

INFORMATION SOCIETY AND THE CHANGED ROLE FOR HUMANITY

Late in the last millennium the pattern of society was concentrated into an information society sharing global television, individual computers, and the Internet. More and more electrical power was required. Global information sharing and foreign trade increased. The changes have come so fast that many customs, mores, ethics, and religious patterns have not adapted. The economic system is large, complex, and changing so rapidly that it is more and more difficult for people to see its energy basis (chapter 9). Few people realize that their prosperity comes from

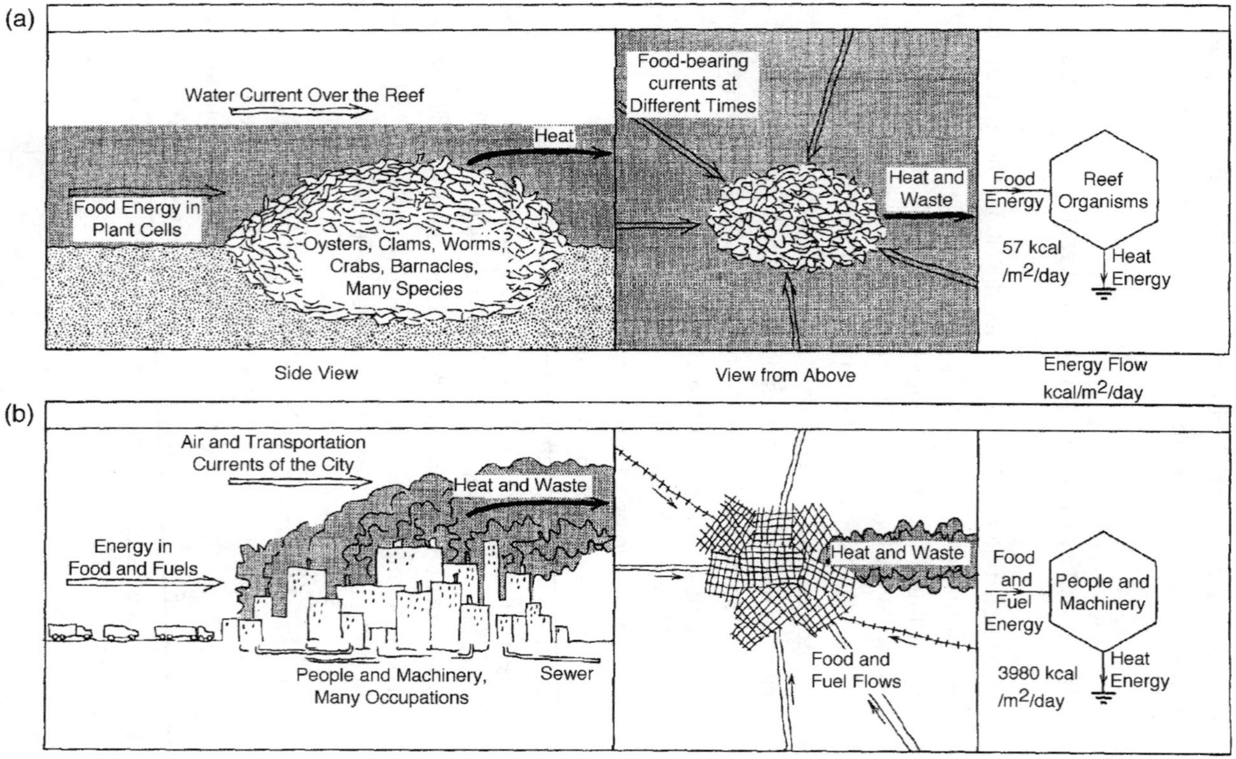

FIGURE 1.5 Comparison of 2 systems of concentrated consumers whose survival depends on strong flows that bring in fuels and oxygen and carry away wastes. **(a)** A reef of oysters and other marine animals characteristic of many estuaries (Copeland and Hoese 1967; Lehman 1974); **(b)** industrialized city based on an early accounting of a city's metabolism (Wolman 1965; Whitfield 1992). (See chapter 10.)

the great flux of fossil fuel energies and not just from human dedication and political design. The principles of energy hierarchy explain much about the organization of society and its population, occupations, health, diversity, institutions, and functions (chapter 11). Useful information and technology are high up in the energy hierarchy, with large energy requirements for their development and maintenance. Less information can be supported in times of less resource (chapter 8).

Because the intense activity and concentrations in urban society are not sustainable without cheap fossil fuels, many raise questions of the ultimate role and survival of *Homo sapiens*. Now scarcities are developing in the global reserves of concentrated fossil fuels on which urban civilization is based (chapter 13). As energy sources decline, how does society return to a lesser position in the earth system without a collapse? Most people are misled about solar technology substituting for fossil fuels (chapter 7).

Our leaders and journalists usually focus on the short-term policies affecting the circulation of money. We now understand many of the relationships between money and energy, the differences between market values and real wealth, and policies to help monetary and fiscal management sustain economies. The energy transformations form a monetary hierarchy (chapter 9).

Critical issues in public and political affairs of human society ultimately have an energetic basis that can be used to select public policy on birth control, private property, people in space, defense, power plants, land zoning, health care, income distribution, trade equity, and environmental impacts (chapter 13). Serving a culture as its programs of energy use, new and old religions can help with the appropriate morality to adapt people to new conditions and times of descent ahead (chapter 11).

Summary

The purpose of this book is to increase our understanding of the system of civilization and its resource basis so as to chart a better future. A macroscopic understanding of environment and society is sought with the principles of general systems, energy hierarchy, and earth metabolism. By accounting for the sequence of society from agrarian landscapes to urban frenzy, we can extend the reasons for history to the future. Even now the environmental resources of the planet are beginning to limit society just as the earth's fossil fuel–based urban civilization is flowering in storms of information. The chapters that follow try to explain the self-organization of energy, materials, money, and information now and in the future.

Bibliography

Beyers, R. J. 1963a. A characteristic diurnal pattern of balanced aquatic microcosms. *Publication of the Institute of Marine Science, Texas*, 9: 19–27.

———. 1963b. The metabolism of twelve aquatic laboratory microecosystems. *Ecological Monographs*, 33: 281–406.

Beyers, R. J and H. T. Odum. 1993. *Ecological Microcosms*. New York: Springer Verlag.

Behrenfeld, M. J., J. T. Randerson, C. R. McClain, G. C. Feldman, S. O. Los, C. J. Tucker, P. G. Falkowski, C. B. Field, R. Frouin, W. E. Esaias, D. D. Kolber, and N. H. Pollack. 2001. Biospheric primary production during an ENSO transition. *Science*, 291(5513): 2594–2597.

Copeland, B. J. and H. D. Hoese. 1967. Growth and mortality of the American oyster *Crassostrea virginica* and high-salinity shallow bays in central Texas. *Publication of the Institute of Marine Science, University of Texas*, 11: 149–158.

Lehman, M. 1974. *Oyster reefs at Crystal River, Florida and their adaptations to thermal plumes*. Master's thesis, University of Florida, Gainesville.

Marino, B., H. T. Odum, and W. J. Mitsch, eds. 1999. *Biosphere 2, Research Past and Present*. Amsterdam: Elsevier.

Odum, H. T., R. J. Beyers, and N. E. Armstrong. 1963. Consequences of small storage capacity in nannoplankton pertinent to measurement of primary production in tropical waters. *Journal of Marine Research*, 21: 191–198.

Odum, H. T. and A. Lugo. 1970. Metabolism of forest floor microcosms. In H. T. Odum and R. F. Pigeon, eds., *A Tropical Rainforest*, I-35–I-56. Oak Ridge, TN: Division of Technical Information and Education, U.S. Atomic Energy Commission.

Vernadsky, V. I. 1926. *Biosphere*. Translated by D. B. Langmuir, revised and annotated by M. A. S. McMenamin (1998). New York: Springer-Verlag.

Whitfield, D. 1992. *Emergy basis for urban land use patterns in Jacksonville, Florida*. Master's thesis, University of Florida, Gainesville.

Wolman, A. 1965. The metabolism of cities. *Scientific American*, 213: 179–190.

Notes

1. Semiclosed aquarium and terrarium systems, such as those found in schoolrooms, are microcosms showing the same energy and metabolic processes of the outdoor world. Beyers and Odum (1993) summarized the hundreds of research papers on microcosms.

2. Microcosms were studied for their carbon dioxide metabolism patterns after they had become stabilized, and their kinetic properties were analyzed as feedback systems (Odum et al. 1963, Odum and Lugo 1970).

CHAPTER 2

SYSTEMS NETWORKS AND METABOLISM

UNDERSTANDING ENVIRONMENT and society as a system means thinking about parts, processes, and connections. To help us understand systems, we draw pictures of networks that show components and relationships. Thereafter, we can carry these system images in our minds. In the process, we learn how energy, materials, and information interact. If we add numerical values for flows and storages, the systems diagrams become quantitative and can be simulated with computers. This chapter introduces a versatile energy systems language for representing verbal concepts with network diagrams. The diagrams explain how photosynthetic production and the respiratory consumption of whole systems are symbiotically coupled and self-regulating.

Ecosystem production (photosynthesis) often is represented as a chemical equation in which carbon dioxide and water are combined to produce organic matter and oxygen (figs. 1.3 and 2.1a). Ecosystem respiration is the reverse, with organic matter and oxygen being used and carbon dioxide, water, and minerals as byproducts.

METABOLISM OF A BALANCED AQUARIUM

The photosynthetic production and respiratory consumption described for the whole earth (fig. 1.3) is present in miniaturized form in a balanced aquarium (fig. 2.1a). Large plants and photosynthetic microorganisms make the food and oxygen that support food chains of microscopic animals, larger aquatic insects, snails, and a few fish. One way to represent these connected processes is with the broad pipe-like pathways shown in figs. 2.1b and 2.1c.

Figure 2.1b shows the inflow and use of energy. Most of the sun's energy inflow on the left is used by plant producers to store food energy in organic matter. Some is reflected. Consumers use the stored organic matter to fuel their work, which usually includes some work controlling the plants, such as distributing seeds and regulating populations (service control pathway in fig. 2.1b). Energy that has done its work passes out of producers and consumers as used energy.

(a) Balanced Aquarium

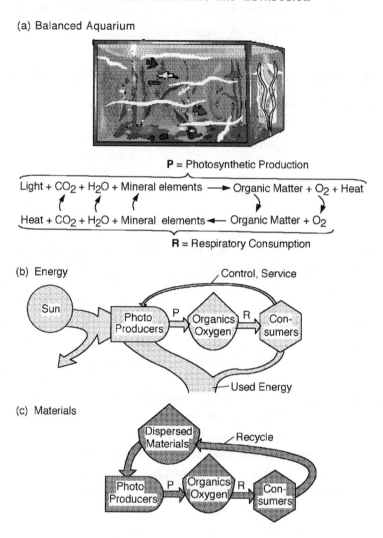

P = Photosynthetic Production

Light + CO_2 + H_2O + Mineral elements ⟶ Organic Matter + O_2 + Heat

Heat + CO_2 + H_2O + Mineral elements ⟵ Organic Matter + O_2

R = Respiratory Consumption

(b) Energy

Control, Service

Sun

Photo Producers

P

Organics Oxygen

R

Con-sumers

Used Energy

(c) Materials

Dispersed Materials

Recycle

Photo Producers

P

Organics Oxygen

R

Con-sumers

FIGURE 2.1 Flows of material and energy in a balanced aquarium system with 2 ways of representing parts and processes. P is gross production, and R is total respiratory consumption. **(a)** Sketch of aquarium ecosystem; **(b)** systems picture with broad pathways to represent energy flows; **(c)** systems picture with broad pathways to represent material flows.

Figure 2.1c shows the material circulation. Plant producers use dispersed material ingredients (e.g., carbon, nitrogen, phosphorus) to make organic matter, which contains these elements in its organic structure. The consumer organisms use up the organic matter and release the raw material nutrients to the environment again, labeled *Recycle* in fig. 2.1c. Circulating materials, such as carbon, are shown as a broad pipe that forms a closed loop. Thus, the diagram shows the biogeochemical cycle of carbon and other necessary materials. In a closed-to-matter system (figs. 2.1 and 2.2), there is no inflow or outflow of matter, and the parts are linked by the internal cycle of materials.

(a) Aquarium System

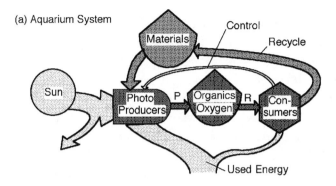

(b) Aquarium System, Using Lines for Flows

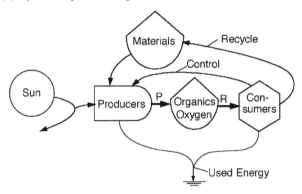

FIGURE 2.2 Combination of the flows of material and energy of the balanced aquarium in fig. 2. 1. **(a)** Combined system with broad pathways; **(b)** combined system with line pathways according to the energy systems language.

WHOLE SYSTEM PICTURES

Although an environmental system can be represented with diagrams of material cycle, such as carbon, only part of the whole environmental system is represented, with only one type of flow showing at a time. Only the processes involved with that material appear, and factors driving the cycles are omitted. When the material flows are diagrammed separately, as in fig. 2.1c, it is easier to keep track of them, but this disconnects the energy pathways and interactive processes that drive the materials. Network pictures must show everything important together (fig. 2.2).

Some of the pathways in a system carry only energy (e.g., light, sound, heat). Other pathways carry materials along with some energy inherent in their concentration. To visualize the whole system, we combine the energy fig. 2.1b and the materials fig. 2.1c into a whole system view (fig. 2.2a). To make the material circulation stand out in a diagram of all flows, shading or color is useful.

A whole system view is one of the tools of the macroscope. By drawing the main parts and connecting the parts with pathways, we learn to think about wholes, parts, and processes at the same time. In figs. 2.1 and 2.2a we used broad arrows like pipes for the flows of energy and materials. Such pictures are easy to understand but hard to draw for complex networks.

In fig. 2.2b, narrow pathway lines were substituted for the broad arrows to make a network diagram that is easier to draw. Network drawings foster thinking about connectivity. But for many people these energy circuit diagrams seem too mechanical, so in this book we represent systems with both styles of pictures.

System Metabolism with Time

Systems diagrams are helpful for thinking about the behavior of metabolism over time. Experiments with closed microcosms show a regular rise and fall of oxygen and carbon dioxide, as light is turned on and off (fig. 2.3a). There is an alternation of production and consumption. With the light on, the production process starts and begins to accumulate organic matter and oxygen while using up the carbon dioxide and nutrients. At this moment, production may exceed respiratory consumption $(P > R)$. Letting your eye follow pathways in fig. 2.2 may help you visualize the way the processes deliver actions from left to right. The more organic matter accumulates, the more the consumers can use it, and their activity increases. The more consumption there is, the more materials recycle.

When the light turns off, photosynthesis stops, but consumption continues and begins to drain down the accumulated organics and oxygen. The recycling of carbon dioxide and nutrients is rapid at first but decreases as the night goes on. Figure 2.3a shows the alternating rise and fall of organic matter storage. Figure 2.3b shows the rates of metabolism (these rates are the slopes of the curves of storage in fig. 2.3a). Net production (P_{net}) of organic matter and oxygen is shown above the zero line in fig. 2.3b; the rate of nighttime consumption of organic matter and oxygen (R_{night}) is shown below the line.

On average, P and R tend to be equal in a system that is closed to matter (has no materials added or removed). In summer, with more light, P tends to exceed R, but in winter, with less light, R tends to exceed P.

Self-regulation by the Loop of Cycling Materials

The closed-to-matter microcosm example shows how the organization of a system determines its responses. The main cycles of materials such as carbon form loops (figs. 2.1c, and 2.2), which have a self-regulatory effect that cushions change. Materials from P, passing to R and returning to P in a loop, are mutually rewarding, with each process stimulating the next. If one of the two processes slows down, so does the

FIGURE 2.3 Diurnal record of photosynthetic production (*P*) and respiratory consumption (*R*) in a microcosm (Odum et al. 1963; Odum and Lugo 1970). (a) Storage of organic matter in the light and discharge in the dark; (b) rates of daytime net photosynthesis (+) and nighttime respiration (–); (c) periods of constant light alternating with a dark period.

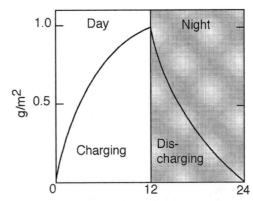
(a) Labile Biomass Stored

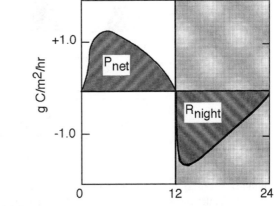
(b) Metabolic Rate

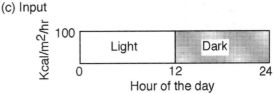
(c) Input

other, until storage concentrations build up and force the process back to its regular rate. Where *P* and *R* are not equal, the metabolism tends to return them to balance.

The systems diagram in fig. 2.4a represents the production, consumption, and recycle process as a simple model. The storage of both inorganic materials and the organic quantities, living and dead, are a means for smoothing out input fluctuations. Flow out of each storage tank is proportional to the quantity stored. Rates increase with storages.[1] After equations are written (appendix fig. A8), the model is simulated on the computer, generating the same rise and fall of storages and metabolic rates found in the microecosystem. Figure 2.4b shows the simulated response of metabolism when the light intensity rises and falls, as in natural daylight.

(a)

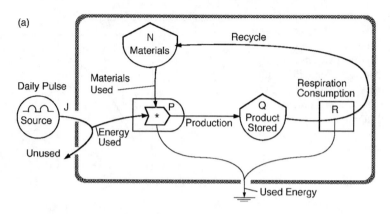

(b) Computer Simulation

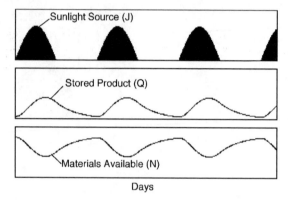

FIGURE 2.4 Model and simulation of gross production (*P*) and total system respiration (*R*) in a closed-to-matter ecosystem using a renewable energy source. **(a)** Energy systems diagram showing *P*, *R*, and recycle; **(b)** computer simulation of the organic storage (*Q*) and material storage (*N*) with an energy input (*J*) resembling sunlight (half sine wave). For equations and simulation program, see appendix fig. A8.

Other examples of daily changes in metabolism in streams, ponds, and estuaries are given in chapters 6 and 12. The circular stimulation is readily observed, where the rate of photosynthesis during the day depends on the amount of respiration of the previous night and vice versa.

P and *R* Diagram

The extent of the balance between the whole system photosynthetic production (*P*) and total respiratory consumption (*R*) can be shown by plotting metabolic rates on a graph (fig. 2.5). When the two processes are in balance, points are on the diagonal line, where the ratio *P*/*R* = 1. Points in the lower left section of the graph have small metabolism and are sometimes called oligotrophic. Points in the upper right section have large metabolism and are sometimes called eutrophic. Because the cycle of materials tends to keep systems in balance, $P = R$. Points tend to move toward the diagonal line in fig. 2.5.

Microcosms adapted to a repeating regime of light and temperature in an experimental chamber develop a quasi–steady state, with only small fluctuations away from a balance of photosynthesis and respiration (fig. 2.6).

FIGURE 2.5 Graph of gross production (P) and total respiration (R) used to compare metabolism of ecosystems. The diagram can also be used to plot the daytime net production and the nighttime respiration.

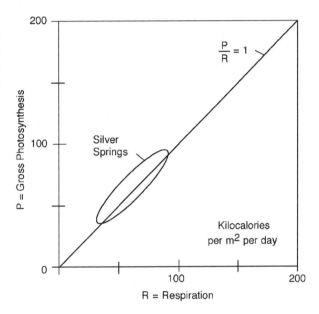

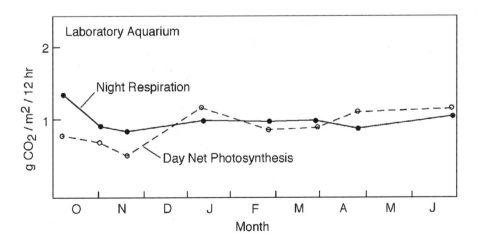

FIGURE 2.6 A metabolism sequence of an ecological system in an aquarium microcosm long adapted to constant conditions of a laboratory experimental chamber (Beyers 1963b).

Metabolic Regulation with Temperature Change

The coupling of P and R in adapted, balanced ecosystems regulates metabolism despite temperature change. Although the rates of biochemical reactions in living organisms (e.g., water fleas, *Daphnia*) tend to increase with temperature, the respiration of steady state ecosystems shown in fig. 2.7 was almost constant when the temperature changed.

In a general way, the earth's biosphere apparently also works by circular self-regulation, even though longer time periods are involved. The carbon dioxide in the air falls when photosynthesis increases in the summer and rises when respiration is excessive in the winter. (See chapter 12, fig. 12.17c.) The summer regime runs on a pulse of light, using the stored raw materials (carbon dioxide, water, and inorganic nutrients). It alternates with a winter regime running on stored food for the consumers. Each regime stores quantities needed for the next stage. For the whole biosphere, oxygen is stabilized by large storages in the air; carbon is stabilized by large storages in the sea and in limestone. From these examples we see that the properties of the whole system determine the outcome of flows and storages, which we observe. A study of systems over time in this way is part of the macroscopic view.

A Table of Circulating Flows

Diagrams of material and energy flows are the main language used in this book to represent systems. Any network diagram can be just as easily represented in a

FIGURE 2.7 Metabolism as a function of temperature, comparing an organism (the water flea, *Daphnia*, fig. 3.8), a community of sewage microbes, and the short-term response of an adapted, balanced ecological microcosm (Beyers 1962, 1963a).

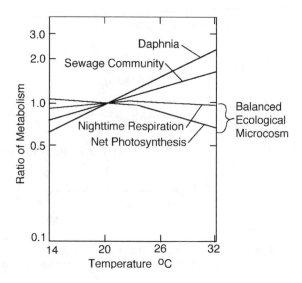

tabular form in which the separate units of the network appear as headings and the numbers in the boxes of the table are the flows. Everyone uses such tables in everyday life. For example, mileage tables have cities as the headings and the miles between the cities indicated in the boxes. An array of numbers in tabular form is also called a matrix.

Figure 2.8 describes the carbon cycle in both network and matrix form. From a microcosm study, there are the same four units already introduced in the closed system diagrams in fig. 2.2, but more of the possible pathways are drawn. Representing systems in this way is a first step in the use of techniques of matrix algebra.[2] A pertinent step after description as a matrix is to count the possible relationships that may have been overlooked. In some systems such as that shown in fig. 2.8b, some pathways are missing, and their place in the table gets a zero. The number of possible boxes and the number of boxes actually used in the system are properties of the complexity of the network (chapters 6 and 7).

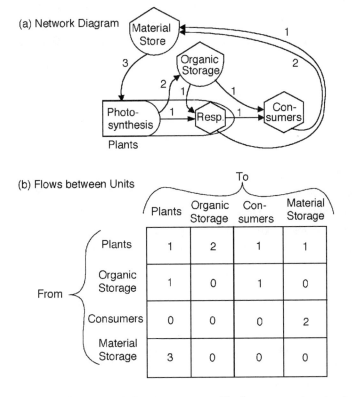

FIGURE 2.8 **(a)** Systems diagram and **(b)** input–output table for representing circulation between production and consumption. Numbers are rates of flow of carbon in grams per square meter per day at steady state. Absence of a pathway is indicated by a zero in the table.

ELEMENTS THAT CYCLE TOGETHER

So far we have used carbon to illustrate the circulation of materials in networks representing metabolism. But living processes require a flow of the other materials of living structure, such as oxygen, water, and mineral elements, given in the equations of fig. 2.1a. When we draw a carbon pathway for a typical ecosystem, we are at the same time implying the flow of other elements, such as phosphorus and nitrogen, that are incorporated into plant production and released by respiratory consumption. They circulate together, each with a characteristic ratio in relation to the other necessary elements. Carbon is the best element to use for overall material calculations (if only one is to be used) because it is the largest fraction. Carbon makes up 40–80% of organic matter and fuels. Another way to represent several different kinds of materials is to use a separate pathway for each kind of material in the network diagram (chapter 3).

The ratio of an element used to carbon processed is called a metabolic quotient. See table 2.1. There are metabolic quotients characteristic of each type of system. The knowledge that elements occur in characteristic ratios to carbon makes it possible to estimate their flows from data on carbon metabolism. By and large, a human economy has a characteristic mix (ratio of elements). The North Atlantic Ocean has a certain mix, as Redfield (1934) showed in extending to nitrogen and phosphorus the physiological concept of respiratory quotients (oxygen to carbon

TABLE 2.1   Chemical Ratios of Regenerative Cycles[a]

ELEMENT	TEMPERATE OCEAN[b]	HUMAN ASSOCIATION[c]	WORLD AGRICULTURE
Total dry weight of food used	2.3	1.8	2[d]
Oxygen consumed in respiring food	1.9	2.9	—
Carbon respired and released in waste	1.0	1.0	1[d]
Nitrogen processes	0.18	0.2	0.029[e]
Phosphorus processes	0.024	0.016	0.011[e]

[a] James Elser and associates (1996) summarize new knowledge on elemental ratios (stoichiometry). Ratios to carbon as one.

[b] Sverdrup et al. (1942).

[c] Per capita wastes. Meyers (1963): carbon, 296 g/day; oxygen, 860 g/day; food 520 g/day. Fair et al. (1968): N, 7.9 g; P, 1 g. Stumm and Morgan (1970), C:N:P 60:12:1.

[d] U.S. Department of Agriculture (1970).

[e] Fertilizer production, 1963: nitrogen, 15.2 million metric tons per year; phosphorus, 5.7 million metric tons per year (Meyers et al. 1966).

ratios). Redfield's innovative principle was that consumers regenerate and recycle elements in the same ratio as assembled together by the producers. Table 2.1 shows ratios for human culture, as taken from its waste products and for an aquatic system. The wastes of the human system have very different elemental ratios from the general environment and can cause the reorganization of ecosystems. Many pollution problems can be generalized as a change in element ratios in environmental cycles through the addition of waste flows from some industry or human settlement. Elser et al. (1996) summarized the ratios for many kinds of processes and organisms. In chemistry, the determination of ratios of elements participating in interactions is called stoichiometry.

Outside Flows

Most systems have inflows and outflows as well as recycling flows. Figure 2.9 shows system metabolism connected to four pathways of outside exchange: inflow of inorganic materials, outflow of inorganic materials, inflow of organic matter, and outflow of organic matter. Included in fig. 2.9 is a rate of change equation for the organic matter and inorganic materials in the system. A system is said to be at steady state when there is no change in storages. A steady state is one that holds a constant pattern of flows, cycles, storages, and structures. The steady state pattern for organic storages in fig. 2.9 is net production (P) plus inflow (I) equals respiration (R, consumption) plus export (E).

Conservation of Matter

An equation is also included in fig. 2.9 for the storage of inorganic raw materials. Whatever flows in (J_i) either goes into the storage, is used by the plants (J_p), or flows out of the system (J_o). Unbound materials are continuously added to the storage of raw materials by the recycle process (J_r). When the material storage is constant and the material flows are in steady state, $J_i + J_r$ equals $J_p + J_o$. The common-sense statement that inputs of matter have to be accounted for in storages or outflows sometimes is called the law of conservation of matter.

A Network Energy Language for All Systems

Many kinds of network diagrams and symbols have been published to represent systems, each with a set of symbols and rules for their use. For example, most people know that the wiring diagrams of electronic equipment use symbols for the hardware components and lines for the wires. Colleges grade applicants for their verbal skills and their quantitative skills, but there is a third skill, the ability to

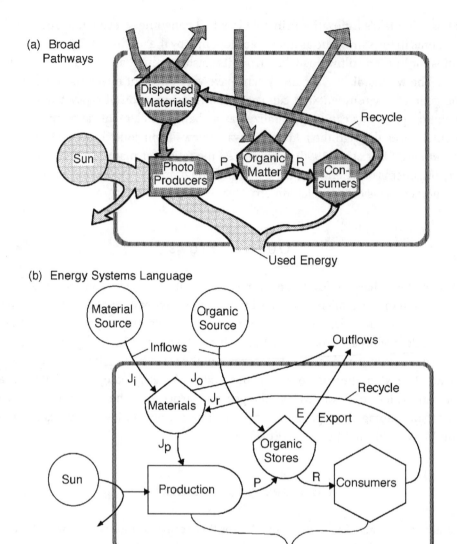

FIGURE 2.9 A system of production, consumption, and recycle that has inflows and outflows. **(a)** Picture drawn with broad pathways; **(b)** drawing using energy systems language.

visualize networks. Appendix fig. A2 shows the way storage is represented in several systems languages.

To understand a whole system and the interaction of its parts, we need a diagrammatic language that could represent everything important. To help the mind understand complexity, a general systems language should follow naturally from verbal thinking while showing system structure, processes, and flows. A systems diagram should help the mind visualize relationships and infer system behavior from the configurations. To evaluate the parts and processes, we use energy, a measure that is found in everything. Energy is part of all storages, and flows of energy (power) accompany all processes. There is some potential energy in any concentration of materials. Numbers are placed on pathways and in storage tank symbols to show energy or material flows (figs. 2.8, 2.9).

First Energy Law: Conservation of Energy

All the energy that enters a unit or system either is stored there or flows out. The fact that energy has to be accounted for is called the law of energy conservation, the first energy law. Sometimes numbers for energy flows and storages are written on the diagrams. Writing the numbers on the pathways and storage symbols of the energy language diagrams helps keep things straight. Energy flows in biological pathways often are estimated from the carbon dioxide or oxygen gases used or released in metabolism.[3]

Energy Systems Symbols

A language of network pictures with a dozen symbols was introduced as an energy systems language in *Environment, Power, and Society* and was widely used. Its network pictures represented energy, material, information, money, system structures, and scale relationships. In a later book, the author provided inter-translations with other kinds of network languages and concepts (Odum 1983). New principles were found by using the network pictures to simplify our view of the real world. The language was designed to represent the parts and relationships of any system. This is possible because there are common properties of all systems at all scales. The symbols and their explanations are given in fig. 2.10. For those interested in the equivalent statements in traditional mathematics, explanations are given in the appendix. Five of the symbols (source, storage, heat sink, producer, and consumer) have already been used in this chapter (figs. 2.1, 2.2, 2.4, 2.8, and 2.9).

Although all systems are interconnected, a frame is drawn to separate the items of interest inside and the connecting items on the outside that become sources, drawn as circles. This artificial boundary helps identify the inflows and outflows. The lines represent the pathways of energy flow and the action of forces. Flows of

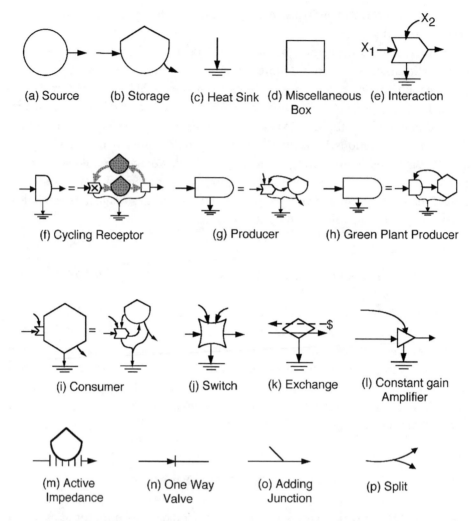

(a) Source   (b) Storage   (c) Heat Sink   (d) Miscellaneous   (e) Interaction
Box

(f) Cycling Receptor   (g) Producer   (h) Green Plant Producer

(i) Consumer   (j) Switch   (k) Exchange   (l) Constant gain
Amplifier

(m) Active   (n) One Way   (o) Adding   (p) Split
Impedance   Valve   Junction

FIGURE 2.10 Modules of the energy systems language. Each of these modules has characteristic equations that may be evaluated in real examples (see the appendix). *First row:* **(a)** Source: This is a source of energy such as the sun, fossil fuel, or the water from a reservoir. A full description of this source would include supplementary description indicating whether the source is constant force, constant flux, or programmed in a particular sequence with, for example, a square wave or sine wave. **(b)** Storage: This represents a form of stored energy such as potatoes in a grocery store or fuel in a tank. **(c)** Heat sink: A sink is needed for all processes that are real and spontaneous, according to the second law of thermodynamics. All processes deliver some potential energy into heat. Heat is the random wandering of molecules that have kinetic energy, and it is this wandering from a less probable to a more probable state that pulls and drives real processes connected to such flows. Energy degradation is also required where storages are dispersing. **(d)** Miscellaneous box: This can be used for any purpose but usually designates an important subsystem. **(e)** Interaction: A force (X_2) interacting with another force (X_1) generates productive output. This action may be as simple as a person turning a valve, or it may be complex, such as the interaction of limiting fertilizer with sunlight in photosynthesis.

Second row: **(f)** Cycling receptor: This represents the reception of pure wave energy such as sound, light, or water waves. In this module energy interacts with some cycling material, producing

(continued)

energy move from the sources, usually drawn on the left along the pathway lines to the right. We refer to these flows as moving from upstream to downstream when they move from the source toward the sink. In the energy systems language, a simple arrow is used to indicate the direction of the driving force. If there are driving forces pushing from both ends, no arrow is used. If backforces push but backflow is prevented, the pathway is a valve. A one-way valve symbol has the arrow symbol with a crossbar, as in fig. 2.10n. A square box (fig. 2.10d) is reserved for any unit for which there is no special symbol; the unit can be identified by a description inside the box.

---

an energy-activated state, which then returns to its deactivated state, passing energy on to the next step in a chain of processes. The kinetics of this module was first discovered in a reaction of an enzyme with its substrate and is called a Michaelis–Menten module. **(g)** Producer: The producer usually contains storages **(b)** and interactions **(e)**. **(h)** Green plant producer: When this symbol is used for plants and vegetation, a cycling receptor **(f)** and consumer unit **(i)** are implied. Energy captured by a cycling receptor unit is passed to a consuming part, which feeds back to the left to keep the cycling receptor machinery working.

Third row: **(i)** Consumer: This represents an interaction of 2 inputs and the autocatalytic coupling of storage that amplifies inflow. The hexagonal consumer symbol usually implies the combination of **(b)** and **(e)** by which stored potential energy is fed back to do work on processing the input. In its simple form this module is said to be autocatalytic. **(j)** Switch: A switch is used for flows. The switch may be a simple on or off, or a flow may be on when 2 or more energy flows are simultaneously on; other flows are on when connecting energy flows are off, and so forth. Many actions of complex organisms and humans are digital switching actions such as voting, reproduction, and starting a car. **(k)** Exchange: This symbol is used for systems in which money is passed in exchange for flows of energy, goods, services, and so on. Money flows in the opposite direction to the flow of real wealth. Human behavior adjusts prices, controlling the ratio of money flow to purchases. For example, a man purchasing groceries at a store receives groceries in one direction while paying money in the opposite direction. The heat sink may be omitted when the work in the transaction is small. **(l)** Constant gain amplifier: Force from the left controls the amount of output and energy flow but contributes only a small energy input. Whatever energy is required comes from the upper inflow as long as there is an adequate source there. The input force from the left causes an output force increased by a constant factor, which is called the gain. For example, a reproducing species inputs effort from the left, generating an output of 10 offspring. Therefore, the module has a gain of 10 as long as the energy supplies are more than adequate.

Fourth row: **(m)** Active impedance: This represents systems that develop a backforce against any input driving force as long as that force is increasing. At the same time, some energy storage is arranged so that when the forcing impetus ceases, the energy unit delivers a forward flow (from some storage or other source) in proportion to the earlier accumulation. Many organisms, human behavior programs, and conservative institutional programs have stubbornness of this form. In electrical systems, such units allow time lags and oscillations to be designed or eliminated. **(n)** One-way valve: This allows flow to pass in one direction only, even though there may be backforce from downstream storage. Symbols **(e–l)** also have the one-way property because of energy losses and interactions with second flows. If there are no backforces from downstream storages, an ordinary barb is used without the bar. **(o)** Adding junction: This joins 2 flows of similar energy type, in which flows are added. The behavior of this symbol may be contrasted with that of the interaction where 2 forces multiply **(e)**. **(p)** Split: One flow splits into 2 of the same kind.

ENERGY DIAGRAM OF A HUMAN HABITATION

To illustrate use of the symbols, fig. 2.11 shows a simplified systems view of a house. Here the energy systems language illustrates and synthesizes the functions and relationships of a familiar system, which was described as an ecological system by Ordish (1960). Some of the flows in a house, such as the heating system, are nonliving processes. There are also living components, such as populations of people, termites, and rats. The diagram (fig. 2.11b) shows force relationships of flows and storages. Note the work done by humans in controlling and making the other flows possible. The people also do work outside the house, for which they receive money, shown in dashed lines. The money received goes back out to pay for the purchased inputs of wood, food, and heating oil. Chapter 10 has more on economic systems.

In the next chapter, the energy systems language is used to represent systems of humanity and nature while introducing some first principles of energetics.

SUMMARY

Using an aquarium microcosm as an example, we used systems diagrams to present an overview of the metabolism of environmental systems. Diagrams of production, consumption, storages, and material recycle were made first with pictures that showed flows with broad, shaded pathways. Then systems networks were shown using the energy systems language, which contains a general set of symbols, pathway lines, and rules introduced to represent all systems. Numbers on the pathways of a network diagram of circulating materials were compared with an equivalent input–output table. The day–night surge of production and consumption was used to show how the closed loop design of basic metabolism is self-regulating. The network symbols were used to represent familiar features of a human house.

FIGURE 2.11 (*opposite page*) Human habitation and an energy circuit diagram of some main parts and processes. **(a)** Sketch; **(b)** systems diagram showing the storages of house structures, pantry and oil furnace, 3 outside energy sources, work by people inside, work and money earned from an outside job, and dissipation of used energy through the heat sink; **(c)** same system showing the inflows of organic carbon and outflow of carbon dioxide.

(a)

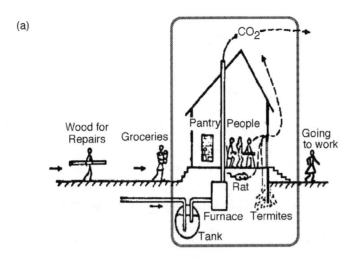

(b)

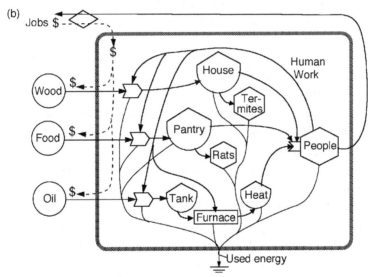

Flows of Carbon

(c)

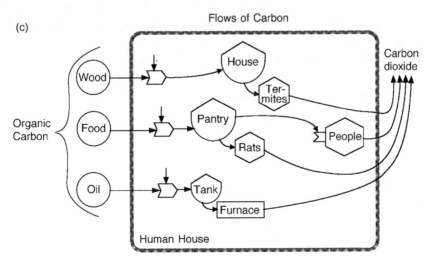

Bibliography

Note: An asterisk denotes additional reading.

Beyers, R. J. 1962. Relationship between temperature and the metabolism of experimental ecosystems. *Science,* 136: 980–982.

——. 1963a. A characteristic diurnal pattern of balanced aquatic microcosms. *Publication of the Institute of Marine Science, Texas,* 9: 19–27.

——. 1963b. The metabolism of twelve aquatic laboratory microecosystems. *Ecological Monographs,* 33: 281–406.

Bullard, C. and R. Herendeen. 1975. The energy costs of goods and services. *Energy Policy,* 3: 268–278.

Elser, J. J., D. R. Dobberfuhl, N. A. MacKay, and J. H. Schampel. 1996. Organism size, life history and N:P stoichiometry. *BioScience,* 46(9): 674–684.

Fair, G. M., J. C. Geyer, and D. A. Okun. 1968. *Water and Wastewater Engineering,* Vol. 2. New York: Wiley.

Hannon, B. 1973. The structure of ecosystems. *Journal of Theoretical Biology,* 41: 535–546.

Herendeen, R. 1981. Energy intensity in ecological and economic systems. *Journal of Theoretical Biology,* 91: 607–620.

Meyers, D. L., T. P. Hignett, and J. R. Douglas. 1966. *Estimated World Fertilizer Production Capacity as Related to Future Needs.* Muscle Shoals, AL: Tennessee Valley Authority, National Fertilized Development Center.

Meyers, J. 1963. Introductory remarks to a symposium on space biology: Ecological aspects. *American Biological Teacher,* 25: 409–411.

Odum, H. T. 1983. *Systems Ecology.* New York: Wiley.

Odum, H. T., R. J. Beyers, and N. E. Armstrong. 1963. Consequences of small storage capacity in nannoplankton pertinent to measurement of primary production in tropical waters. *Journal of Marine Research,* 21(3): 191–198.

Odum, H. T., A. Lugo, and L. Burns. 1970. Metabolism of forest floor microcosms. In H. T. Odum and R. F. Pigeon, eds., *A Tropical Rain Forest.* Oak Ridge, TN: Division of Technical Information and Education, U.S. Atomic Energy Commission.

*Odum, H. T. and E. C. Odum. 2000. *Modeling for All Scales, An Introduction to Simulation.* San Diego, CA: Academic Press.

Ordish, G. 1960. *Living House.* Philadelphia: Lippincott.

Redfield, A. C. 1934. On the proportions of organic derivatives in sea water and their relation to the composition of plankton. In *James Johnstone Memorial Edition,* 176–192. Liverpool: University Press of Liverpool.

Stumm, W. and J. J. Morgan. 1970. *Aquatic Chemistry.* New York: Wiley.

Sverdrup, H. U., W. Johnson, and R. H. Fleming. 1942. *The Oceans.* New York: Prentice Hall.

U.S. Department of Agriculture. 1970. *The World Food Budget.* Foreign Agricultural Economic Report No. 19, 1964. Washington, DC: USDA.

NOTES

1. See Chapter 3. A flow is delivered from a storage in proportion to the driving force and this in turn is proportional to the stored quantity.

After fig. 2.8 was published, many papers followed using the input–output way of representing energy, materials, and money for ecosystems and the economy (Hannon 1973; Herendeen 1981; Bullard and Herendeen 1975). Methods of matrix algebra were used to make group calculations with large data tables.

2. The power flow through an organism is approximated by measuring the oxygen consumed in the cold fire of the body's metabolism. About 4.5 kcal is processed per gram of oxygen consumed or per gram (dry weight) of food metabolized.

CHAPTER 3

ENERGY LAWS AND MAXIMUM POWER

SYSTEMS DEPEND on power, which they use to develop structure and functions that self-organize according to laws of energy transformation and use. As suggested by Alfred Lotka in 1922, maximum power results from self-organization according to the natural selection of systems designs. This chapter explains energy laws, including the maximum power principle and its control of production, growth, competition, succession, energy storage, diversity, and the oscillatory pulsing of all systems.

POWER

In human affairs the word *power* often refers to the effectiveness of action or the capability of action. Great military power implies large military bodies involving many people and machines and exerting a directing force over large areas. Great political power suggests command of large numbers of votes and a wide influence on government systems and on the actions of many people. Great economic power implies control of large amounts of money and of influences that can be bought with volume spending. Almost everyone understands in a qualitative way what power means in human affairs, but few equate general concepts of power with scientific measures.

In science and engineering, *power* is defined precisely in terms of measurable units as the rate of flow of useful energy. Scientific definitions of energy and power are quantitative. Energy can be measured in such units as the calorie, the joule, the British thermal unit (Btu), and the erg. The flow of energy (power) is a rate measured in time units such as calories per day, watts (joules per second), or horsepower (l hp = 10.688 kilocalories (kcal)/min; 1 Btu = 0.252 kcal; 1 kcal = 4,186 joules).

The ability of a machine to accomplish a function is determined by its power rating, with a large number indicating a large role. For example, the functions provided by an air conditioner, a heating system, or a locomotive are described by their power deliveries. Although nearly everyone is familiar with power ratings of household appliances and automobiles, our educational system has rarely emphasized that the affairs of people also have quantitative power ratings and that the important issues of human existence and survival are as fully regulated by the laws of energetics as are the machines. It is possible to put calories-per-day values on human institutions, on the flows of energy in cities, on the power needs and delivery of activities of nations, or on the relative influences exerted by humans and their environmental systems. However, different kinds of energy and power are not comparable in their abilities to do work until they are put on a common basis as empower (chapter 4).

Most people think that society has progressed in the modern industrial era because human knowledge and ingenuity have no limits, a dangerous partial truth. Twentieth-century progress was power enriched. Progress evaporates whenever and wherever power is limited. Knowledge and ingenuity are the means for applying power subsidies when they are available, and the development and retention of knowledge also depend on power (chapter 9).

Heat and the Principle of Energy Degradation

The most important reality about energy flow is the principle of energy degradation, illustrated by the heat sinks in energy flow diagrams. In fig. 3.1a water in the mountains contains potential energy, which drives flow and the work of pumping water into the tank. For any process, some of the available potential energy must be dispersed along the way as unusable heat. In other words, the potential energy of water in the mountains can be connected to some process if the arrangement allows part of this potential energy to go to waste in the form of useless dispersed heat. In textbooks on energetics this principle is called the second law of thermodynamics. A general waste dispersal is shown with the heat sink symbols. In terms of further use, the energy has gone "down the drain" as heat. Potential energy is also called available energy.

Heat is the state of energy consisting of the random motions and vibrations of the molecules. An object is hot when its molecules have increased vibration and motion. These motions tend to spread from a hot body to a cooler one by random wandering and bumping that transfers and disperses the energies of these motions. It is this random dispersal that ultimately is responsible for spontaneous energy transformations. The wandering pulls other processes coupled to it. Energy can flow only if some of it is dispersed by heat flow into the sink. Without such energy dispersal, the downstream energy storages would push back with as much force as the upstream source. If we try to arrange an energy flow to operate so that the heat drain is eliminated, the process simply will not work. The heat drain sometimes is called the energy tax necessary for operation.

The energy that goes to the sink is not actually destroyed as energy but ends up as the speeding, bouncing, shaking velocities of billions of molecules, each going in a different direction. This motion is disorganized, however, and cannot

(a)

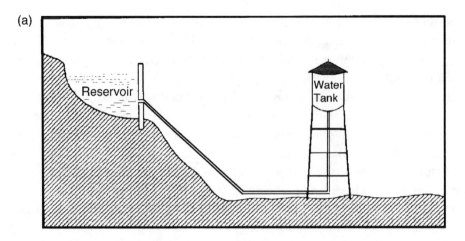

(b) Energy Systems Diagram:

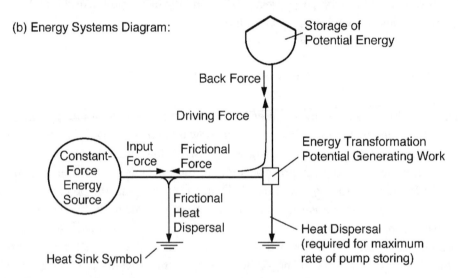

FIGURE 3.1 Example of an energy flow (power circuit) from a source through pathways that have frictional backforces to a point where there is a storage of the energy, doing work against potential generating backforces. (a) Schematic picture of water system from the mountains, with energy being restored ("pump storage"); (b) energy network diagram using symbols defined in fig. 2.10.

be harnessed except when one object has more of this energy than another next to it. If the energy that flows down the heat tax drain (fig. 3.1b) is evenly dispersed in the environment as random heat energy, it can be said to be lost as potential energy, although it still exists as molecular motion.

The first law of energetics states that energy in processes not involving interconversion of energy and matter is neither created nor destroyed. It is conserved. Thus, in fig. 3.1 all the calories flowing in from the potential energy source on the left must be accounted for in the storage and two outflows. However, the second law states that the potential energy, which is the energy available to carry out additional processes, is used up. It is degraded from a form of energy capable of driving phenomena into a form that is not capable of doing so. In recent years the available potential energy has been called *exergy*.

Thus we may restate energy laws: Energy is neither created nor destroyed, but in any real process the availability of potential energy is lost. In our culture these ideas are implied in some favorite expressions of our common sense: "You can't get something for nothing," and "Perpetual motion is impossible."

Absence of Heat, Third Energy Law

If something has no heat, then its atoms and molecules are at rest, with no motion. This condition is called absolute zero, and the Kelvin scale of temperature has absolute zero as its lowest value. (On the Celsius–centigrade scale absolute zero is –273 degrees). No place has been found that is this cold. In the depths of space of the universe the temperature of matter is about 3 degrees Kelvin. In chapter 7 of this book are measures of complexity. One of these, entropy, measures the complexity of molecular states, which depend on the heat content. Where there is no heat, the molecules fall into simple arrangements like crystals. As a third energy law, it is said that the entropy at absolute zero is zero.

The One-Way Direction of Power Delivery

Because of the power drain needed for a process to function, any operation involves a one-way processing of energy with the availability of the potential energy lost as the energy is dispersed into unavailable form. Thus, any procedure is unidirectional, and use and reuse of potential energy are not possible in the processes on Earth.[1] The unidirectional nature of energy processes is reflected in such common expressions as "Time and tide wait for no man" and "Use it or lose it."

What Fraction of the Power Goes Down the Heat Sink?

What fraction of the power flow must go down the compulsory tax drain? What law governs the quantity of this loss? Let us consider the arrangements of weights in the Atwoods machine in fig. 3.2, where weights falling on the left raise them on the right. The falling of a weight is an inflow of power from the potential energy of the weight's high starting position. The energy contributed is a product of the mass of the weight times the velocity of the weight. The raising of a weight is a flow of power into potential energy storage as the weight is lifted. The falling weight does the work of lifting the other weight. The ratio of energy stored to the energy input used is called the efficiency.

When the weights are equal, the system, if it operated, would convert all its potential energy inflow into useful, reusable, stored potential energy. The efficiency of this arrangement is 100%. However, for such an arrangement to result in motion and energy transfer is contrary to the principle of energy degradation (second law).

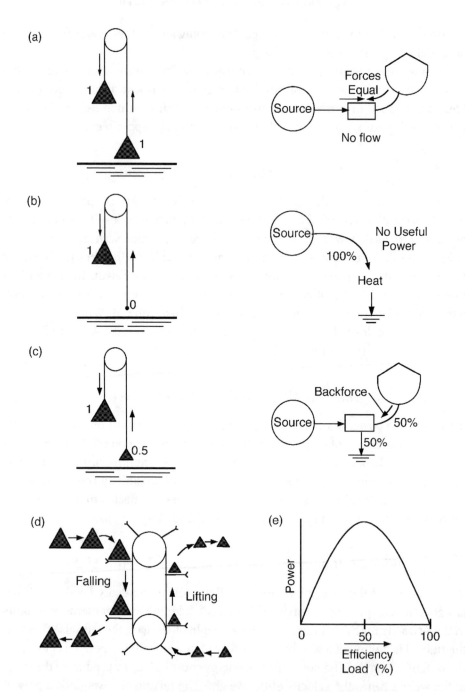

FIGURE 3.2 Atwoods machine: The fall of one weight supplies the force to raise another. Diagrams of energy transformations with 3 different conditions of loading of input and output weights. Energy flow diagrams are shown for each. **(a)** Reversible stall when backforce and input forces are equal; there is no flow, no power delivery. As the loading approaches the stall, efficiency approaches 100%. **(b)** With no load, there is a free drop at maximum speed, discharging heat to the ground. There is no output power for load-lifting work. **(c)** Maximum rate of power storage with intermediate loading. **(d)** Arrangement for a series of weights dropping and lifting in steady state. **(e)** Useful output power as a function of the load and efficiency based on equations in appendix A3 from Odum and Pinkerton (1955).

Common sense also tells us that the two weights, being balanced, remain still, and nothing happens. Situations with balanced forces of opposing energy storages are said to be reversible.

In the middle diagram (fig. 3.2b), the weight on the right is zero. Now the weight on the left can drop without drag from any weight on the right. The potential energy stored in the weight flows rapidly out of its storage state into the kinetic energy of the falling weight. When the weight stops at the bottom, all the concentrated energy of the falling object goes into heating the object and the surface it hits. Very shortly, that heat is dispersed evenly in the environment in the invisible motions of the molecules. Because there was no weight on the right-hand side, all the potential energy went into the heat drain and none into the useful storage process of elevating a weight. The process in fig. 3.2b is thus very fast, dispersing all the potential energy and storing none because it does not lift a weight.

Figure 3.2c is an intermediate arrangement. The weight on the left is about twice that on the right. In this example the weight on the left falls slowly, and part of the energy is stored as available and useful potential energy for further processes. The rest of the energy is dispersed into the heat drain, unavailable for further work. The loading arrangement in fig. 3.2c is the optimum one for processing the greatest amount of useful power in lifting work. If the loading weight is greater than one-half, the process of storage is slower; any loading weight that is less wastes too much in the heat drain.

If one weight falls, the efficiency of energy transfer is about 63%. With the arrangement in fig. 3.2d, a series of weights fall one after another in a kind of steady state, dragging a belt that raises weights one after another on the other side. For a steady state arrangement the optimum efficiency that maximizes the output work is 50%. Figure 3.2d summarizes the relationship of output power and efficiency of transfer.[2]

Maximum Power, Darwin, Lotka Energy Law, and Fourth Energy Law

Whenever it is necessary to transform and restore the greatest amount of energy at the fastest possible rate, 50% of it must go into the drain (Odum and Pinkerton 1955). Nature and society both have energy storages as part of their operations, and when power storage is important, it is maximized by adjusting loads, as demonstrated in fig. 3.2. In the 19th century Darwin popularized the concept of natural selection, and early in the 20th century Lotka (1922) indicated that the maximization of power for useful purposes was the criterion for natural selection.[3] Darwin's evolutionary law that applied to organisms thus developed into a general energy law that applied also to selection of design relationships. In other words, systems that prevail are those with loading adjusted to operate at the peak of the power efficiency curve of fig. 3.2e. During self-organization, these systems reinforce (choose) pathways with the optimum load for maximum output.

Power Transformation for Varying Energy Sources

The explanation of the effect of load on efficiency in fig. 3.2 considered situations in which the potential energy available from the source was constant. However, in many situations the available source of energy varies, but the loading of the energy-using system is set to be useful on the average. In this case the efficiency is inverse to the energy concentration. As the energy availability increases, the efficiency decreases, as shown in appendix fig. A4.

Photosynthetic Production and Varying Sunlight

The sunlight energizing the biosphere and society varies daily, seasonally, and annually. Plants have adapted by developing green chlorophyll structures to receive this energy and transform it into growth and other useful work.

In the first step of photosynthesis, photons separate electrons from organic structure, which gains a plus charge as a result. In other words, the green pigmented structures of plants are natural photovoltaic cells. As shown in fig. 3.3a, the plus charges of most plants interact with water to make oxygen, and the minus charges drive the load of organic production biochemistry.

In laboratory biophysical research, the output of isolated chloroplasts was measured for different reaction loads. The result was the power–efficiency curve in fig. 3.3b, quite like the theoretical curve in fig. 3.2e. In other words, there is a thermodynamic optimum efficiency for maximum production at intermediate light.

If the load is constant but the input energy is varied, fig. 3.3c results. The efficiency decreases with increasing input power. As predicted from the equations (appendix figs. A3, A4), the efficiency is inverse to the light intensity (fig. 3.3c).

Given a longer time, plants adapt to light levels by adding or decreasing the chlorophyll so as to maximize the energy conversion. Ireland and Oregon are dark green in winter because oceanic clouds shade the light, so plants need more chlorophyll to catch enough energy from the scarce light.

Photoelectric Ecosystem

In very shallow (10 cm [2.5 inches] deep or less) fertile waters, temperatures, dissolved oxygen, and acidity (pH) have wide diurnal ranges to which most plants are not adapted. A carpet of matted blue-green algae develops, often with brilliant color in summer sunlight. The ecosystem develops a 0.5-volt difference between the positively charged upper surface and negative charges under the mat. The steep voltage drop pulls negatively charged nutrients released by respiration under the mat up to the photosynthesizing algae. The properties of this photoelectric ecosystem

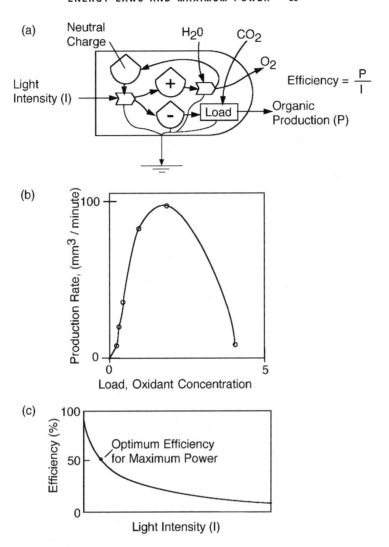

FIGURE 3.3 Efficiency of plant cells using chlorophyll to transform solar energy into organic matter. **(a)** Energy systems diagram showing the first step, a photovoltaic process receiving light photons; **(b)** efficiency of light transformation where light intensity is constant and physiological load is varied (Clendenning and Ehrmantraut 1950); **(c)** photosynthetic efficiency as a function of varying light intensity using equations in appendix fig. A4.

were studied by transferring cores of the algae mats to tubes (fig. 3.4a), which delivered electric power in response to changing load, not unlike a silicon photovoltaic cell (appendix fig. A6). The parabolic curve of power output is another example of the inverse power efficiency principle that guides the adaptation to solar energy.

Tropical Forest

The principle that efficiency of photosynthetic production decreases with light intensity applies to large ecosystems as well as to isolated chloroplasts and algal mats.

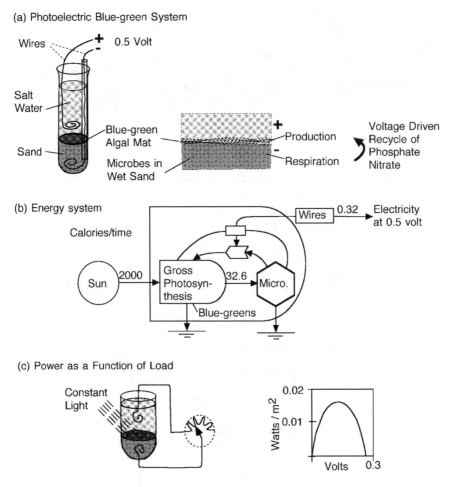

FIGURE 3.4 Blue-green mat ecosystem and its photoelectric characteristics. **(a)** Blue-green algal mat growing in a test tube made into a photoelectric cell by connecting platinum wires to oxidized waters above and reduced organic sediments below; **(b)** energy systems diagram; **(c)** power delivered as a function of load while light is constant. See maximum power discussion in fig. 3.2 and appendix fig. A3. Micro., microbes.

Recent studies measure the photosynthetic production of landscapes from the rate of turbulent exchange of carbon dioxide eddying down from the atmosphere into the ecosystems. Figure 3.5 is a graph of photosynthetic efficiency as a function of light intensity in an area of tropical forest measured with this free air method (Ryan 1990).

KINDS OF WORK: STORING, PROCESSING, AND ACCELERATING

When power from a potential energy source is flowing through and driving a useful process, we are accustomed to describing the process and its output as work. Thus

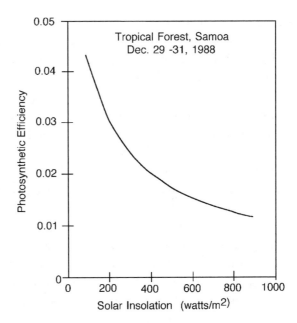

FIGURE 3.5 Efficiency of forest photosynthesis as a function of light intensity in a tropical forest in Samoa (Ryan 1990).

mechanical work is done when weights are lifted or objects are arranged.[4] However, not all work is mechanical. Work is the output of any useful energy transformation.

Let's distinguish between three kinds of work. The first, *storing work,* has already been illustrated in fig. 3.2. Maximum power flow in storing energy at maximum speed requires a 50% drain. Such power involves the force of one potential energy source directed against the backforce from another potential energy storage. The energy of the process can be measured either by the energy stored or by the total power flow, which is twice the storage rate when the process operates in optimal fashion. Most energy transformations also involve a spatial concentration of energy, which uses up some of the available energy, causing efficiencies to be less than what is required just for storage (chapter 4).

In *processing work,* power flows from a potential energy source and passes through the system, arranging matter but affecting no storage and no final acceleration. All the energy eventually disperses into the heat drain, but first useful work is accomplished. For example, if we arrange books in a library or trucks in a parking lot, power is delivered into temporary motions, which are then stopped, with energy being transformed by the various frictions of the disordering and stopping motions. Arrangement of structure is a principal task in preserving or establishing any kind of order, whether it involves the maintenance of a living being or the operation of an industry. Thus work done against frictional forces to accomplish necessary and useful ends without storage disperses all the available energy. The energy of the process is not measured by the energy stored, for none is stored, but it may be measured by the power delivered to accomplish the work. Processes of this kind represent much of the work of human beings, such as the typing and filing of a secretary or the handiwork of a carpenter. In industrial planning it is customary to measure work of this type by the time required to do the job when efficient procedures are used. To convert data on work time into power figures, we may multiply time by the energy flow through the human being doing the work for a full day. This flow includes the actual work plus the necessary maintenance of the human being's biological systems that allows the work to be done.[5]

A third kind of work is *acceleration of matter*. Potential energy from a source increases the speed of objects. Thus we may throw a ball and pass energy from body storage into the kinetic energy of the ball. If the ball is thrown in a place where there is little or no friction, as in outer space, 100% of the energy may go into kinetic form. At first glance, this conversion may seem an exception to the principle of energy degradation because no heat is dispersed into a drain. But this class of energy flow is only a relative energy transfer and not a typical energy transformation. In 1902 Einstein indicated that a process of acceleration, such as throwing a ball on one planet, would look like deceleration when viewed from a second planet traveling by in a direction opposite that in which the ball was thrown. Any process whose direction depends on the place from which we are looking, when viewed from the universe, has no unique direction.

Acceleration is work against an inertial force, but inertial force is defined, in turn, as a backforce against acceleration. This circular reasoning started by Newton made it possible to state that a force always has an opposing force. Work done in acceleration requires no heat sink dispersal, and the energy stored in kinetic form is still available to drive other kinds of work. Therefore it does not fall within the second principle of energetics, which applies to processes with real power flows that exist regardless of the position from which we observe them. A pendulum operating in a vacuum is an example of energy changing its relative form from potential to kinetic and back without involving the second principle of energetics.[6]

Thus energy flows involving accelerations and work against inertial forces are not energetic transformations in the same sense as those involving loss of potential energy. When we accelerate matter, it is not like the true power delivery for useful processes that entail heat drain. When matter is accelerated, potential energy is changed from a resting form to one with a velocity relative to the observer. Thus kinetic energy and the energy in the magnetic fields that form around flowing electric charges really are interchangeable forms of potential energy. However, if we stop the motion with frictional force, energetic degradation results. Thus the many accelerations and frictional decelerations involved in most operations constitute processing work. In a real way, frictional forces are necessary to the processes of work that support society. If there were no friction, there would be endless pendulum motions.

Energetic Determinism and Causal Force

In the affairs of forests, seas, cities, and human beings, potential energy sources flow through each process, doing and driving useful work of one of the types mentioned. The availability of power sources determines the amount of work activity that can exist, and control of these power flows determines the power in human affairs and in their relative influence on nature. Every process and activity on Earth is an energy manifestation measurable in energy units. We may say that phenomena are energetically determined. Many people prefer to call the role of energy a constraint because they believe factors other than energy are causal.

The common use of the cause concept[7] is another manifestation of energetic determinism. As we have already stated, power delivery from a potential energy source involves the expression of a force X against some opposing force. Power (P) is the product of the force X and flow J; see equation 3.1. The amount of force can be scientifically defined by dividing the power JX delivered by the velocity of motion J.

Important processes are unidirectional because of the energetic degradation that makes them irreversible. We may designate the input force X as the causal force because it goes in the same direction as the flows.[8] Flows J induce friction, but if there were no frictional heat dispersal the flow would not go spontaneously.

$$\text{Power} = \text{Flow} \times \text{Force},$$

$$P = J \times X. \quad (3.1)$$

Frictional forces develop to balance input force and exist only against a flow. Because force X is proportional to the power delivery JX, most everyday ideas of causal force are also statements of the ability of the energy source to deliver power and are adequately measured by energetic data. Units either of force X or of power JX can be used to measure causal action.

When we apply the concept of force to social groups, movements, and influences, we must consider program sequences and groups of associated physical forces involving power flows through a population. Whether we examine a simple process such as the fall of an apple or a complex cluster of millions of related processes in a military campaign, we can measure the causal actions as power delivered from a power source through the action of forces. For the complex world of environmental systems and the economy, it is usually easier to measure the energies in fuels involved than to determine all the component and transient physical forces exerted.

The existing power budget in an unchanging system does not necessarily measure the power available to focus on a new causal action, although powerful systems often are capable of flexibility in the use of their flows or in tapping new sources as needed. Power to do new things depends on new power development or a shift in the use of power.

All the flows of available energy in the environment are in use by our life support system already. When we set up a water wheel, a windmill, or a solar collector to use energy for a new purpose, we divert the energy from its existing use.

We should not do this until we are sure that the new use contributes more to the economy than the old indirect use.

ENERGY ON SYSTEMS DIAGRAMS

The energy network language given in fig. 2.10 represents the networks of energy flows in a useful way and includes the requirements and limitations of energetic laws. Here a pathway of potential energy flow is represented by a line, and if it is inherently unidirectional, incapable of reversal, it is marked with an arrow symbol (fig. 2.10n). The potential energy storages are marked with the tank symbol (fig. 2.10b), which indicates a source of causal force along the pathway lines. The force is in proportion to the storage whenever the storage function is one of stacking up units of similar calorie content. Most ecological and civilization systems have such storage.

We uphold the second energetic principle by drawing the network so that any process has some potential energy diverted into the dispersed random motion of molecules (heat). The downward flow into the heat sink is symbolized by an arrow directed into the ground (fig. 2.10c). The first energy law is upheld by having all inflows balance outflows into storages, into the heat sink, or into exports. The many kinds of work done against frictional forces are illustrated by arrows that flow from the potential energy storage tanks into the heat sink. Whenever a work process is necessary for a second flow, facilitating the other flow by the work done on it, the arrow interacts with the second flow (fig. 2.10e; appendix fig. A5). For example, the work of people on a farm facilitates the flow of sunlight energy into food storage, although the energy of the workers' food, which sustains them during the period of work, is converted into heat by their metabolism (fig. 2.10i).

Whenever an energy flow is transformed and restored into a tank (fig. 3.1) at the optimum rate for energy storage, part of it must go into the heat drain. If pathways of a system must pass over hills or valleys, over which the flow must be pumped, temporary additions of energy are needed. These pathways are called energy barriers. An energy carrier that crosses the barrier must increase in potential energy temporarily and hence must lose available energy to the heat sink. An ordinary arrow symbol barb is placed on a line if the pathway receives no backforces from downstream storages. In this case the only backforces are frictional.

Flow in Proportion to the Populations of Forces

Each line of the energy network diagram represents a population of forces acting against frictional forces to cause a flow of energy either alone (as in heat flows, wave energy flows, or light) or associated with a flow of materials (such as minerals or organic matter). For the energy diagrams of population phenomena given in this book, when driving forces balance friction, flow J is proportional to population of active forces N:

$$J \propto N, \quad (3.2)$$

where the forces usually are from populations of events working together or in succession. Equation 3.2 states that the rate of flow of such stored energy packages as minerals, money, or work is in proportion to the number of forces acting and to their individual magnitude. For example, the total body metabolism, money expenditures, or nitrogen excretion of a population of people is in proportion to the number of people involved.

For a circuit of particular conducting tendency, the proportionality in equation 3.2 can be measured by L, a constant characterizing the friction in the pathway, sometimes called conductivity:

$$J = LN. \quad (3.3)$$

The reciprocal of conductivity L is the resistance R:

$$R = 1/L. \quad (3.4)$$

An example from electrical systems is Ohm's law, which states that the flow of electrical current is proportional to the driving voltage (electrical force) according to the resistance R of the wire. Another example from groundwater geology states that the flow of water through the ground is proportional to the water level head according to the conductivity of the aquifer, which depends on its flow through porous earth (Darcy's law). Chemical reactions depend on the population of molecules acting along pathways and on resisting energy barriers and temperature. The energy systems language has helped apply concepts of flows and forces to ecological and social systems also.

Constant Current and Constant Force Sources

Energy flows may be further classified according to the kinds of driving energy sources and the properties that result as we use them for work. *Constant flow sources* keep the current constant. For example, an old-fashioned waterwheel dipping into a steady river flow may take varying degrees of energy from the river, but there is no change in source flow. Another example is the steady fall of leaves in some tropical forests, providing energy to the soil organisms. If earthworms use all of the organic matter, it does not increase the litterfall. Sunlight is another example. There is nothing the plants can do to bring more sunlight to their leaves.

Constant force sources deliver a constant force to the user, even though the energy consumer uses more and more. Because no energy source has infinite resource, it cannot provide a constant force beyond its limit. For example, if a dam has a very large reservoir holding the water pressure constant, we can open various turbine tunnels to obtain the same pressure in each tunnel. As long as we do not draw enough water to change the water level appreciably, the force exerted will be constant. The energy source in this instance is a constant force source. Small drains on large batteries are pushed with constant force. A bird colony would have a constant force applied to bird reproduction if the zookeeper kept the food concentration constant regardless of the number of birds eating.

Autocatalytic Growth

When resources are not limiting, systems grow. This happens when the concentration of available energy is fairly constant. The resource exerts a nearly constant input force, at least for a while. Consistent with the maximum power law, the products of growth are used to accelerate the capture of more energy so that growth goes faster and faster (fig. 3.6). The systems diagram for this situation (fig. 3.6c) shows the products of production (in storage tank symbol) being fed back (to the left) to amplify capture of more energy. This kind of system was called *autocatalytic* in chemistry, but the name is now in general use for any system of this design. The mathematics of this kind of growth generates exponential equations and growth graphs, and the name *exponential* often is used for such growth acceleration. Such circular stimulation is called positive feedback and mathematically produces exponential-shaped growth curves (fig. 3.6d), as Malthus showed long ago.

When available energy levels are large enough, the system develops a self-interaction to accelerate even faster, a superacceleration (fig. 3.7). The feedback of products amplifies in proportion to the self-interactions, which mathematically is the square of the storage. Heinz Von Foerster et al. (1960) used superacceleration to model the growth of the U.S. economy. The economy was in superacceleration until 1973, after which availability of energy was not as constant. The American pattern of rapid weedy growth involved high levels of cooperation in the economic and cultural frenzy to exploit. For example, the mining law of 1872 gave priority rights for mining resources over surface use of land. Later, there was an oil depletion allowance that encouraged oil development by tax exemption.

Competition for Energy Sources

When energy supplies support accelerating growth, users with faster growth rates outgrow and displace others. It is inherent in the mathematics of the exponential growth process that the differences in growth rates increase. The competitor (species or business) with the better reproductive performance is self-amplifying, and the discrepancy between the two stocks increases. Competitive overgrowth often is shown in laboratory experiments if other controls are removed (Gause 1934;

(a) Sheep Supported on Grass

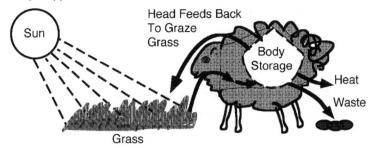

(b) Town Using Hydroelectric Power

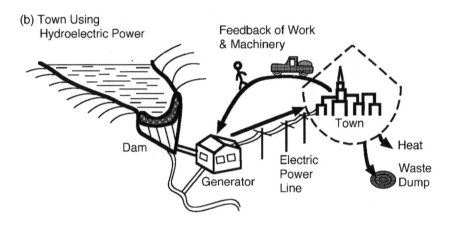

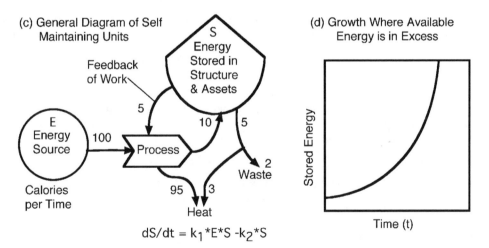

(c) General Diagram of Self Maintaining Units

Feedback of Work

S Energy Stored in Structure & Assets

E Energy Source

100

Process

Calories per Time

5

10

5

2 Waste

95

3

Heat

$$dS/dt = k_1 *E*S - k_2 *S$$

(d) Growth Where Available Energy is in Excess

Stored Energy

Time (t)

FIGURE 3.6 Autocatalytic growth where available resources are in excess. **(a)** Sheep living off grass; **(b)** village living off hydroelectric power; **(c)** autocatalytic feedback of storage to amplify inflow; **(d)** graph of accelerating exponential growth. The rate equation implied by the diagram in **(c)** is $dS/dt = (k_1 * E * S) - (k_2 * S)$.

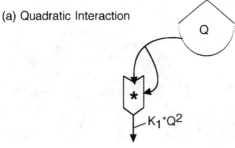

(a) Quadratic Interaction

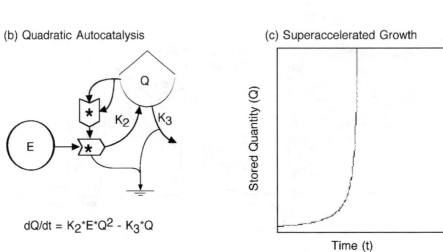

(b) Quadratic Autocatalysis

$dQ/dt = K_2{}^*E{}^*Q^2 - K_3{}^*Q$

(c) Superaccelerated Growth

Stored Quantity (Q)

Time (t)

FIGURE 3.7 Superaccelerated growth from quadratic interactions. **(a)** Quadratic output of population interaction; **(b)** growth with quadratic autocatalytic feedback; (c) graph of superaccelerated growth.

DeWit 1960). For example, one of the two species of water fleas growing in fig. 3.8 overgrew the other during the period when energy supplies were in excess.

AVAILABLE ENERGY IN WATER
AND OTHER CHEMICAL SUBSTANCES

We are not used to thinking of water, fertilizer, and air as available energy, but they are forms of potential energy. Any inflowing reactant to a process makes an energy contribution to that reaction. When calculations are made,[9] the potential energy value of the water to cause cleaning processes (its chemical potential energy) is ten times the value of the water used for hydroelectric purposes (in areas of ordinary topography). Water can be more valuable as a chemical energy than as a hydroelectric energy source. The available energy in freshwater reaching the sea drives oceanic currents along coastlines.

(a) Simple Competition

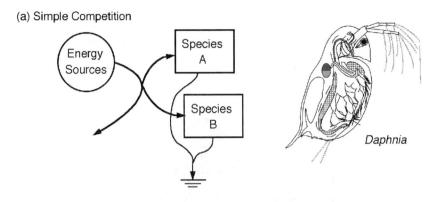

(b) Competitive Exclusion

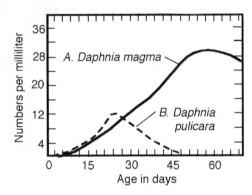

FIGURE 3.8 Competitive exclusion where 2 species of *Daphnia* compete for available resources (Frank 1957).

FACTOR CONTROL AND THE PRODUCTION PROCESS

The production processes that support growth usually need more than one kind of input. Each input involves a flow driven by a force. For example, fast-growing cattails in a wetland in fig. 3.9 catch the sunlight with their leaves and inorganic mineral materials such as phosphorus with their roots. The input of sunlight is pure energy. An input of phosphorus is a chemical substance, but it has the energy content of its chemical concentration.[10] Both are required to make this weed's organic matter, much of which is buried as peat. The systems diagram (fig. 3.9a) is drawn to show only the factors that may become limiting and thus control growth. Photosynthesis requires other inputs (fig. 1.3g), but they are not limiting to the cattails in this situation. The wind, which supplies carbon dioxide to the leaves, and the water needed for photosynthesis are in excess.

A work process that generates a second flow in proportion to its necessary activity is mathematically a product function and therefore is indicated by a pointed box symbol containing an × or * to indicate multiplication (fig. 2.10e). Such a box

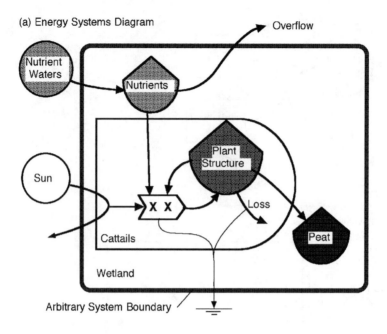

(a) Energy Systems Diagram

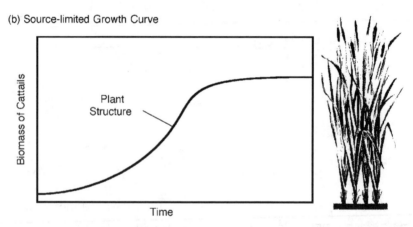

(b) Source-limited Growth Curve

FIGURE 3.9 Fast-growing, weedy cattails in a wetland. **(a)** Energy systems diagram showing the net production and deposition driven by sunlight and nutrients; **(b)** the S-shaped growth curve that results with an externally limited flow source.

indicates a controlling role of the flow in short supply. Such pathways are called limiting factors, and the work can be thought of as a control valve on the other flow (fig. 3.10, appendix fig. A5). Some of these interactions derive their driving potential energies from their own storages and thus are autocatalytic, multiplicative positive feedbacks (fig. 3.9a, appendix fig. A7).

The interaction symbol is used most often for inputs that have multiplicative actions when they join. If a water valve is turned on, the water flows according to the turning action's effect on the water pressure. The flow is a product of the two input factors. We often think of the one with less energy but higher quality as the control.

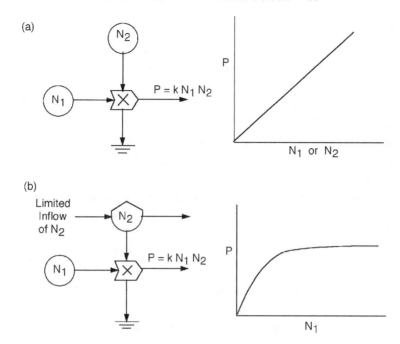

FIGURE 3.10 Production (P) by an interaction process that is in proportion to the product of 2 forces, N_1 and N_2. **(a)** Production where conditions are maintaining local forces or concentrations constant; increasing either increases production and input flows. **(b)** Increase of force N_1 increases production but with diminishing response because the inflow of N_2 is limiting. See equations in appendix fig. A5.

Work is done in the interaction, and the process is controlled in the same way that a gate controls water in irrigation ditches. We used the term *work gate* in *Environment, Power, and Society*. The interaction symbol usually has a multiplicative interaction of two different forces to generate a product. In many fields this is called a production function. Used potential energy is dispersed into the heat sink.

In the chemical industry the reaction between two inflowing chemicals is proportional to the probability of the molecules encountering each other, and this probability depends on their concentrations. The output of the reaction is a function of the product of the concentrations of the two chemicals at the reaction site. This is an example of the interaction in a production process. As shown in fig. 3.10a, varying either of the two reactants increases the output proportionately.

However, if one of the inflowing reactants is limited at its source, either from outside the system or from some restrictions in recycling within the system, then increases in the other concentration cause the output to increase, but at a diminishing rate, until the output is dependent solely on the limitation of the reactant in short supply. The characteristic response of such curves, called a rectangular hyperbola, is shown in fig. 3.10b (appendix fig. A5). The limiting factor concept, by which the amount of some small limiting nutrient such as phosphorus determines the flow of some process such as agricultural production, has become well established since the time of Liebig. Responses of plant production to nutrients follow the limiting factor hyperbola in fig. 3.10b (appendix fig. A5). Thus the symbol implies a limiting factor control action when the flows are limited.

Many energy sources important to the processes of civilization are also limiting factors, although few people dealing with public resources have given them their appropriate calorie values. Large quantities of energy go into processing water for economic use, but many substances such as water and minerals have been regarded as free goods, partly because their real wealth values were not realized, and no dollar values were assigned until human work was involved. In chapter 4 we evaluate all resources on a single basis according to nature's work in generating them. Then the real wealth measures are compared with market values in chapter 9.

Energy in Production Processes

All inputs and outputs of an interaction or production process contain available energy, but each is a different kind of energy. In fig. 3.9 solar energy is large in quantity but low in quality, whereas the chemical potential energy in the nutrient input is small in quantity but of high quality. The output of organic matter is intermediate in quantity and quality. For example, Alkire (1965) evaluated the energy flow of fish used by Pacific atoll inhabitants. For about 125 kcal that humans spent in fishing, the production yielded about 720 kcal of organic matter, and the fish required even larger energy inputs of food. As we explain in the next chapter, calories of different kinds should not be considered equal.

Limiting Nutrients and Linear Programming

Many production processes need inputs of several kinds of necessary materials, any one of which can limit output. Biological production requires several kinds of nutrient materials: nitrogen, phosphorus, potassium, and others in appropriate ratios (table 2.1).

Where there are two or more kinds of producers and two or more kinds of materials used, there is an optimum ratio of producers that uses and recycles materials for maximum output. The system may be analyzed by a mathematical approach called linear programming in order to determine the optimum ratios of the units. For example, in fig. 3.11a there are two parallel agricultural producers consuming phosphorus and nitrogen fertilizer. Each crop has its characteristic ratio of use of two nutrients, and the system of fertilizer supply is providing them in a nitrogen/phosphorus ratio of 7:1. The model also applies to two species of phytoplankton algae in the sea, each drawing from the nutrient supply upwelling from deepwater reserves in a ratio of nitrogen to phosphorus of 7:1.

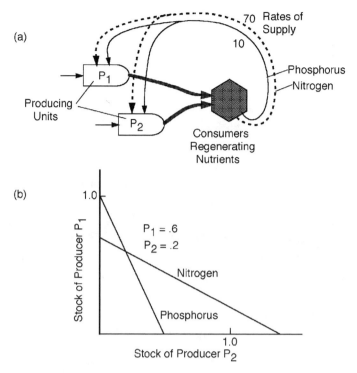

FIGURE 3.11 Linear programming of a system of 2 producing plant units, P_1 and P_2, each with characteristic rates of phosphorus and nitrogen use. Rate of supply of the fertilizer units is 7:1. Equations for use of 10 units of phosphorus and 70 units of nitrogen inflowing per day are as follows: phosphorus, $10\ P_1 + 20\ P_2 = 10$; nitrogen, $100\ P_1 + 50\ P_2 = 70$. If selection works to adjust the ratios of P_1 and P_2 so that total phosphorus and nitrogen use leaves no accumulations or shortages, then an optimum is located by the intersection of the lines. Reasonable figures for plankton are in milligrams per liter.

The effects of each unit on each process are expressed by terms in the two equations (see figure legend), one for phosphorus and one for nitrogen. Each line is plotted on a graph that relates the two producers, P_1 and P_2. Where the two lines intersect in fig. 3.11b, use of the incoming fertilizer nutrients is complete, without wastes or shortages. During self-organization, selection by the closed loops of recycling nutrients adjusts the ratio of the weights of P_1 and P_2 so that there is no leftover resource (where the lines in fig. 3.11b intersect).

GROWTH ON LIMITED SOURCES

When its main energy source is flow limited, growth of a dependent unit soon levels off because it cannot increase its energy use beyond what is inflowing. The weeds growing in a wetland can accelerate at first when there is plenty of light and soil nutrients, but growth levels off (fig. 3.9b) after the ground is completely covered with green foliage absorbing as much light as possible and drawing into its roots as much of the nutrients as possible. Thereafter, its production process (interaction) is source limited, as shown in the energy systems diagram (fig. 3.9a).

CONCEPT OF SUCCESSION

A concept of ecological succession was developed by R. Clements at the end of the 19th century from studies of the sequence of plants, animals, and physical changes that occurred when lands were abandoned to natural processes and seeded with organisms from surrounding forests. Often there was rapidly accelerating growth by a few weedy, rapid colonizing species, soon displaced by a higher diversity of longer-lasting plants and animals. Figure 3.12 contrasts the simple, early stages of growth with the later, more complex ecosystem that puts its energy into sustaining larger structures and relationships instead of growth. The mature stage was called a climax. The ecological model of succession and climax has now been applied to national policy under the name *sustainability*. But seeking a constant level of civilization is a false ideal contrary to energy laws.

THE PULSING PARADIGM

A century of studies in ecology, and in many other fields from molecules to stars, shows that systems don't level off for long. They pulse. Apparently the pattern that maximizes power on each scale in the long run is a pulsed consumption of mature structures that resets succession to repeat again. There are many mechanisms, such as epidemic insects eating a forest, regular fires in grasslands, locusts in the desert, volcanic eruptions in geologic succession, oscillating chemical reactions, and exploding stars in the cosmos. Systems that develop pulsing mechanisms prevail. Figure 3.12 includes the downturn for reset that follows ecological climax. In the long run there is no steady state.

Figure 3.13 shows how growth and pulsing over a longer period can generate repeating patterns. By averaging over many cycles you can calculate average storages and rates of energy processing. The model generates pulsing with two connected units. One, a producer, grows and accumulates storages; the second, a consumer, goes into a frenzy of consumption and generates a pulse of activity based on its use of the accumulations.[11] Our universe is full of such pulsing pairs on every scale of size and time. Rises and declines in the Mayan civilization of Mexico have been attributed to the gradual net accumulation of environmental resources, soils, forests, and fisheries, followed by a frenzied period of net consumption. The pyramids and temples are symbols of unsustainable achievement (fig. 3.13c).

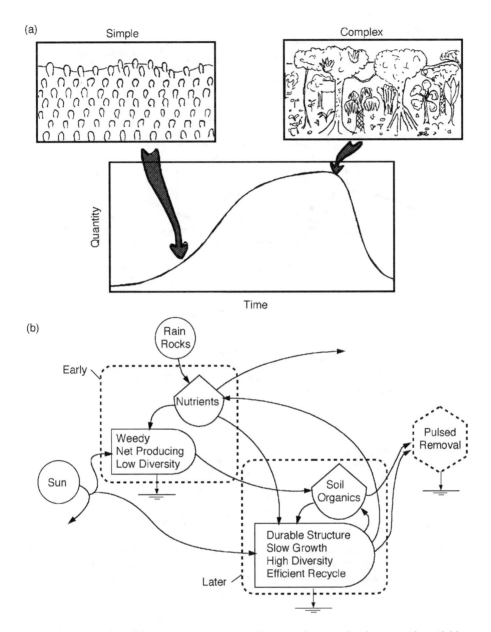

FIGURE 3.12 Growth and diversity in succession, climax, and restart that begins with available energy and nutrients to be colonized. Power is maximized at first by low-diversity overgrowth and net production and later by durable structures, high-diversity division of labor, and efficient recycle. **(a)** Growth curve and diversity sketches; **(b)** simplified energy systems diagram with dashed frames to indicate the parts of the system that are important in the 3 regimes.

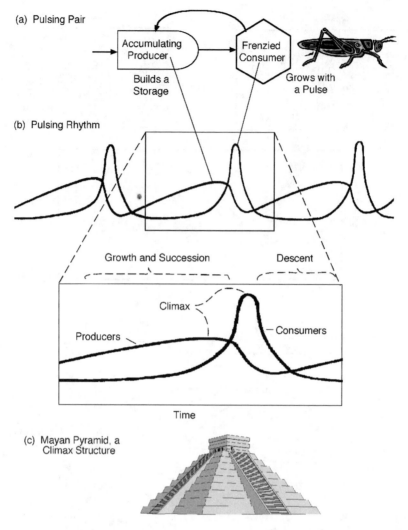

FIGURE 3.13 Pattern of repeating pulses produced by accumulating production followed by frenzied consumption. **(a)** Energy systems overview; **(b)** result of simulating a simplified model to explain the pulsing paradigm. Appendix fig. A10 has details and equations. **(c)** Mayan pyramid, a climax structure of human civilization in the past.

Perhaps a consensus has emerged that ecological succession is best regarded as a time span within the pulsing cycle (Holling 1986; Odum 1983; Odum et al. 1995). For more, see chapter 6 and fig. 6.14.

OVERGROWTH AND DIVERSITY

The maximum power principle explains why early succession minimizes diversity and later succession maximizes diversity. The first priority of a system is to maximize energy intake, to cover the area with energy receivers quickly, with units adapted for most rapid growth (Yount 1956; Kent et al. 2000). The system reinforces the uncontrolled expression of the ecological principle of competitive exclusion (Gause 1934), which states that one of two populations of self-duplicating organisms, when given excess resources to expand with Malthusian growth, will overgrow and displace the other. Figure 3.8 diagrams this kind of competition with an example.

The second priority is to maximize efficiency in its energy processing. When there are no excess, unused resources to be found, a high diversity of cooperating units develops, with better efficiency and division of labor. Succession such as that in a rainforest develops a higher diversity of species and repair mechanisms that give the system resilience to disturbances such as wind damage and atomic radiation (Odum and Pigeon 1970). Chapter 7 explains the relationship of energy to diversity and other kinds of complexity.

When resources are in excess, maximum power is achieved by the uncontrolled overgrowth of a few species specialized for quick capture of energy and materials. They do it by throwing up flimsy structures quickly. In human society we have the example of the American colonization of North America with weedy structures, fanatic elimination of diversity, capitalistic elimination of competitors, and laws written to facilitate growth. Religious reasons were used for displacing native people, a process that maximized the growth and power of the colonists.

It is a well-known property of growth acceleration that the competitor that starts first wins out. Thus in capitalism enterprises that begin by borrowing money to get a quick start win out as long as resources are not limiting. Later, after all sources are in use, they are replaced with more diversity, more controls, and longer-lasting structures. But many, if not most, people believe humans are somehow above the limits of energy resources. Ignorance about energy develops during times of accelerating growth.

UNDESIRABLE WORK, SHORT CIRCUITS

As power levels increase, problems with excess power develop. Potential energy either is stored or drives a work process. Energy does not flow quietly if available energy is above the threshold for autocatalytic development. When storages are large, there is an immense concentration of ability to drive processes that will dissipate the energy accumulation. If no special protection for the stored energy has been arranged, opportunistic circuits may be connected and the energies discharged in disorder. Storage packages in nature need special protections, and most animals and plants divert energies into such protection, into special skin, bark, or individuals specialized to protect the group. In human systems the police perform one such function by protecting storages. The costs are the necessary part of having such a network.

A transformation of potential energy stalls if it is overloaded by efforts to accomplish too much potential generating work, but if underloaded it tends to develop additional work. If we arrange for a flow of concentrated energy to disperse the energy directly into the environment, the heat releases will become so concentrated that localized heat gradients will develop to do work while the energy is dissipating. The electric short circuit, the forest fire, and the waterfall are examples of potential energies released into heat with resulting eddies, structural damage, or work done on the environment.

A related example is found in human populations, when support is provided without work being required in return. If an idle population does not have to engage in a regular work process, it tends to set up various unorganized activities. If no organized outlet for human contributions is arranged, unrest, mob actions, irresponsible reproduction, and social eddies can result, at least until a new closed-loop system evolves by selection. The giveaway of support in human affairs can be an energy short circuit.

CANCER AT MANY LEVELS

Whereas the overgrowing acceleration of early succession has its place, there is always a possibility for runaway competitive exclusion of one part of a network using another part as excess resource. Pathological overgrowth is a fearsome, ever-present danger against which all surviving systems must be protected by organizing influences.

When a well-organized system is disrupted and its controls are destroyed, the parts may go into Malthusian competitive exclusion and in the process destroy the remnants of the system. When this happens with cells in humans, we call it cancer. In biological systems there are many kinds of cancer.

The exact causes of cancer have been difficult to identify primarily because so many influences are capable of disordering the normal control systems. Radiation, chemicals, and senescence can interfere with the controls and release the dreadful competitive exclusion growth. These disturbing agents may be additive. Once started, the cancer gains control of the system's energy sources.

Our environmental crisis, with its many stresses and disorders, is producing cancer at many levels. Its chemicals are causing cell cancers. The disordering of ecosystems is turning well-controlled networks into widely exploding, erratic growths of species that seem as cancerous to the ecosystem as are cancerous cells in the body. Cancerous capitalism is explained in chapter 9.

The biggest cancer of them all is the human population itself. Removed from its normal controls by modern medicine, global population has accelerated past 6 billion. Using fossil fuel, some human populations have gone into a mode of competitive exclusion, consuming resources, setting up subcompetitions in their cultures and races, and generally draining the operating capital of the world toward its collapse. How will our system restore the controls and cut off the fuel supplies to the runaway components?

Maximum Power and the Explosion of Humanity

Two centuries ago human societies were still agrarian, based mostly on the environmental energy flows from the solar-based climate, oceanic resources, and earth processes. Cultures and civilizations grew and dispersed, first in one place and then in another, as energy resources were accumulated and consumed in frenzied pulses of local, momentary growth. Like ecosystems, human societies alternated between the stages of simple overgrowth and climax periods of complexity and diversity. These patterns of society apparently maximized power in both the short run and the long run, like those of other self-organizing systems.

Then ways of using fossil fuels were discovered, and a huge quantity of highly concentrated potential energy became available. Society was newly driven with an energy source of the constant force type. For the last two centuries available energy has not been source limited. Growth has been possible on a much larger scale of space and time. As can be predicted from the maximum power principle, humanity and its economy surged into global growth that is still under way. It has been a time of early successional overgrowth, competition, wars of conquest, capitalism, and uniformity. But to understand better, let's postpone consideration of the power basis for society until we introduce additional energy principles in chapters 4, 5, and 6.

Summary

The universal dependence of all systems, including human society and its economy, on available energy and energy laws was introduced with the help of energy systems diagramming. Production, growth, and consumption require that available energy (exergy, potential energy) be used up in work processes and the energy dispersed in degraded form. The natural processes of design selection during self-organization cause the systems to develop that maximize power intake and useful consumption. In each energy transformation an optimal intermediate efficiency is selected that maximizes power.

Production processes depend on the necessary inflowing ingredients, which can be limiting. When potential energy and material resources are available in excess, overgrowth of simple units prevails with autocatalytic growth and competitive exclusion. Later in succession, when available inflows are fully in use, the designs that maximize power have high-diversity units with more structure and storages, symbiotic interactions, efficiency, and recycle of materials to eliminate material limitation. But each period of accumulation is ended with pulsed consumption and restart, a pattern that apparently maximizes power in the long run. The surge of growth and reorganization of humanity in the last two centuries on rich fossil fuels has also produced pathological short circuits, cancerous drains, and disruption of the global life support system.[12]

BIBLIOGRAPHY

Alexander, J. 1978. *Energy basis of disasters and cycles of order and disorder.* Ph.D. dissertation, University of Florida, Gainesville.

Alkire, E. W. H. 1965. Lamotrek Atoll and the inter-island socioeconomic ties. In *Illinois Studies in Anthropology.* Urbana: University of Illinois Press.

Clendenning, K. A. and H. C. Ehrmantraut. 1950. Photosynthesis and hill reactions by whole chlorella cells in continuous and flashing light. *Archives of Biochemistry,* 29(2): 387–403.

Curzon, F. I. and B. Ahlborn. 1975. Efficiency of a Carnot engine at maximum power output. *American Journal of Physics,* 43: 22–24.

DeWit, C. T. 1960. On competition. *Verslagen van Landboukundig Onderzoekingen Netherlands,* 66: 1–82.

Fairen, V. and J. Ross. 1981. On the efficiency of thermal engines with power output. *Journal of Chemical Physics,* 75(11): 5490–5496.

Frank, P. W. 1957. Coactions in laboratory populations of two species of *Daphnia. Ecology,* 38: 510–519.

Gause, G. F. 1934. *The Struggle for Existence.* New York: Hafner.

Holling, C. S. 1986. Resilience of ecosystems: Local surprise and global change. In W. C. Clark and R. E. Munn, eds., *Sustainable Development of the Biosphere,* 292–317. Cambridge, UK: Cambridge University Press.

Jammer, M. 1957. *Concepts of Force.* New York: Harper & Brothers.

Kent, R., H. T. Odum, and F. N. Scatena. 2000. Eutrophic overgrowth in the self-organization of tropical wetlands illustrated with a study of swine wastes in rainforest plots. *Ecological Engineering,* 16(2000): 255–269.

Lotka, A. J. 1922. Contribution to the energetics of evolution. *Proceedings of the National Academy of Sciences,* 8: 147–151; Natural selection as a physical principle. *Proceedings of the National Academy of Sciences,* 8: 151–155.

Odum, H. T. 1983. *Systems Ecology.* New York: Wiley.

Odum, H. T. and R. F. Pigeon, eds. 1970. *A Tropical Rain Forest.* Oak Ridge, TN: AEC Division Technical Information.

Odum, H. T. and R. C. Pinkerton. 1955. Time's speed regulator: The optimum efficiency for maximum power output in physical and biological systems. *American Scientist,* 43: 331–343.

Odum, W. E., E. P. Odum, and H. T. Odum. 1995. Nature's pulsing paradigm. *Estuaries,* 18(4): 547–555.

Patten, B. 1985. Energy cycling in the ecosystem. *Ecological Modeling,* 28: 1–71.

Rabinowitch, E. T. 1951. *Photosynthesis,* Vol. 2. New York: Interscience.

Richardson, J. R. 1988. *Spatial patterns and maximum power in ecosystems.* Ph.D. dissertation, University of Florida, Gainesville.

Ryan, S. 1990. Diurnal CO_2 exchange and photosynthesis of the Samoa tropical forest. *Global Biogeochemical Cycles,* 4(1): 69–84.

Von Foerster, H., P. M. Mora, and L. W. Amiot. 1960. Doomsday: Friday 13 Nov. A.D. 2026. *Science,* 132: 1291–1295.

Yount, J. L. 1956. Factors that control species numbers in Silver Springs, Florida. *Limnology and Oceanography,* 1: 286–295.

NOTES

1. When energy is transformed in successive stages in circular pathways, the available energy decreases, but it is possible to trace some energy through many loops, as studied by Patten (1985).

2. The curve for power as a function of loading and efficiency is a parabola (appendix A3 from Odum and Pinkerton 1955). Later derivations were published by Curzon and Ahlborn (1975) and Fairen and Ross (1981).

3. Because every real process requires power, the maximum and most economical collection, transmission, and use of power must be one of the primary selective criteria. The criterion for natural selection has often been considered simply the maximum possible reproduction and survival of the species concerned, a principle that, operating alone, would make shambles of any multispecies system through destructive competition.

4. The scientific, quantitative definition of mechanical work characterizes it as the product of a force times the distance over which that force operates. Any process involves a force and an opposing force, and any process involves work done in operating the forces for a distance. The force that is used to overcome resisting backforces flows from a potential energy source and is the means for delivering power. The opposing forces may be from other potential energy sources, from the friction that opposes any velocity, or from the inertial force that opposes any acceleration in the absence of other forces.

5. See chapter 2, note 3.

6. In electric transformers, energy is transformed from one electric coil to another by the surges of electric current, which cause temporary magnetic fields created by one coil to induce a current flow in the other coil. In coils of different properties, the voltage of the electrical power may be changed as desired. When side effects are eliminated, such transformers pass power approaching 100% efficiency. One may wonder whether such energy transfers are exceptions to the second principle of energetics. Transfer of electric energy by transformer is yet another kind of acceleration of matter (charged electrons in wires), once again a relative process. As viewed from one position, energy changes its apparent form from electric potential to magnetic field energy. Viewed from a different position that is moving in the opposite direction, the transformation appears to be from magnetic field energy to electric potential energy. The energy transformation is only a relative one.

7. See Jammer (1957) for a review of force causality concepts.

8. In the past some physics textbooks, in trying to emphasize the concept of balance of forces, made the point that there is no cause or effect. This half truth applies to problems of static force balance and accelerations not involving heat-dispersing work processes and energy degradation.

9. The potential energy in chemical concentrations of gases and solutions in the biosphere where the pressure is nearly constant is Gibbs free energy (ΔF), a logarithmic function of the concentration. For example, there is more available energy in rainwater than in seawater. The presence of 3.5% salt in seawater reduces the concentration of the water to 96.5%:

$$\Delta F = \Delta F_o + nRT \ln(C_2/C_1);$$
$$\Delta F = 0 + (1/18 \text{ mole/g})(1.99 \text{ cal/deg mole})(300°K)[\ln(96.5/100)].$$

The result is 1.18 cal/g, or 4.6 kcal/gallon. Additional energy may be included in water, depending on the kinds of molecules dissolved in the water as it is used in industrial and living activities.

10. The Gibbs free energy in phosphorus concentration can be calculated using the concentration relative to average concentration in the biosphere and the Gibbs equation. The result is about 83.5 cal/g.

11. The model (fig. 3.9a) for cattail growth implies the following equations:

$$dN/dt = J_n - (k_1 * N) - (k_2 * R_s * N * S);$$
$$R_s = J_s / [1 + (k_3 * N * S)];$$
$$dS/dt = (k_4 * R_s * N * S) - (k_5 * S) - (k_6 * S);$$
$$dP_t/dt = k_6 * S,$$

where S is plant structure, N is nutrient concentration, J_n is inflow of nutrient waters, J_s is inflow of sunlight, R_s is residual unused sunlight in plant tissue, and P_t is peat accumulation.

12. This model was discovered by John Alexander (1978) and its spatial and temporal properties simulated by John Richardson (1988). For details, see appendix fig. A10.

WHEREAS THE classical energy laws (chapters 1–3) explain much of what happens in single processes, energy hierarchy principles relate processes between scales. This chapter explains how the energy hierarchy accounts for spatial organization and distribution of pulse intensity, stored quantities, and material concentrations. The fundamental principle offered in this chapter is not only that energy organizes hierarchical patterns but that energy is itself a hierarchy because the many calories flowing through numerous small units in a network converge to form units with fewer total calories but with larger size and territory. The energy hierarchy provides a measure *emergy*, spelled with an *m*, for evaluating work on all scales in units of one scale, a concept of natural value.

Energy transformations form a series in which the output of one is the input to the next. Available energy decreases through each transformation, but the energy quality increases, with increased ability to reinforce energy interactions upscale and downscale. For systems organized on many scales from small to large territory of influence, a maximum power design develops in which each scale is symbiotically connected by feedback loops with the next. Energy transformations organize that converge energy, materials, and money to centers. With increasing scale, storages increase, depreciation decreases, and pulses are stronger but less frequent.

ENERGY TRANSFORMATION AND QUALITY

An energy transformation (fig. 4.1) is a conversion of one kind of energy to another kind. For example, coal fuel is burned in a power plant to generate an output of electric power. Or a person transforms food to generate an output of human services. As required by the second law, the input energy with available potential to do work is partly degraded in the process of generating a lesser quantity of the output energy.

FIGURE 4.1 Energy transformations with steady state energy flows. **(a)** Coal transformation into electricity; **(b)** transformation of food into human services.

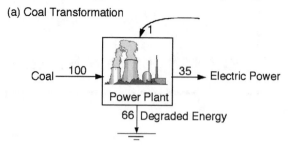

(a) Coal Transformation

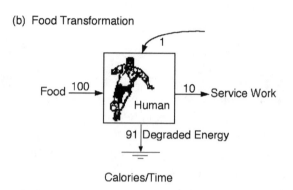

(b) Food Transformation

Calories/Time

Efficiency

Efficiency is the traditional measure used to represent energy transformations. It is the percentage of input energies that is output energy. See examples in fig. 4.1.[1]

Control Inputs

Most energy transformations are controlled by small inputs of high-quality energy, as shown on top of the boxes in fig. 4.1. In most energy texts these energy flows are ignored because they are small, but their effect is major. The few calories of the control inputs have as much effect on the output power as the many calories of lower-quality intake, as explained with fig. 3.10.

In one sense, the output energy usually is regarded as an increase in quality because more available energy was used to make it. There is some consensus that such high-quality outputs can accomplish more useful work if coupled to amplify other lower-quality energy flows according to their potential. This coupling of high- and low-quality energies to maximize useful output is the energy matching principle. In these examples a calorie of electric power can do more than a calorie of coal. A calorie of human service contributes more than a calorie of food. As explained in chapter 3, transformed energies can maximize power by feedback interactions that capture more resources. Recall the example of autocatalytic reinforcement in fig. 3.6.

Although one can imagine an energy transformation that does not generate something that aids useful work, it is reasoned that such transformations will not be selected by the natural self-organizational processes that reinforce the individuals and connections that generate more useful power.

As we explained in chapter 1, for purposes of evaluating energy budgets, we can aggregate or disaggregate depending on our scale of interest. For example, in fig. 4.1b we show human service work as one aggregate process, but we could have shown the human as a series of connected transformations by the organs, tissues, and cells within the body.

Energy Transformations in a Series

When energy transformations are connected in a series, the outputs of one transformation are the inputs to the next, except that some of the outputs go back to the input process, interacting and controlling the input. Figure 4.2 shows energy flows in a series with three transformations. Each calorie of energy flow is represented as a separate flow line in fig. 4.2a to help readers visualize the way available energy decreases through the transformations and disperses into degraded heat when its ability to do work is used up. The stepwise decrease in power flow between transformation units is shown as a bar graph in fig. 4.2b. The transformation units are represented as an energy systems diagram in fig. 4.2c.

A Law of Energy Hierarchy

When many units of one kind combine to support another kind of unit that feeds back control, we describe the organization as a hierarchy. As shown in the typical energy transformation system in fig. 4.2a, energy flows decrease in steps to the right. Many calories of lower quality on the left support fewer calories to the right. In other words, energy transformations are hierarchical.

To explain why energy organization is a hierarchy, consider a military organization. We call the organization of an army a hierarchy because many privates are required to support the action of sergeants, and many of these to support lieutenants, and so on. Conversely, high-quality control actions feed back from lieutenants to sergeants to privates. The military organization is a special case of the energy hierarchy (fig. 4.3).

The author proposed the universal hierarchical self-organization of energy systems as a fifth energy law (Odum 1987, 1988): *All the energy transformations known can be connected in a series network according to the quantity of one kind of energy required for the next* (fig. 4.4). Note the reinforcing feedbacks that appear to be universal (but often ignored) by which each transformed power flow feeds forward or backward so that its special properties can have amplifier actions.

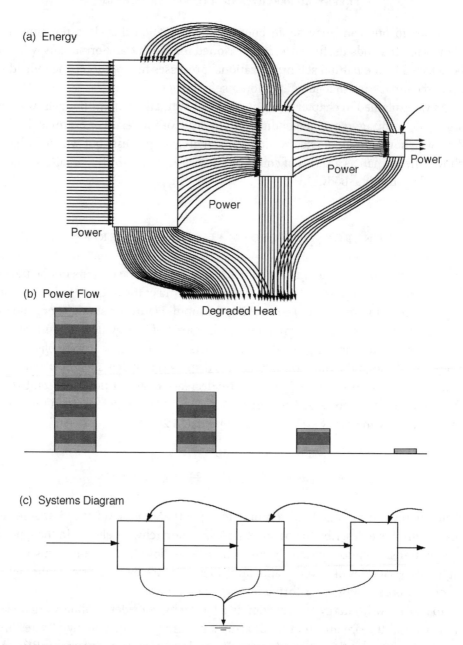

FIGURE 4.2 Hierarchy of converging energy flows and transformations. **(a)** Picture of power flows with each line representing a flow of 1 calorie per time; **(b)** useful power output of each transformation process (flows out of each box toward the right); **(c)** energy systems diagram.

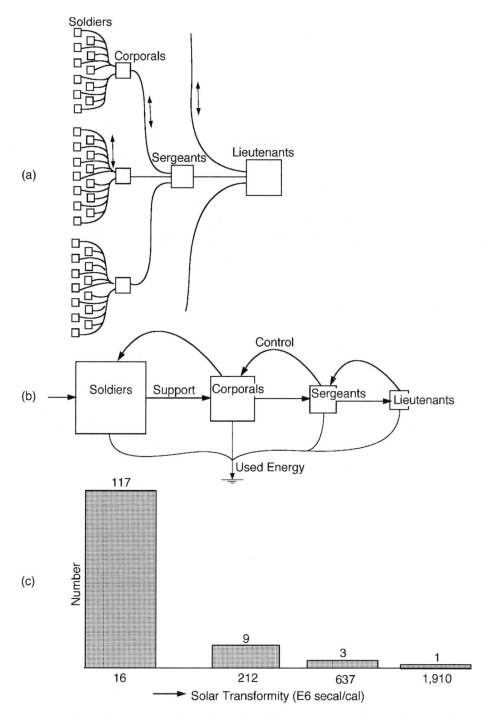

FIGURE 4.3 Flows of energy in a military hierarchy (Odum 1987). **(a)** A schematic diagram of military organization, **(b)** an energy systems diagram including control actions, and **(c)** the number of units of different transformity.

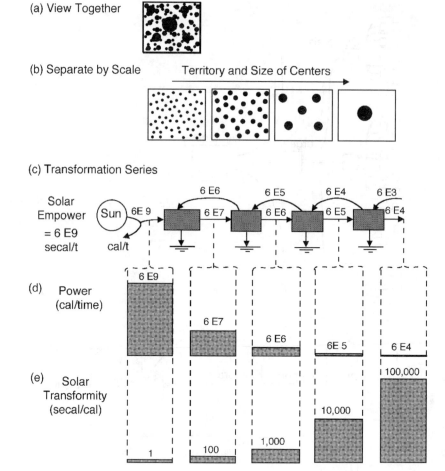

FIGURE 4.4 Energy transformations connected in a series. **(a, b)** Size and territory of the units composing each; **(c)** energy systems diagram of transformations with values of energy flow on pathways; **(d)** useful power flowing between transformations; **(e)** transformities.

EMERGY AND EXERGY

In recent years the terms *available energy* and *exergy* (spelled with an *x*) have been used for energy that still has the potential to do work. For one kind of work process, the more exergy, the more work. For example, using more electrical exergy causes more water to be pumped with an electric pump.

Unfortunately, different kinds of available energy (exergy) often have been added as if they are equivalent measures of work. In fig. 4.1b it is incorrect to say a calorie of food is equivalent to a calorie of human service. It makes no sense to say a calorie of sunlight is equivalent in work done to a calorie of electric power.

Where energies of different kinds from different levels in the universal energy hierarchy are to be compared or their effects combined, they need to be expressed in units of one kind of energy. A new measure was defined: *Emergy* (spelled with an *m*) *is the available energy of one kind previously used up directly and indirectly to make a product or service*. The energy required for the transformations is no longer in the product or service, but emergy carries the memory of the availability that was used up. Emergy is a new kind of state variable. The units of emergy[2] are defined with the prefix *em-* (e.g., *emcalories, emjoules, emBtus*). In the energy transformation series in fig. 4.4, the solar emergy is the same for all the flows in the chain. In this book we use solar emergy as the common measure, although in the past coal emergy and electrical emergy have been used for some comparisons. The abbreviations used for quantities of solar energy are as follows: solar emcalories (secal), solar emkilocalories (sekcal), and solar emjoules (sej).

Figure 4.5 is an example of an energy transformation series (solar energy–wood–coal–electric power–night lights). Abundant energy from the left supports one unit of electric light power on the right. This series uses the three kinds of energy that drive the geobiosphere: the sun, the tide, and deep earth heat. These are combined in the figure into one source expressed as solar emergy units (2.4 E6 solar emcalories per time).[3] The useful power decreases sharply with each step.

Empower

The flows of energy through a network carry their emergy, a property of their nature relative to the universal energy hierarchy. Whereas power is useful energy flow per time, empower is emergy flow per time. In the energy chain in fig. 4.5, the empower flow is 2.4 E6 solar emcalories per time. The energy that was required is no longer a part of the product, but emergy is a property that represents its history and implies its importance. It remains with the product and the product's products until the available energy is gone. Emergy disappears when the available energy is used up (degraded).

Comparisons

In *Environment, Power, and Society,* tables summarized the power in various processes of our civilization and the globe to show the importance of energy. However, to see how important an energy flow is, we need to express it as empower. In table 4.1 the energy content (kcal/cm³) being consumed by four different processes is expressed as solar emergy per unit volume (sekcal/cm³), and the power flows of the processes (kcal/cm³/day) have been expressed as solar empower (sekcal/cm³/day).

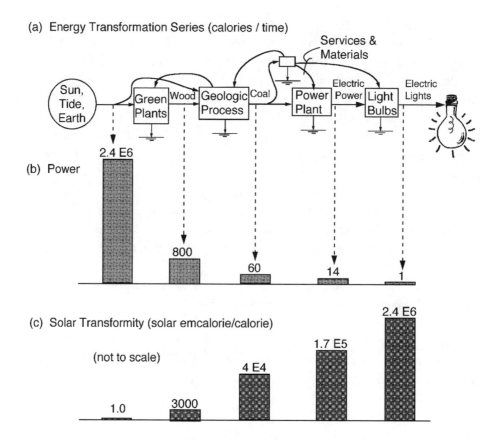

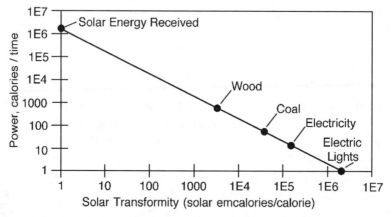

FIGURE 4.5 Example of an energy transformation series from solar energy through biomass to electrical lighting. **(a)** Energy flows and transformations; **(b)** bar graph representing the decrease of available energy at each step; **(c)** transformities increasing at higher levels to the right; **(d)** graph showing power as a function of transformity.

TABLE 4.1   Comparison of Some Fuels and Consumers

System	Fuel Content		Consumption Rate	
	Energy (kcal/cm³)	Solar Emergy (sekcal/cm³)	Energy (kcal/cm³/day)	Solar Emergy (sekcal/cm³/day)
Green plant using sunlight (covering 1 m² of ground)	6 E-16	1 E11	4.0 E3	6.7 E7
Waterfall (10 m high, 10^5 gallons/day) consuming energy of elevated water	2 E-5	1	9.0 E3	4.2 E8
Human consuming food	4.5	7.6 E5	3.0 E3	5.0 E8
Small automobile	10	1.1 E6	9.0 E5	1.0 E11

TABLE 4.2   Comparison of Power Flows

System	Power Density (kcal/m²/day)	Empower Density (sekcal/m²/day)
Incoming energy (dilute type)		
Sunlight absorbed by biosphere[a]	5,110	5,110
Sunlight reaching green plant level[a]	3,400	3,400
Maximum conversion[b]	170	170
Organic production as in fig. 1.3		
Production of a rainforest[c]	131	655,000
World primary production[d]	6	10,000
World agriculture on 8.9% of the world's area[e]	0.26	44,000
World agriculture contribution to biosphere[e]	0.024	4,000
Production removed by farming[f]	3.6	605,000
Consumer systems		
Respiration of a rainforest[c]	131	655,000
A village without machines, 100 m²/person	30	4,080,000
Fossil fuel consumption in the whole biosphere[g]	0.135	15,000
Total consumption of biosphere (production and fossil fuel)	6.1	25,000
Fossil fuel consumption in the United States (per U.S. area)	3	330,000

(continued)

TABLE 4.2   Comparison of Power Flows (*continued*)

System	Power Density (kcal/m²/day)	Empower Density (sekcal/m²/day)
Animal city, Texas oyster reef[h]	57	9,600,000
Fossil fuel consumption in a large city[i]	4,000	444,000,000

[a] Sellers (1965): 3.67 E18 kcal/day; 5.1 E14 m²/earth, 29% reflected.

[b] At high light intensities 10% of visible light is converted to organic matter; visible is 50% of total sunlight.

[c] Odum and Pigeon (1970).

[d] Hutchinson (1954). This figure is defined as the sum of daytime net photosynthesis measured by metabolic substances released or absorbed by the medium of plant cells (such as oxygen) plus the nighttime respiration. The actual gross photosynthetic process is much greater than can be measured in this way because there is fast internal cycling (making and using) in the day, stimulated by photorespiration and by the impetus of storages in the various sites of reaction in the cells of the plants. Adding night respiration is an objective procedure but one that apparently underestimates the amount of the daytime cycling. Therefore the figure given is much too small as a measure of gross production. In recent years maps of the world's primary production have been published using much lower values from radiocarbon uptake studies in bottles of seawater from the tropical oceans. These world averages are one-tenth of the figure in the table. In one respect, most of these radiocarbon measurements were mostly erroneously calculated so that they represent neither the basic gross inflow of carbon into the cells nor the net accumulation but something in between. However, as a measure of the food available to other food chains they are useful, showing how little of the gross primary production gets outside of the plant cells over much of the earth. The actual net production is even less than these maps show.

[e] U.S. Department of Agriculture (1970). Production is first calculated per unit of agricultural land (4.55 E13 m²) and then per area of the biosphere (5.1 E14 m²).

[f] [(40 kcal/m²/day)(Agricultural lands)]/(Area of earth).

[g] Revelle et al. (1965).

[h] Copeland and Hoese (1967).

[i] 0.025 people/m²; 1.6 E5 kcal/person/day.

Power density is the flow of energy per unit area per unit time (kcal/m²/day). It is a measure of the intensity of energy use per area. Table 4.2 summarizes power (energy flows) per square meter for various levels of the global system of humans and environment. Note that the power density decreases from sunlight to fossil fuel consumption. However, their importance is evident when power density (kcal/m²/day) is converted to empower density (sekcal/m²/day).

INTERBRANCHING NETWORKS

Although we often isolate a simple chain in our minds or in drawings to think about a system (for example, figs. 4.2, 4.3, and 4.4), real systems form more complex networks, as we showed with the house in fig. 2.11. If each energy transformation is connected to the ones that supply inputs and use outputs, an interbranching network results (fig. 4.6a). Like the single chain, the flow of power decreases from left to right while the quality increases. Perhaps it is time to recognize the energy hierarchy as a fundamental property of the universe in which each kind of energy has its place. To put it more simply, large quantities of low-quality energy are the basis for but controlled by small quantities of high-quality energy.

Similar phenomena on different scales were called isomorphic structures by Len Troncale (1978, 1988). Gerrit Feekes (1986) wrote a book titled *The Hierarchy of Energy Systems from Atoms to Society* in which he described the way energy is associated in comparable ways with structures and processes on different scales of living organisms and their psychological organization.[4]

TRANSFORMITY

One can relate all the forms of energy in a series of energy transformations to one form of energy by calculating their transformities. *Transformity is defined as the calories of available energy of one form previously required directly and indirectly to generate one calorie of another form of energy.*[5] The units of transformity are emjoules per joule, or emcalories per calorie. For example, in fig. 4.5 the solar transformity of electric power is 1.7 E5 solar emcalories/calorie. Transformity is the emergy divided by the energy. In this book we use calories as the measure of energy, solar emcalories for emergy.

$$\text{Solar transformity} = (\text{Solar emergy/Time})/(\text{Energy/Time}),$$

which is the same as

$$\text{Solar transformity} = (\text{Solar empower/Power}),$$

with the units solar emcalories/calorie.

Although we use joules in other books, transformity is numerically the same:

$$\text{Solar emcalorie/Calories} = \text{Solar emjoules/Joule}.$$

In the energy transformation chains in figs. 4.4, 4.5 and 4.6, the energy decreases at each step, but the emergy is the same. Hence the solar transformity increases. In fig. 4.5 solar transformities rise from 1 for sunlight to 2.4 million solar emcalories per calorie for electric lights. Transformity increases with each energy transformation. It is a measure of the increasing quality of energy as it is passed through successive transformations. *Transformity measures the position of each kind of energy in the universal energy hierarchy.*

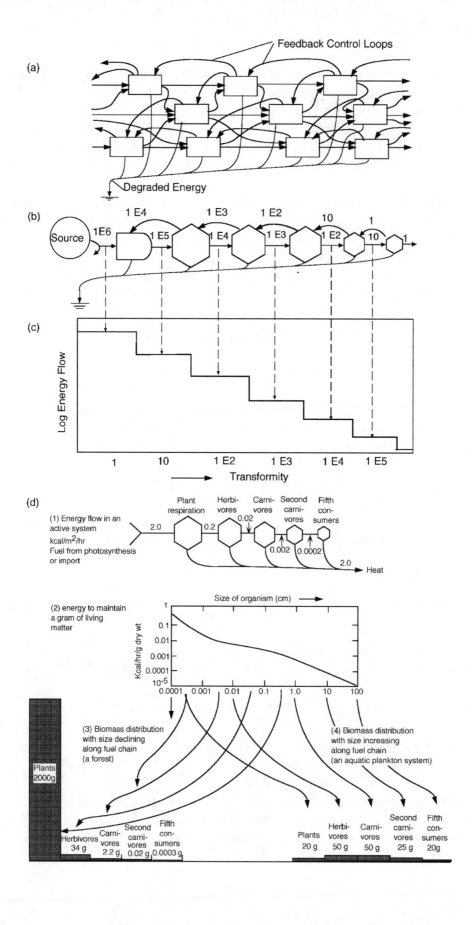

(a)

Feedback Control Loops

Degraded Energy

(b)

Source

1E6   1 E4   1 E3   1 E2   10   1

1 E5   1 E4   1 E3   1 E2   10   1

(c)

Log Energy Flow

1   10   1 E2   1 E3   1 E4   1 E5

Transformity

(d)

(1) Energy flow in an active system

kcal/m²/hr
Fuel from photosynthesis or import

Plant respiration   Herbi-vores   Carni-vores   Second carni-vores   Fifth con-sumers

2.0   0.2   0.02   0.002   0.0002   2.0   Heat

(2) energy to maintain a gram of living matter

Size of organism (cm)

Kcal/hr/g dry wt

1
0.1
0.01
0.001
0.0001
10⁻⁵

0.0001   0.001   0.01   0.1   1.0   10   100

(3) Biomass distribution with size declining along fuel chain (a forest)

(4) Biomass distribution with size increasing along fuel chain (an aquatic plankton system)

Plants 2000g

Herbivores 34 g   Carni-vores 2.2 g   Second carni-vores 0.02 g   Fifth con-sumers 0.0003 g

Plants 20 g   Herbi-vores 50 g   Carni-vores 50 g   Second carni-vores 25 g   Fifth con-sumers 20g

Graph of Power and Transformity

A useful way to represent energy in relation to the hierarchy is the plot of energy (power or storage) as a function of the transformity (fig. 4.5c). Because transformity is a measure of energy quality, we sometimes call this a quantity–quality graph. In a series of transformations the energy flow steps down with each transformation, and transformity increases accordingly, causing a plot on regular (arithmetic) coordinates to be very hollow and cover only a small range.

Usually we use logarithmic scales, where each tick represents a tenfold increase. The horizontal scale has transformities beginning with sunlight and increasing to the right 1, 10, 100, 1000, 10,000, etc. The quantity of energy flow (power) on the vertical axis is also plotted on a logarithmic scale. When the graph represents a series of energy transformations with the same emergy, the energy flow and transformity are inverse (fig. 4.6c). The following definitions are obtained by rearranging the definition that follows from the hierarchy law:

$$\text{Power} = \text{Empower}/\text{Transformity}.$$

Sometimes we call a power–transformity plot an energy transformity spectrum:

$$\text{Energy flow} = \text{Emergy flow}/\text{Transformity}.$$

On the graph with logarithmic scales (fig. 4.6c), the relationship is a straight line. For example, in fig. 4.5d the line connecting points for the solar–coal–electricity–light series is straight.

Typical Transformities

As explained further in chapter 5, the earth receives its main energies from sunlight, tide, and the hot interior of the earth. Almost everything else on Earth is derived from these flows through the complex web of energy transformations in atmospheric, oceanic, geologic, biological, and societal systems. The

FIGURE 4.6 (*opposite page*) Patterns of the energy hierarchy. (a) Web of energy transformations in series; (b) network aggregated into a web; (c) power as a function of transformity on double logarithmic scales; (d) relationship between a steady energy flow and the mass of living structure maintained. (1) Diagram of a representative energy flow between compartments of the food chain. (2) Graph of the respiratory metabolism of organisms as a function of their individual size (based on data in Zeuthen 1953). (3, 4) Two distributions of mass estimated by dividing the energy flow by the energy per gram maintenance requirements given in (2). Different sizes cause masses to vary, even though the energy flow is the same.

more transformations that occur at successive levels in the global hierarchy, the higher the transformity and the less the actual energy, although the small energies at high levels have great impact (e.g., birds of prey, earthquakes, information storms). Figure 4.7 summarizes some of the transformities of environment and society.

Spatial Dimensions of Energy Hierarchy

The self-organization of energy into hierarchies explains how and why phenomena over the surface of the landscape form spatial centers of energy transformation on all scales. Small centers converge their outputs to larger centers that represent the larger scale. These converge to even larger centers at the next level, as suggested by the sketch in fig. 4.8a. The hierarchy of towns, cities, and villages, so vivid in maps, is an old principle of geography. The hierarchy of small spatial centers of human settlements converging to larger centers was described by Al-Muqaddasi in 986 AD and further documented by Christaller (1966).

Ecologists have documented the pattern of small centers to larger centers in the relationship of leaves, branches, and tree trunks and in such animal cities as nests of ants and termites. Extensive studies on geomorphology, stream branching, hydrology, and energetics of watersheds were made by Luna Leopold (Leopold and Langbein 1962) and others (Jarvis and Woldenberg 1984). The flow of water self-organizes hierarchies of water channels converging on larger streams.[6]

So by what mechanisms are energy hierarchies and spatial organization of centers linked? Because energy decreases with each transformation, the downstream products have less energy to feed back and amplify. However, if systems converge the transformation products to centers spatially, they concentrate these flows so

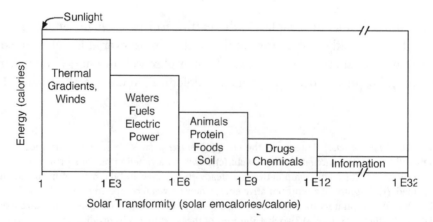

FIGURE 4.7 Some transformities of environment and society (Odum 1996).

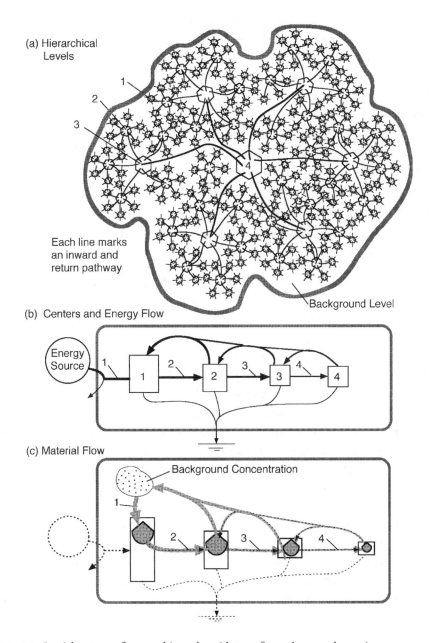

(a) Hierarchical Levels

1

2

3

Each line marks an inward and return pathway

Background Level

(b) Centers and Energy Flow

Energy Source

1   1   2   2   3   3   4   4

(c) Material Flow

Background Concentration

1   2   3   4

FIGURE 4.8 Spatial pattern of energy hierarchy with transformed energy becoming more concentrated in centers. **(a)** Spatial pattern of converging pathways; **(b)** energy systems diagram showing transformations and transformity increasing to the right; **(c)** diagram of energy-driven processes concentrating materials to the right and recycle to the left.

that the feedback out from the centers is concentrated enough to have a strong effect by spreading its useful work over the contributing area. The pathway arrows inward and outward in fig. 4.8 show the manner by which an area is both a center's support and its territory of controlling influence. Systems prevail that form hierarchical centers spatially, a design that helps maximize power and empower.

The cascade of similar structural patterns on different scales can be generated by fractal models according to Mandelbrot (1983), a well-known way to understand hierarchical snowflakes, tree branches, and flowers. The energy hierarchy is a fractal cascade with fractions of energy inflow transmitted through each energy transformation. That fractal forms are so abundant is another evidence of the energy hierarchy law.

Pulsing Intensity in the Energy Hierarchy, the Temporal Dimension of Energy Hierarchy

The accumulation–pulsing sequence (fig. 3.13) is observed on all scales, but the pulsing appears to be more concentrated in time at the higher levels (to the right in our diagrams). With less energy with which to feed back and reinforce, the higher levels can achieve concentration not only through spatial concentration but also by storing for a longer period and discharging in a shorter delivery time. Figure 4.9 shows how the larger the scale, the longer the accumulation time, and the more concentrated in time is the feedback to the supporting territory. Jorgensen and associates (1998) found that ten ecological phenomena, including photosynthesis and avalanches, fit the straight-line, logarithmic plot of frequency as a function of magnitude of change (fig. 4.9). Many examples were found by Daeseok Kang (1998) in plant and animal populations in lakes and oceans. Strength of floods is proportional to the scale of contribution and inverse to their frequency. Doxiadis (1977) found many people taking small trips and a few people with longer trips, expressing their areas of influence. Pulsing follows from the hierarchical self-organization of energy. In other words, the higher levels can do more with less by accumulating for longer times and delivering sharper pulsing.

Pulse Adaptation

Pulses from a larger scale often are called catastrophes, although they are inputs of high-transformity energy with the ability to cause large responses. Systems that prevail have to draw benefit and adapt to catastrophic pulses. A recent summary of anthropological views (Hoffman and Oliver-Smith 2002) recognizes the human responses to catastrophic pulses such as terrorist attacks as a normal part of the culture and genetics that have been self-organized into human society.

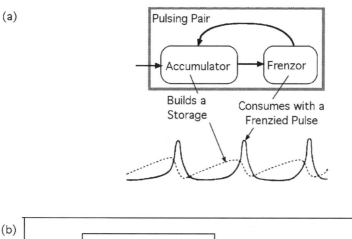

(a) Pulsing Pair

Accumulator → Frenzor

Builds a Storage

Consumes with a Frenzied Pulse

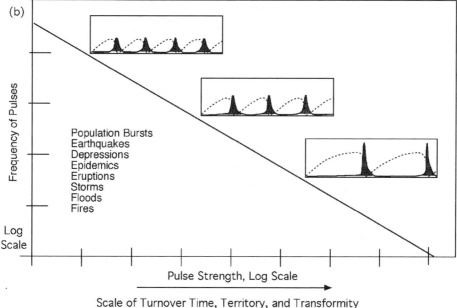

(b)

Frequency of Pulses

Population Bursts
Earthquakes
Depressions
Epidemics
Eruptions
Storms
Floods
Fires

Log Scale

Pulse Strength, Log Scale

Scale of Turnover Time, Territory, and Transformity

FIGURE 4.9 Pulsing patterns of energy hierarchy. **(a)** Accumulator–frenzor design that generates performance-increasing pulses. **(b)** Inverse relation between number of pulses and their strength. Example: earthquake frequency and intensity on the Richter scale (log of energy).

HIERARCHICAL STORAGE OF ASSETS AND TURNOVER TIME

Usually each energy transformation process has an associated storage from which the autocatalytic feedbacks originate. There are feedbacks within the transformation unit and longer feedbacks that go to other energy transformation units (fig. 4.10). These storages fill and discharge as part of the accumulations and frenzied pulses. The size of the storages *per area* that is maintained increases along an energy transformation series. This is a necessary consequence of the selection for longer periods of accumulation at higher levels. To have a longer period of accumulation for

(a)

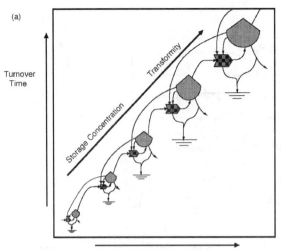

Turnover
Time

Territory of Support and Influence

FIGURE 4.10 Energy transformations on a scale diagram of territory and turnover. Storage concentration increases with scale. **(a)** Scale diagram showing the increase of transformity and storages. **(b)** Biomass of animals related to the territories of support and influence (composite of figures from Holling 1992). **(c)** Building structure of human civilization related to zones surrounding hierarchical centers (modified from Doxiadis 1977).

(b)

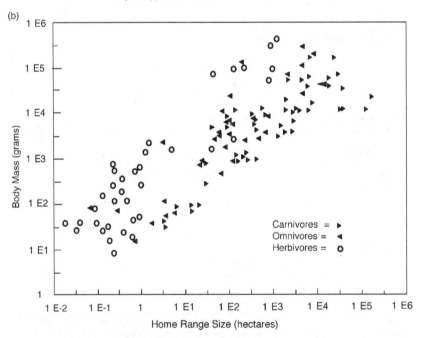

Body Mass (grams)

Carnivores = ▶
Omnivores = ◀
Herbivores = ○

Home Range Size (hectares)

(c)

Increasing Storage, Replacement time, Territory of influence, Pulsing intensity & Transformity

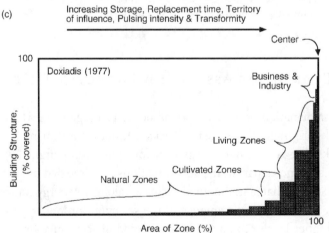

Center

Doxiadis (1977)

Business & Industry

Living Zones

Cultivated Zones

Natural Zones

Building Structure, (% covered)

Area of Zone (%)

levels with less energy flow requires a larger storage. Larger storage concentrations also have larger volume–surface ratios and longer depreciation times. Figure 4.10a shows larger storages with position in the energy hierarchy. It is well known that the size of many animals increases along the food chain even as the total energy flows diminish.[7] For example, data from C. S. Holling in fig. 4.10b show the biomass of animals increasing with their territories of support and action (home range).

EMERGY STORAGE

As explained in chapter 2, growth and succession on any scale require and are accompanied by development of the storage necessary to maximize the energy intake with feedback pumping. When resources from transformations are stored, both energy and emergy accumulate. The emergy is the sum of the inflowing empower required to generate the storage, overcoming any necessary depreciation. Emergy storage is the product of the energy going into storage and its transformity. Some emergy storage may be lost by transport away for other use; some emergy is lost if the energy or material content of the storage is allowed to disperse.

Ecologists Ramon Margalef (1958) and Sven Jorgensen (1997) have suggested that maximum power occurs through the building of biomass storages. In a population of animals the biomass often is the hierarchical center. For example, visualize the spider at the center of its web. Troncale (1988) found the mass of biological entities increasing as the square of the size.

Everyone who has observed skyscrapers and skylines is familiar with the accumulations of structural mass that develop in the center of cities. Figure 4.10c shows the relationships of building mass concentration in urban centers, first explored by Doxiadis (1977).

TERRITORY, TURNOVER, AND TRANSFORMITY

In recent years it has been customary in many fields to show that territory and turnover time (replacement time) are correlated and useful for registering the scale of phenomena. We can summarize some of the principles of energy hierarchy by relating territory, turnover time, and transformity. An energy transformation series has been plotted as the diagonal of fig. 4.10a, which has the usual coordinates of territory and turnover time. As the figure suggests, a series of energy transformations really is a passage between scales of size and time in which transformity becomes a third measure of scale. Transformity measures the position of something in the universal energy hierarchy. Veizer (1988) did the fundamental work to show how the sectors and cycles of geologic process form such a diagonal series.[8]

If it was not clear before, by representing systems with series of energy transformation modules the energy systems language (fig. 2.10) automatically arranges items of a system in order of the scale properties (turnover time, territory, and transformity). When we diagram a new problem, our first task is to arrange the sources and components from left to right in order of transformity.

Role of the Energy Hierarchy in Concentrating Materials

According to the second law, concentrations tend to disperse spontaneously (fig. 4.11a). For example, a concentration of salt in water diffuses under the influence of movements and oscillations on smaller scales. To concentrate something or maintain a concentration, available energy has to be used up in some transformation process that increases concentration. See the pumping relationship in fig. 4.11b. The potential energy of a chemical concentration is its Gibbs free energy.[9] When a substance is concentrated, its Gibbs free energy is increased somewhat, but much more available energy is used up concentrating it at the optimum efficiency for maximum power (chapter 3). Note the relationship of driving energy to the concentrating process in the energy systems diagram (fig. 4.11b).

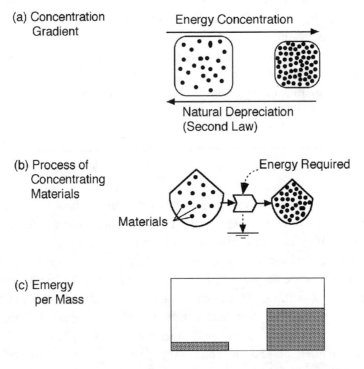

FIGURE 4.11 Use of available energy to concentrate substances. **(a)** Concentration and the second law; **(b)** energy availability required to concentrate materials and energy; **(c)** increase of stored emergy with concentration.

Because available energy is used up in each transformation step, there is less and less available energy to do further concentrating. This means that there has to be an inverse relationship between the quantity of materials concentrated and the number of steps (levels) of concentration. *Just as the quantity of concentrated energy has to decrease with scale, so does the quantity of concentrated materials, also because of the nature of the energy transformation hierarchy.* Figure 4.12b shows the inverse relationship between the quantity of materials concentrated and the degree of concentration for an energy transformation series with one emergy source.

(a) Hierarchical Concentration of Materials

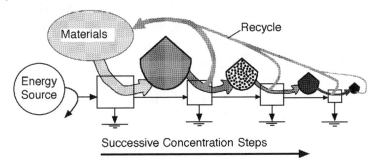

(b) Inverse Relation of Quantity and Concentration

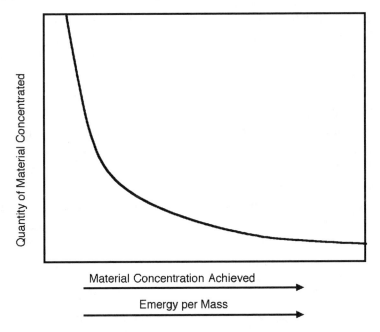

FIGURE 4.12 The distribution of materials caused by the transformations of the energy hierarchy. **(a)** Systems diagram showing decrease of materials with each step up in concentration; **(b)** materials concentrated as a function of concentration.

Specific Emergy for Evaluating
Biogeochemical Cycles

Whereas materials by their concentration have available energy and emergy relative to their surroundings, it is often more convenient to represent materials according to the emergy per unit mass, which Sergio Ulgiati (2001) called *specific emergy*. The materials of the universe can be arranged in a series according to their values of emergy per mass (fig. 4.12b).

It has been customary for a century to draw chemical cycles of ecosystems, organisms, and the earth, but there has been no accepted pattern for arranging components and flows. The energy-based law of material hierarchy suggests that material diagrams can be made more thermodynamic and uniform by using the energy systems convention that arranges components and flows of materials from left to right in order of their emergy per mass values. Flows to the right represent increases in concentration, whereas cycles back to the left indicate dispersals of material to larger area, with loss of emergy concentration (fig. 4.12a). For example, figs. 2.8 and 2.9 contain material flows for metabolism, and fig. 2.11c shows material circulation in a house.

Energy Hierarchy and Molecular Energy

This chapter started by explaining energy hierarchy concepts involving a series of transformation steps using ecological food webs and biochemical networks as examples. A food chain of energy transformations is on the scale of human observation and therefore in clear view. People can see that successive steps are generating products of increased order, significance, and unit importance such as whales, tigers, and creative people.

But much of the science of energy has come from physical scientists thinking about the molecular scale. The submicroscopic world was out of sight, and the energy of molecules was assumed to be unorganized and disorderly, to be represented as a statistical distribution.

Figure 4.13a is the distribution of kinetic energy used by Maxwell and Boltzmann to represent molecular energy at equilibrium (i.e., without any available energy incoming or degrading). On the left are molecules of lower velocity and kinetic energy bumping each other about. But some of the collisions generate a smaller number of faster molecules with more kinetic energy, and these are on the right of the figure. At the far right the energy level is high enough to cause small pulsing observable in the microscope as Brownian motion. Energy moving upscale to the right equals that dispersing downscale to the left. Note the sketch of equilibrium in fig. 4.13a. Even at equilibrium, self-organization generates an energy hierarchy. No assumptions of randomness are needed.[10]

(a) Energy Circulation at Equilibrium

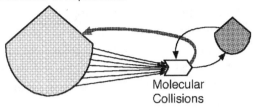

Molecular
Collisions

Low Energy Molecules   High Energy Molecules

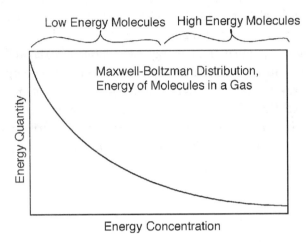

Maxwell-Boltzman Distribution,
Energy of Molecules in a Gas

Energy Quantity

Energy Concentration

(b) Self Organization of Molecular Energy Hierarchy

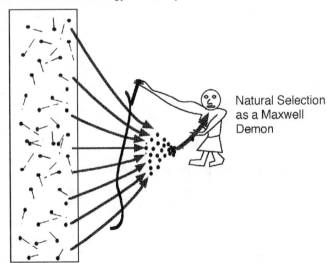

Natural Selection
as a Maxwell
Demon

FIGURE 4.13 Energy hierarchy in a gas at thermodynamic equilibrium. **(a)** Energy systems diagrams of molecular collisions used to explain the hierarchical distribution of energized molecules; **(b)** cartoon of natural selection as a Maxwell demon with a broad base of choices and support.

A favorite image for discussing molecules in motion is the Maxwell demon (fig. 4.13b). The idea is that a tiny molecular-sized being might concentrate energy by opening a door to let molecules go in one direction while closing the door to those that would pass back out. Arguments were made that the demon's work was impossible because the energy gathered by choosing a higher-energy molecule was less than that required to operate the door. Yet fig. 4.13a shows that, in a larger system, energy could converge from a broader base of many inputs to support selective action at a higher level. The natural self-design process of developing a hierarchy accumulates a few molecules of higher energy (higher velocity). In other words, natural selection in a population of energy interactions is a Maxwell demon.

Now, if a small inflow of available energy is added to the energy distribution, more of the energy moves to the right, and the system is now energetically open but without much change in the energy hierarchy. Figure 4.14a shows the molecular energy distribution and diagrams the continuous distribution of energy in different energy concentrations with weak input. This continuous pattern is found

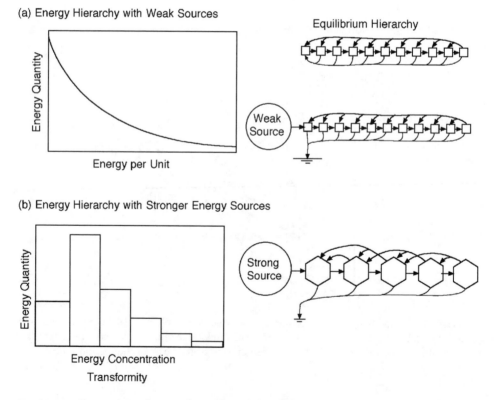

FIGURE 4.14 Comparison of energy hierarchies with strong and weak energy sources. **(a)** A more continuous energy hierarchy with weak energy source; **(b)** an energy hierarchy of discrete autocatalytic units.

in the thermal energy of molecules near equilibrium and in fluids with weak turbulence. There is an increase in transformity along the series (left to right).

To measure availability of energy, a quantity *entropy* was defined as the quotient of heat added divided by the temperature (on the Kelvin scale). In chapter 8 entropy is found to measure molecular complexity (logarithm of the possible combinations and connections) and used as a measure of order and disorder. At absolute zero (−273°C), where there is no heat, the molecules don't move and have only one arrangement (as in a crystal). The entropy is small, and the molecules are orderly. Adding heat increases entropy and complexity.

Because it was assumed in physical–chemical sciences that the molecular scale was unorganized, it was assumed that entropy measured complexity disorder. Many people outside science used the word *entropy* as a synonym for *disorder*. Yet the same measure (logarithm of possible combinations and connections) was applied to the larger scale of ecosystems, society, and information to measure complex organizational order (not disorder). If the energy hierarchy law applies to all scales, then the molecular distribution of energy (fig. 4.14a) is indicating hierarchical organization of molecules (not disorder). Each year, advances in molecular science are finding more and more organized complexity at the molecular level. (See chapter 8.)

When there are large available energy flows from outside, the scale of self-organizational processes increases. The continuous energy distribution breaks up into autocatalytic units (fig. 4.14b). Each autocatalytic unit occupies a zone of the energy spectrum on its scale, and other units form in adjacent scales. Recently, papers in the systems science and ecological literature have studied the observed width of the energy spectrum occupied by species of animals and the gap in energy concentration before the next species in the network.[11]

If a surge of input energy of one kind is added to a system, it creates a bulge in the energy spectrum, causing energy to be propagated upscale and downscale. For example, the average distribution of energy in water waves in the ocean is like fig. 4.9, with many waves of small energy and few of larger energy.[12] When a storm passes, it generates waves with energy in the middle of the spectrum, causing a bulge in the spectral graph. Some waves interact to form larger waves, but most lose energy to friction, moving downscale to waves of lesser energy and heat.

INFORMATION IN THE ENERGY HIERARCHY

On earth, information has the highest transformities of the energy hierarchy. Here *information is defined as the parts and relationships of something that take less resources to copy than to generate anew.* Examples are the thoughts on a subject, the text of a book, the DNA code of living organisms, a computer program, a roadmap, the conditioned responses of an animal, and the set of species developed in ecological organization. Each of these takes emergy to make and maintain.

Information is carried in energy flows and storages. For example, information passes between people in sound waves of their voices, over telephone wires, and through radio waves. Biological information in genes passes by means of the seeds of plants and eggs and larvae of fishes. Information is stored in libraries, computer disks, human memories, and the archaeological remnants of history. All these information carriers and memory devices contain available energy. However, the energy of the carrier (e.g., calories in paper, brain cell functions, computer disks) is small compared with the emergy involved in creating the information. The emergy/energy ratio (transformity) of information generally is quite high.

Because information has to be carried by structures, it is lost when the carriers disperse (second energy law). Therefore, emergy is required to maintain information. Information is maintained by copies made faster than they are lost or become nonfunctional. But copying from one original is not enough because errors develop (second law), and copying doesn't make corrections.

So in the long run, maintaining information requires a population operating an *information copy and selection circle* like that in fig. 4.15a. The information copies must be tested for their utility. Variation occurs in application and use because of local differences and errors. Then the alternatives that perform best are selected, and the information of the successful systems is extracted again. Many copies are made so that the information is broadly shared and used again, completing the loop. In the process, errors are eliminated, and improvements may be added in response to the adaptation to local variations.

For example, a wiring diagram is used to make radios and is widely used with some variations. The radios that work best are selected and the information extracted into plans again. These successful plans are copied and widely shared and reapplied.

The life cycles of organisms are good examples. Seeds are planted to make many individuals with variation caused by local conditions and errors; the best performers are selected by nature or by humans. These are used to make seeds, which are broadly shared and available to replant for the next generation.

The position of information stages in the energy hierarchy is shown in fig. 4.15b by the positions from left to right. Making the first copy takes large emergy. Examples are writing a book or evolving a new species. The first copy contains the emergy of its formation.

If 1,000 copies are made, the emergy in each copy has 1/1,000 of the formation emergy plus that used to make the copy. The emergy of the set of copies is increased by the copying. When the copies are distributed so that information is broadly shared, and if more emergy is used by advertisers or receivers to incorporate the copies into their functions, then the set of information that is now shared has the highest emergy, with a high transformity (fig. 4.15b). With many copies in use, the shared information has a large territory and a slower depreciation turnover time.

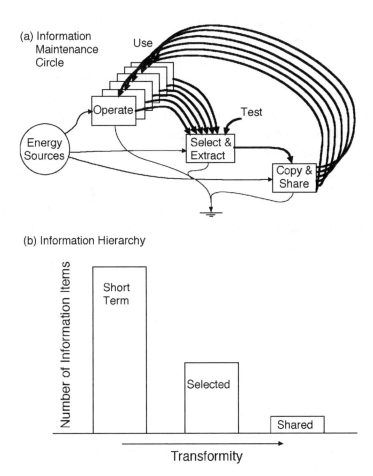

FIGURE 4.15 Information and the energy hierarchy. **(a)** Information maintenance cycle; **(b)** information classes and transformity.

When shared information is used to make and operate systems over the territory of application, the emergy of the information is absorbed into energy transformations, and the emergy concentration is returned to its lower transformity. An information circle resembles the mineral cycle somewhat. However, in a material cycle matter is conserved. In an information maintenance circle, the information is increased with copying and decreased with selections and depreciation, but a successful circle maintains enough copies to exceed depreciation and destruction rates.

MAXIMUM EMPOWER AND SCALE

When we consider systems containing several scales, the statement of the maximum power principle (chapter 3) becomes ambiguous. Maximizing power could imply that the lower part (lower quality) of the energy network gets priority because that is where energy flows are greatest. The energy hierarchy principle provides that each level in the energy hierarchy self-organizes the maximum power possible for its part of the spectrum at the same time. In other words, no level has priority over the others, but more power comes from collective development of all scales.

To discuss and compare many levels, we express each energy transformation in *empower* units (which puts all values in terms of energy of one type and scale). Thus, we amend Lotka's principle and propose that *self-organization develops designs to maximize empower of each scale at the same time.* Although the energy flows are of different magnitude, empower of small, intense information storms at the top of the energy chain of society may contain a similar range of empower values as higher-energy flows of environmental life support. The hypothesis is that the importance of an exchange between scales is in proportion to the empower involved.

Human society occupies the middle scales of the earth's energy hierarchy, where we can see so much confusing detail that most people don't even look for principles. Some suggest that the middle levels are more indeterminate, with more randomness and fewer scientific principles, a belief that fits ideas of human glory, independence, and control by free market choices. But according to the principles of this chapter, patterns of society and environment fit the energy hierarchy because the trial and error of human social behavior is constrained and adapted to fit.

NATURAL VALUE

By measuring work from any scale in comparable units, emergy measures the real wealth generated by nature and by humans. By measuring what is required to make something, emergy measures the real contribution to the system of environment and society. As explained in chapter 9, emergy measures real wealth on any scale of time, whereas money measures the human services needed to buy it in the short term.

ENERGY HIERARCHY IN THE COSMOS

If the energy hierarchy concepts are general laws, they should apply to the universe. We can examine the observations of astronomy with the expectations of energy hierarchy summarized in fig. 4.4. Many students of astronomy have suggested that the stars and galaxies of the universe are hierarchically organized (Charlier 1908; Lerner 1991), although the majority have doubts. The idea is that planets are organized around stars, stars are organized in galaxies, galaxies are organized in clusters, and clusters are organized in superclusters. The organization on each scale of view is part of a larger organization on the next larger scale of view.[13]

Let's consider how energy laws may account for the structure and functions of the universe. In the vast realm of space, stars and other units that self-organize are gravity produced, as described in astronomy textbooks (Carroll and Ostlie 1996; Chaisson and McMillan 1998). Under the pull of their own gravity, units of matter fall together, concentrating mass and energy and developing structure. The resulting increase of mass at the center increases gravity and captures more material. The pull of gravity between two bodies is the product of their masses divided by the square of their distance apart. The self-organization of a concentrated mass in space according to the gravity equation fits the autocatalytic energy diagram of self-organization for the maximum power principle (fig. 4.16).

The potential energy of mass falling inward together is concentrated and transformed into heat and kinetic energy of rotation. When the gravity and temperature are high enough, fusion reactions start that convert the mass of hydrogen into energy, turning such units into light-emitting stars. Subsequently, there are sequences of structural change not unlike succession in ecosystems (fig. 3.12). All units send out radiant energy, which disperses, losing concentration and thus degrading, consistent with the second law. The points of light that we see in the night sky are the energized centers of this pattern of energy and matter.

According to the energy hierarchy theory, self-organization of separate units should result in a connected energy hierarchy pattern forming an energy transformation series, drawn from left to right like that in fig. 4.4. The distribution of energy in the universe as a function of wavelength, shown in fig. 4.17, is decreasing to the right. The shorter the wavelength, the more energy there is per photon of light. Most of the universe's energy is in microwave radiation that radiates back and forth in equilibrium, with the dilute ordinary matter at 2.7°K, where zero is

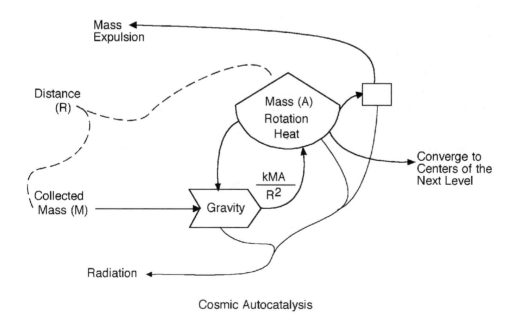

Cosmic Autocatalysis

FIGURE 4.16 Energy systems diagram of the autocatalytic action of gravity in developing a star.

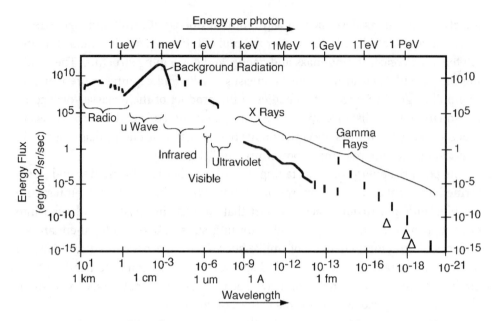

FIGURE 4.17 Distribution of radiant energy in the universe (Kolb and Turner 1986; Hoyle et al. 2000). Energy flow by wavelength. sr, steradian.

absolute zero (defined by the third energy law for conditions without the molecular motions of heat). This uniformly distributed energy is called *background radiation* (fig. 4.16). Heavenly bodies, mass aggregates, stars, and so on, with more concentrated energy, radiate shorter-wavelength radiation, some visible and some the high-energy X-rays and gamma rays. Thus abundant low-quality energy on the left appears to be connected through the series of transformations to centers that have more concentrated energy on the right.

Mass in the universe also fits the energy hierarchy. Most of the mass of the universe is ordinary matter evenly dispersed or in small aggregates. Going from smaller to larger scale, the mass of stars increases as their numbers decrease. Plots of mass distribution are hierarchical (many small stars and fewer massive stars). Plots on logarithmic coordinates show a nearly straight line relationship of star numbers and star mass (Carroll and Ostlie 1996). In ecosystems where biomass storage increases upscale, animals are larger. In the cosmos, where mass storage increases upscale, the stars get smaller because the intense gravity of large masses packs the matter into tiny volumes. The territories of gravitational influence increase, with increasing distances between massive centers. The turnover times increase.

All the bodies recognized in space are known to accumulate storages and then emit pulses of radiant energy and expulsions of mass. Their growth cycles and life history behavior are not unlike those of populations in ecosystems. Mechanisms involve sequences of nuclear reaction, centrifugal force, concentration by gravity, and thermal expansion that accumulate the conditions to cause explosive mass expulsion. For example, there are supernova explosions of old stars. Periods of energy accumulation followed by pulsing outflows are known or have been proposed for most of the universe's objects, ranging from proto-star aggregates and stars to the quasars and black holes of galaxies. At the centers of larger-scale organization, the energy hierarchy concepts predict the most concentrated bodies at their centers, with the most intense pulses and the longest accumulation periods between pulses.

Chemically, most of the ordinary matter of the universe is hydrogen, widely dispersed as background matter. Nuclear reactions in the stars generate chemical atoms of higher atomic weight. Figure 4.18 shows the cosmic abundance that results, another example of the energy-driven hierarchy of materials described with fig. 4.12.

As in earlier studies of other realms, drawing a hierarchical energy systems diagram is a useful method for organizing facts about energy and matter while considering theories. Figure 4.19 is an energy systems diagram that models the universe, showing principal structures, energy, and mass flows. The diagram separates populations of different scale and position in the energy transformation hier-

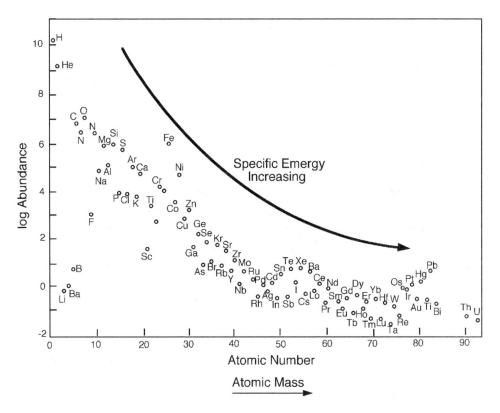

FIGURE 4.18 Distribution of matter in the universe by atomic weight, following Brownlee (1992) using the data of Anders and Grevesse (1989). Transformations using the mass–energy of the abundant hydrogen on the left generate heavier atoms in stars higher in the energy hierarchy (to the right).

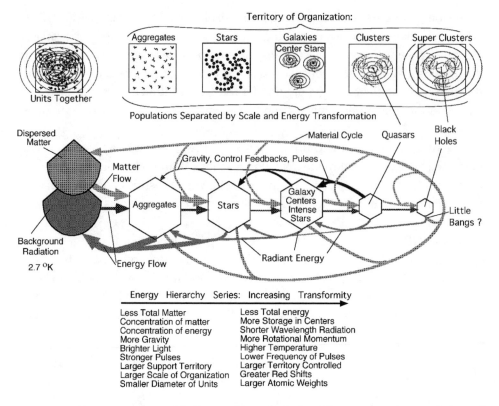

FIGURE 4.19 Energy systems model of energy hierarchy in the universe, separating populations of structures and processes according to their scale and energy transformations, increasing from left to right (Odum 2002).

archy from left to right. The background radiation and dilute matter are the lowest concentrations on the far left. These contribute energy and gravity-concentrated mass into matter aggregates. The aggregates condense into higher-energy, visible units to the right. For each level of increased scale (e.g., galaxies, galaxy clusters), there is a more intensive center (smaller and denser because of the increased gravity). Many stars are organized around larger gravitational centers, and on a larger scale these are organized around even larger gravitational centers. From left to right, nuclear reactions generate elements of higher atomic weight but in lesser quantities. The larger the territory of organization, the higher in the energy hierarchy is its center unit. To the far right are black holes with enough gravitational mass to draw in light energy and inhibit outbound radiation. Mechanisms are known or proposed to generate high-energy radiation pulses and emissions even by black holes.

The model of fig. 4.19 is not unlike models of systems on Earth, which operate in quasi–steady state. Each level of hierarchy has its cycle of transformation, growth, and pulsing but in the long term maintains an average presence and function overall. Radiant energy has mass and delivers light pressure, which helps recycle matter to the background state while being absorbed and reradiated at longer wavelengths.

Energy hierarchy concepts help explain how the fairly uniform distribution of background energy and hydrogen matter (mass energy) helps generate and sustain the fantastic forms and variety in the heavens. The dilute hydrogen mass is transformed to energy as it concentrates. To illustrate the energy hierarchy concept, the author calculated a transformity relating the energy of the sun and the earth to the sun's share of the universe's background energy and mass energy (fig. 4.20).[14]

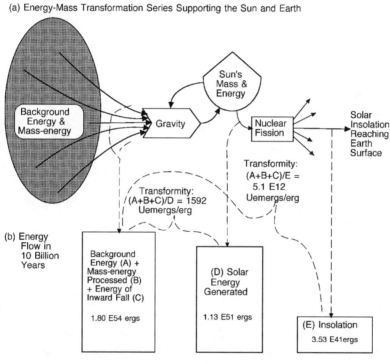

(a) Energy-Mass Transformation Series Supporting the Sun and Earth

A = Background energy in the volume occupied by the background mass used
   = (2.0 E33 g of sun)(4 E-13 ergs/cm^3)(5 E30 cm^3/g) = 4.0 E51 ergs

B = Mass-energy equivalent of background mass used (E = mC^2).
   = (2 E33 g/sun)(3 E10 cm/sec)^2 = 1.8 E54 ergs

C = Gravitational potential energy from the collapse of matter
   = 6.6 E48 ergs (Allen, 1981)

D = Solar Energy emitted in the 9 billion year life of the sun
   = (4 x 10^{33} erg/sec)(3.15 x 10^7 sec/yr)(9 x 10^9 yrs) = 1.13 x 10^{51}ergs

E = Solar energy reaching the earth's surface
   = (3.93 E24 J/yr)(1E7 erg/J)(9 E9 yrs) = 3.53 E41 ergs

FIGURE 4.20 Calculations of the sun and earth's share of background energy (Odum 2002). Uemergs, universal emergy ergs. Analogous to the calculation of transformities in solar emjoules per joule (or solar emergs per erg), the uemerg is used here to express everything in emergy of the most abundant, lowest-quality energy, the background radiation of the universe.

The Maxwell–Boltzmann distribution (fig. 4.13) shows how a smooth, uniform distribution of matter and energy can make a hierarchical, "lumpy" distribution of matter and energy. Once there is a mass center, background radiation can help raise its temperature because of the spherical geometry of incoming radiation. Once there is some mass concentration, its fields of gravity can evolve into autocatalytic units (fig. 4.16) and grow (the classic explanation of star formation). Self-organization may reinforce and select spiral organization and rotation, a design that helps maximize power transformations and feedback efficiency. Once there are units of higher mass and energy, these can be self-organized to converge into higher-level centers that have higher concentration and territory of influence, as in any energy consumption chain (figs. 4.4, 4.8). Hoyle et al. (2000) suggest that quasars at the center of clusters are important in pulsing recycle of energy and matter.

Steady State Universe

The big bang concept of cosmology that is most popular among astronomers is based largely on the *red shift* of light. When a person drives a car away from a bell tower, the sounds reaching the driver shift to a lower pitch because the delayed arrival of the sound waves mimics sound waves of longer wavelength (called the Doppler effect). Similarly, when light of known wavelength comes from a star that is traveling away from Earth, the light is received with a longer wavelength, one shifted toward the red. Nearly all the stars we can see are red shifted, and stars further away are red shifted more, as if they were accelerating away faster.

However, Zwicky (1929), following Einstein, suggested that light could be red shifted if some influence in space was taking energy out of the light (tired light theory). Because the speed of light in space is constant (186,000 miles per second), reducing light energy doesn't slow the light down but instead increases its wavelength, which reduces the energy of its photons. Light is known to be pulled by gravity. The more massive the star, the more its gravitational pull can red shift its own light emissions. According to the energy hierarchy concept, the most massive stars are centers of larger systems and bright enough to be seen at great distance. Thus, larger red shifts are further away but not necessarily traveling away. As the model in fig. 4.19 suggests, radiant energy dispersing from stars loses concentrations by diverging. Perhaps it is in the nature of diverging light in space for some red shift as part of energy recycle. Thus, several factors may contribute to observed red shifts.

When viewed as a whole with a long scale of time, the universe could have a constant average pattern of structure, mass, and energy flows (fig. 4.19). The energy hierarchy concepts seem to make the big bang theory unnecessary. At the top of the energy hierarchy (right side of fig. 4.19) are the most intense gravity centers, the black holes, which may be expected to have little bang explosions at long intervals. Some kind of little bang with intense radiations was reported recently in *Science*

(Schilling 1999). If the intense packing in black holes regenerates hydrogen, as has been suggested, the pulsed bang of black holes could help recycle the hydrogen used up in the nuclear fusion of stars.

The kind of steady state universe described in fig. 4.19 views the heavens as a perpetual Christmas tree with flashing lights. This kind of steady state universe should not be confused with the disappear-and-reform idea or the *big bang and crunch* theories of steady state (Steinhardt and Turok 2002). In the latter, a big bang blows the whole universe apart, followed by big crunches in which everything falls together again into a tiny mass before banging apart again. Advocates might find the bang and crunch idea to be consistent with the theory of energy hierarchy pulsing.

Summary

The network of all energy transformations forms a hierarchical series that explains the capabilities of available energy of different kinds to do work. The energy hierarchy accounts for ecological food webs, spatial organization of centers, the hierarchy of material concentrations, the increase of storage concentration with scale, and the inverse relationship of the frequency and intensity of pulses. Energy hierarchy is found in molecular populations, even at equilibrium. Many of the important facts of astronomy and astrophysics seem to be explained with a dynamic but steady state model of the cosmos based on the energy hierarchy.

Transformity indicates the position of each kind of energy in the hierarchy and the ability of that energy to amplify processes when organized for maximum power. Emergy and transformity measure the work from all scales on a common basis, a measure of natural value. In energy evaluations each exergy value should be multiplied by its transformity so that totals can be expressed in emergy units, a measure independent of scale. Because of the energy hierarchy, materials are also organized hierarchically in space and in concentration series. Self-organization develops systems and relationships that maximize production and the use of real wealth (empower). Emergy and transformity provide quantitative measures to guide public policy to maximize real wealth of the environment and society. At the top of the energy hierarchy is information, which depends on a copying cycle. Widely shared information is the highest of all transformities. More explanations and examples of energy hierarchy are available elsewhere (Odum 1996).

Bibliography

Note: An asterisk denotes additional reading.

Allen, C. W. 1981. *Astrophysical Quantities*. London: Athlone.

Anders, E. and N. Grevesse. 1989. Abundance of the elements. Meteorite and solar. *Geochimica et Cosmochimica Acta*, 53: 197–214.

Brownlee, D. E. 1992. The origin and early evolution of the earth. In S. S. Butchers, R. J. Charlson, G. H. Orians, and G. V. Wolfe, eds., *Global Biogeochemical Cycles*, 20. New York: Academic Press.

Carroll, B. W. and D. A. Ostlie. 1996. *An Introduction to Modern Astrophysics*. Reading, MA: Addison-Wesley.

Chaisson, E. and S. McMillan. 1998. *Astronomy*, 3rd ed. Upper Saddle River, NJ: Prentice Hall.

Charlier, C. V. L. 1908. Das planeturische Rotations. *Arkiv for Mathematik, Astronomi, och Fysik*, 4(1).

Christaller, W. 1966. *Central Places in Southern Germany*. Translated by C. W. Baskin. Englewood Cliffs, NJ: Prentice Hall.

Copeland, B. J. and H. D. Hoese. 1967. Growth and mortality of the American oyster *Crasostrea virginica* and high salinity shallow bays in central Texas. *Publications of the Institute of Marine Sciences*, 11: 149–158.

Doxiadis, C. A. 1977. *Ecology and Ekistics*. Boulder, CO: Westview.

Feekes, G. 1986. *The Hierarchy of Energy Systems from Atoms to Society*. Elmsford, NY: Pergamon.

*Haar, T., H. Vonder, and V. E. Suomi. 1969. Satellite observations of the earth's radiation budget. *Science*, 163: 667–668.

Hoffman, S. M. and A. Oliver-Smith, eds. 2002. *Catastrophe and Culture, the Anthropology of Disaster. School of American Research, Santa Fe, New Mexico*. New York: Oxford University Press.

Holling, C. S. 1992. Cross scale morphology, geometry and dynamics of ecosystems. *Ecological Monographs*, 62(4): 447–502.

Hoyle, F., G. Burbidge, and J. V. Narlikar. 2000. *A Different Approach to Cosmology*. New York: Cambridge University Press.

Hutchinson, G. E. 1954. The biochemistry of the terrestrial atmosphere. In G. P. Kuiper, ed., *The Earth as a Planet*, 371–433. Chicago: University of Chicago Press.

Jarvis, R. S. and M. J. Woldenberg. 1984. *River Networks, Benchmark Papers in Geology #80*. Stroudsburg, PA: Hutchinson Ross.

Jorgensen, S. E. 1997. *Integration of Ecosystem Theories: A Pattern*, 2nd ed. Dordrecht, The Netherlands: Kluwer.

Jorgensen, S. E., H. Mejer, and S. N. Nielsen. 1998. Ecosystems as self organizing critical systems. *Ecological Modeling*, 111: 261–268.

Kang, D. 1998. *Pulsing and self organization*. Ph.D. dissertation, University of Florida, Gainesville.

Kolb, E. W. and M. S. Turner. 1986. *The Early Universe*. Reading, MA: Addison-Wesley.

Leopold, L. B. and W. B. Langbein. 1962. The concept of entropy in landscape evolution. U.S. Geological Survey Professional Paper 500-A. Washington, DC: U.S. Government Printing Office.

Lerner, E. J. 1991. *The Big Bang Never Happened*. New York: Times Books.

Mandelbrot, B. B. 1983. *The Fractal Geometry of Nature*. San Francisco: W.H. Freeman.

Margalef, R. 1958. Temporal succession and spatial heterogeneity in ecology. In A. A. Buzatti-Traverso, ed., *Perspectives in Marine Biology*, 323–349. Berkeley: University of California Press.

Odum, H. T. 1976. Energy quality and carrying capacity of the earth. *Tropical Ecology*, 16(1): 1–8.

——. 1987. Living with complexity. In *Crafoord Prize in the Biosciences*, 19–85. Stockholm: Crafoord Lectures, Royal Swedish Academy of Sciences.

——. 1988. Self organization, transformity, and information. *Science*, 242: 1132–1139.

——. 1996. *Environmental Accounting, Emergy and Decision Making*. New York: Wiley.

——. 2002. Energy hierarchy and transformity in the universe. In *Proceedings of the Emergy Conference, Sept. 20, 2001*, 1–14. Gainesville: Center for Environmental Policy, University of Florida.

Odum, H. T. and E. C. Odum. 1983. *Energy Analysis Overview of Nations*, with sections by G. Bosch, L. Braat, W. Dunn, G. de R. Innes, J. R. Richardson, D. M. Scienceman, J. P. Sendzimir, D. J. Smith, and M. V. Thomas. Working Paper #WP-83–82. Laxenburg, Austria: International Institute of Applied Systems Analysis.

Odum, H. T. and R. F. Pigeon, eds. 1970. *A Tropical Rain Forest*. Oak Ridge, TN: AEC Division Technical Information.

Revelle, R., W. Broecker, H. Craig, C. D. Keeling, and J. Smagorinsky. 1965. Atmospheric carbon dioxide. In *Restoring the Quality of the Environment*, 111–133. Washington, DC: President's Science Advisory Committee.

Schilling, G. 1999. Watching the universe's second biggest bang. *Science*, 283: 2003–2004.

Scienceman, D. 1987. Energy and emergy. In G. Pillet and T. Murota, eds., *Environmental Economics*, 257–276. Geneva: Roland Leimgruber.

Sellers, W. D. 1965. *Physical Chemistry*. Chicago: University of Chicago Press.

Sendzimir, J. P. 1998. *Patterns of animal size and landscape complexity: Correspondence within and across scales*. Ph.D. dissertation, University of Florida, Gainesville.

Steinhardt, P. J. and N. Turok. 2002. A cyclic model of the universe. *Science*, 296: 1436–1439.

Straskraba, M. 1966. Taxonomical studies on Czechoslovak Conchostraca III. Family Leptestheriidae, with some remarks on the variability and distribution of Conchostraca and a key to the Middle-European species. *Hydrobiologia*, 27: 571–589.

Troncale, L. R. 1978. Linkage propositions between fifty principal systems concepts. In G. J. Klir, ed., *Applied General Systems Research*, 29–52. New York: Plenum.

——. 1988. The new field of systems allometry: Discovery of empirical evidence for invariant proportions across diverse systems. In R. Trappl, ed., *Cybernetics and Systems '88*, Part I, 123–130. Dordrecht, The Netherlands: Kluwer.

Ulgiati, S. 2001. Energy, emergy and embodied exergy: Diverging or converging approaches? In M. T. Brown, S. Brandt-Williams, D. Tilley, and S. Ulgiati, eds., *Emergy Synthesis. Theory and Applications of Emergy Methodology*, 15–32. Gainesville: Center for Environmental Policy, University of Florida.

U.S. Department of Agriculture. 1970. *The World Food Budget*. Foreign Agricultural Economic Report No. 19. Washington, DC: USDA.

Veizer, J. 1988. The earth and its life: System perspective. *Origins of Life and Evolution of the Biosphere,* 18: 13–39.

Zeuthen, E. (1953). Oxygen uptake as related to body size in organisms. *Quarterly Review of Biology,* 28: 1–12.

Zwicky, F. 1929. On the red shift of spectral lines through interstellar space. *Proceedings of the National Academy of Sciences,* 15: 773–779.

NOTES

1. Efficiency of transformation is the output energy divided by the inputs. In fig. 4.1a the efficiency is 34.7% and in fig. 4.1b 9.9%.

2. Although our embodied energy concept had been in use since 1967 and was used in *Environment, Power, and Society,* emergy units were defined in 1983 to clarify the confusion that arose from use of the same units for both embodied energy and energy. We purposely avoided the confusing practice of taking over a common word in general use for a quantitative measure. Instead, we sought a new word. David Scienceman (1987), who collaborated in our energy analysis studies for two decades, after library scholarship searches suggested the word *emergy,* which implies *energy memory.* Emergy records the available energy previously used up, expressed in units of one kind but carried as a property of the available energy of continuing outputs.

3. For convenience, in text and tables we use the computer notation E1, E2, E3, etc., which means to multiply by 10^1, 10^2, 10^3, etc., which is the same as multiplying by 10, 100, 1,000, etc., moving the decimal to the right. The notation E-1, E-2, E-3, etc., means to multiply by 10^{-1}, 10^{-2}, 10^{-3}, etc., which is the same as dividing by 10, 100, 1,000, etc., or multiplying by 0.1, 0.01, 0.001, etc. The notation moves the decimal to the left.

4. With biological and societal examples, Feekes (1986) sought similarities in energy activity at each level by identifying phenomena on each scale with a verbal model diagrammed with four boxes. The four categories were ability, performance, impulse, and motivation.

5. This quotient was proposed as a measure of energy quality (Odum 1976) and called the energy quality ratio and the energy transformation ratio, but it was renamed *transformity* in 1983 (Odum and Odum 1983). Because emergy and its intensive measure, transformity, refer to a property of its surroundings, it is a new kind of dimension. It does not have the dimensions of energy. It is not a dimensionless ratio. Emergy calculation in practice involves measuring observed energy transformations, but the best (lowest) transformity that is compatible with the maximum empower of open systems operation appears to be a thermodynamic property of the universal energy hierarchy. Processes operated wastefully have higher transformities than the minimum best possible.

6. Luna Leopold discussed the energetics of watersheds in relation to Ilya Prigogine's early theory that systems organize to *minimize power.* As explained in chapter 3 and illustrated in chapter 5, the opposite is true. Watersheds self-organize for maximum power, as do chemical reaction systems, which Prigogine finally recognized later, reversing himself, with his dissipative structures theory.

7. Many publications find the sizes of organisms increasing with their territories and position in food chains as the total population numbers and energy flows decrease (Straskraba 1966).

8. To describe the hierarchy of the universe from molecules to the stars, Veizer (1988) plotted stored mass representing spatial organization and half life representing turnover time.

9. A logarithmic expression for Gibbs free energy of a difference of concentration (C_1, C_2) comes from integration of the product of pressure and volume (which is energy), using the pressure and volume relationship of the gas law, which includes absolute temperature T. R is a constant, and n is the number of molecules (expressed in moles):

$$DF = nRT \ln C_2/C_1.$$

For an example, see note 9 in chapter 3.

10. The idea that randomness was inherent in molecular populations was encouraged by the fact that a hierarchical distribution of molecular velocities is Gaussian (bell shaped, normal), a pattern found in devices that generate symmetric variation. Because the kinetic energy of molecules is a square of their velocity, a bell-shaped velocity distribution has an exponential distribution of energy (fig. 4.13a). A reverse interpretation is that the energy of molecules is distributed exponentially because of the energy hierarchy principle (self-organization for maximum interlevel energy flows). The expression for the plot of the Maxwell–Boltzmann distribution says that there is a constant percentage decrease of energy quantity with increase of molecular energy flow (fig. 4.13a), and thus the distribution of molecular energy flows is exponential. As a result, the velocity distribution is normal. No assumptions of inherent randomness are needed.

11. The energy spectrum in living systems (energy quantity and concentration in fig. 4.14b or energy flow and transformity in fig. 4.3) is not continuous but is made up of populations connected in food webs (Holling 1992; Sendzimir 1998).

12. Ocean wave climate is represented in manuals for wave forecasting showing energy spectra. The distribution of ocean waves of different energy on average has the typical energy hierarchy spectra and frequencies.

13. Lerner (1991) provided a table showing hierarchical characteristics of heavenly bodies and groups.

14. In fig. 4.20 the sun's share of the universe's radiant energy and mass energy is used to estimate a transformity of the sun and earth in terms of the background energy required. 1,592 joules of background energy and mass energy is the universe's low-quality contribution to generate each joule of the sun's nuclear-based solar irradiation. 3.2 E9 joules of solar irradiation are required for each joule of solar energy received at the earth's surface.

CHAPTER 5

ENERGY AND PLANET EARTH

T HE ENERGY hierarchy principles from chapter 4 invite us to look at the planet
Earth as a diversity of units organized to maximize empower. The self-organizing units of air, ocean, and land are intimately coupled by the circulation of material and the automatic planetary cogeneration of nature's giant heat engines. The
earth's multiple energy sources process materials through energy transformation
networks, developing patterns of concentration and scarcity within watersheds and
seascapes. In the latest of the earth's episodic pulses, fuel use by civilization is
enriching the earth's emergy and changing climate. This chapter explains how the
energy hierarchy distributes the resources on which life and civilization depend.

GLOBAL HEAT ENGINES

The practical operational definition of energy is something that can be converted
100% into heat, where heat is the motion energy of invisible molecules, particles
and structures at the molecular scale (chapter 3). A heat engine uses differences
of heat concentration to drive machinery. The mechanical work that can be obtained depends on the percentage of the heat that is more concentrated than the
surroundings. Temperature measures the concentration of heat. If the Kelvin temperature scale is used, the heat is proportional to the temperature starting with the
absence of heat at 0°K. The work that can be obtained from a flow of heat is shown
in fig. 5.1a as the fraction of the heat that is at higher temperature than the surroundings. This fraction is called the Carnot ratio.[1] In fig. 5.1a it is 27/300 = 0.09,
which means that a maximum of 9% of the heat can be converted into work. A heat
engine causes fluids to circulate. The heat from absorbed sunlight causes circulation of the atmosphere and the oceans, and heat deep in the earth under pressure
is enough to make the earth flow, very slowly, like cold molasses.

   When engineers build a heat engine, they use pipes to control the movement of
the hot fluid to the cold background so as to drive pistons, turbines, and jets (locomotive in fig. 5.2a). The potential energy inherent in temperature differences does
mechanical work. In order to use all the available potential energy, some power

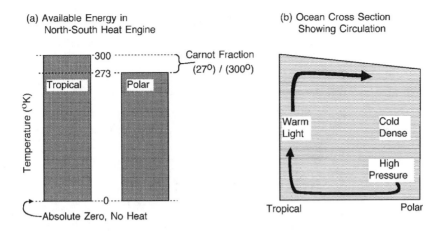

FIGURE 5.1 Concept of heat engines. **(a)** Carnot fraction; **(b)** ocean cross-section.

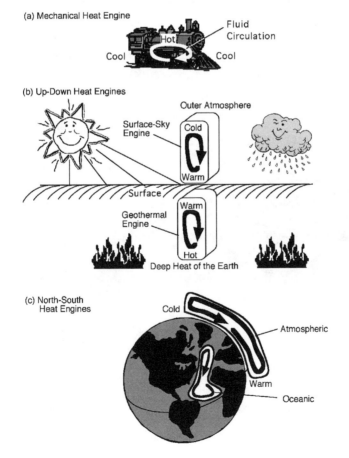

FIGURE 5.2 Heat engines. **(a)** In a locomotive; **(b)** operating between the warm earth surface and cold top of the atmosphere; **(c)** operating between warm tropics and cold polar region.

plants use *cogeneration,* in which the higher temperature range runs electric generators, whereas the smaller remaining differences in temperature provide heating and air conditioning for comfort.

The geobiosphere of the earth operates heat engines, drawing its power from solar heating and from the heat underneath the earth's surface (fig. 5.2). The geobiosphere has few pipes but builds its own fluid structures, changing potential energy of heat to the kinetic energy of circulating air, water, or plastic earth.

NONLIVING WHIRLCELLS

The heat engines feed whirlcells. These are self-maintaining units that are nonliving but have the same autocatalytic patterns (chapter 3, fig. 3.6c). They transform, concentrate, and store potential energy, using that availability to pump in more inputs and circulate the fluid materials.[2] For example, whirlcells are readily visible in the famous Bernard bowl (Swenson 1989) (fig. 5.3a), where a dish of viscous fluid is heated from below, causing the fluid to circulate and self-organize into hexagonal cells.

Each whirlcell of the atmosphere (fig. 5.3b) has input energy from the smaller scale, where energy is stored and material concentrated in potential and kinetic energy of fluid structure (natural capital). Autocatalytic feedback of winds ampli-

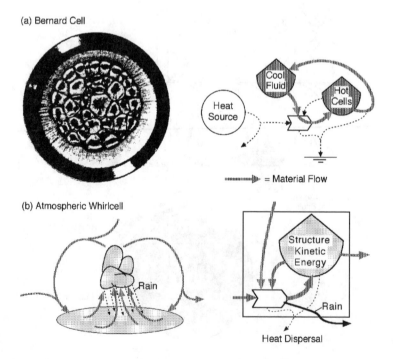

FIGURE 5.3 Nonliving, autocatalytic fluid cells units. **(a)** Bernard cell (Swenson 1989); **(b)** whirlcells of atmosphere.

fies the input transformations, diverges, and recycles material to the surroundings, with control actions from the next larger scale.

From satellite view, populations of whirlcells are visible in the patterns of cumulus clouds over the solar-heated landscape on a summer day. Each cloud marks the hierarchical center of a whirlcell with air rising in the center, spreading out, and descending in the surrounding area, to be warmed by the solar-heated ground, to converge and rise again in the center (fig. 5.3b). The system collects and converges the energy of the sun into the kinetic energy, fluid structure, and clouds. Highest transformity is in the cloud center.

Operations for Maximum Power

In *Environment, Power, and Society,* the autocatalytic nature of storms was cited as another example of a design for maximum power. Several authors, starting with G. W. Partridge in 1975, found evidence that the atmosphere organizes to maximize the rate of generation of entropy (rate of use of available energy).[3] If convection cells at each scale organize for maximum power at that scale, the combined system of many scales is maximizing empower (chapter 4). *Environment, Power, and Society* used the flows and storages in electrical circuit models as analogs to overview earth systems. Partridge found that a pulsing electrical circuit analog for the atmosphere self-organized its operation for maximum power.

The Gathering of Sunbeams

After several billion years of self-organization, the earth spreads out a broad solar collector over most of the earth and the ocean, with the property of catching both visible light and invisible infrared sunbeams, which heat the top layer of the sea. Winds evaporate water from the sea, carrying a half kilocalorie of heat energy per gram of water vapor. Each cubic meter of air at sea level carries 25 g of water vapor and a kilogram of air. The water vapor of sea air is not far from equilibrium with the water and thus has no potential energy there. But when that moisture-laden air is moved elsewhere and is manipulated by storms to condense as clouds and rain, the latent heat is released where surroundings are colder. Condensing vapor becomes concentrated heat that makes parcels of air rise like balloons, driving circulation. The temperature differences drive atmospheric heat engines that circulate air. Arriving sunbeams, when concentrated and transformed, become higher quality, the kinetic energy of flowing fluid.

The eons of self-organization appear to have optimized the proportions of earth in sea surface for catching light, in clouds for transforming energy, and in land for using the rains and winds. Seventy percent of the earth's surface is deep, dark, blue-black ocean, adapted for solar energy absorption and conversion into atmospheric water vapor (analogous to the boiler of a steam engine). Because the oceanic ecosystems keep the concentrations of nutrients and organic matter low, the water is clear. But transforming energy into winds also generates clouds, snow, and ice, which reflect some of the sunlight (32%) back into space unused (called the *albedo*). Too many clouds cause too much sun to be reflected unused. Too few clouds means not enough solar heating of the sea is being converted to drive the winds and waters. There is a self-regulating optimum cloudiness for maximizing empower.[4]

PLANETARY HEAT ENGINES

The differences in temperature maintained in air, water, and land operate several kinds of heat engines that are coupled to each other.

Surface–Sky Engine

Most of the sun's energy penetrates the atmosphere, making the earth's surface warm (about 15°C). The top of the atmosphere is cold where energy radiates out to space, balancing the solar input on average. The earth sends out infrared heat radiation, based on an overall outside earth temperature of −18°C (Wallace and Hobbs 1977:290). Therefore the temperature difference from surface to effective outside atmosphere is around 33°C. As illustrated in fig. 5.1, the available energy to drive the vertical heat engine is proportional to this difference in the concentration of heat energy.

North–South Atmospheric Engine

The cold air masses bursting out from polar areas engage the tropical air flowing toward the poles with large temperature differences. Cyclonic storms convert the potential energy in heat differences and water vapor to kinetic energy of wind circulation.

Oceanic Engine

Over half of the earth, the sun's heat makes a layer of warm water on the ocean surface. The tropical sea is 25°C warmer than the icy waters of the polar oceans. Part of the circulation of the ocean is generated by this difference in temperature (the rest by the wind stress, tide, and differences in salinity). The warm water is lighter and tends to overflow the colder waters; on the sea bottom, dense colder water tends to underflow the warmer water (fig. 5.1b). Cold polar waters from both poles move toward the equator with a deep dive to the floor of the sea, making the deep sea bottom environment near freezing.[5] The heat engine of the sea operates partly with the north–south temperature difference and partly with the up–down temperature difference (fig. 5.2).

Ocean thermal energy conversion (OTEC) technology tries to use the up–down temperature differences by substituting pipes and turbines. Such systems have been built, but the efficiencies are low because the temperature difference is small (compared with fuel-fired power plants). The net yield is small and not competitive with renewable or nonrenewable fuels on land.[6]

Up–Down Geothermal Earth Engine

On average, the temperature of the land increases downward toward the center of the earth by about 30–35°C per kilometer (Palmerini 1993). But much higher temperature gradients are found at geologic centers of convergence such as volcanic mountains. The inside of the earth is hot for four reasons. First, the earth was heated by the kinetic energy of space materials as they fell together by mutual gravitational attraction in the early history of the solar system (fig. 4.16). Much of this heat remains in the earth's center. Second, the solid earth generates heat as its radioactive elements decompose. Third, sediments at river deltas pile up, compressing the sediments underneath and generating heat. Fourth, sedimentation buries chemically reduced substances, such as organic matter, with chemically oxidized sediments, such as ferric iron, in a ready-to-react mix. Under pressure, these react, generating heat.

Between the hot interior of the earth and the cooler earth surface, a natural geothermal heat engine operates. Under high pressures the earth is plastic, and its mass circulates as convection in the mantle layer below the crust, not so different in principle from the water circulating in the steam engine.

Geothermal Technology

Geothermal technology uses the difference in temperature within the earth to circulate hot steam to operate machinery or generate electricity. A few successful systems operate in the vicinity of large temperature differences in volcanic regions (e.g., in New Zealand, Iceland, Italy, and California).[7] But over the rest of the earth, where temperature gradients are ordinary (32°C/km), yields are small and not economical.

Planetary Life Support or Economic Development

People often look to planetary heat engines as power sources for civilization in the future, as if these energies were unused resources. Actually, the heat engines are already in full use, converting heat to land forms and moving and purifying atmospheres, rivers, glaciers, and land surfaces. To tap these sources is to take power away from their roles in supporting civilization indirectly. Society depends on the environmental systems that run on planetary power. The structures and stores of environmental units are called *natural capital* (e.g., soils, forests, mountains, reefs). Natural capital, represented by the storage symbols in fig. 5.4, is an essential part of renewable power sources that keep the biosphere livable.

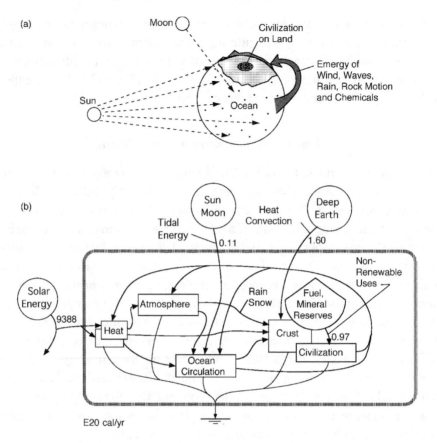

FIGURE 5.4 Energy system of the geobiosphere. (a) Sketch; (b) energy systems diagram with annual empower.

Planetary Cogeneration

The planet's several main heat engines are coupled to each other in an amazing example of cogeneration. All these engines interact as they operate so that available energy in excess for one process is passed to another. This is an example of one of the self-organizing mechanisms used to maximize empower (chapter 3). The tropical hurricane, embedded in air masses without much lateral temperature difference, is a surface–sky engine. The northeaster that develops off New England in winter is a north–south engine. The slips and bursts of earthquakes and volcanoes are part of the under-surface engine.

The circulation of the crust of the earth is driven from below by the geothermal engine and from above by the rivers and glaciers fueled by the sun-driven atmospheric engine spreading rain and snow. Sediments are eroded by rivers and glaciers, deposited underwater at the seashore, compressed into rock, and uplifted to form land again. This circulation of solid matter is called the *sedimentary cycle*.

THE UNIFIED SYSTEM OF ATMOSPHERE, OCEAN, AND LAND

The global energy systems model (fig. 5.4) helps us understand how potential energies are flexibly shared in earth processes. It shows how each of the earth engines reinforces the other engines. Figure 5.4b has the main pathways arranged from left to right in order of transformity and scale. The energy flows operating the earth are included, but each flow is of a different kind and quality. Circulation and turnover are rapid in air flows on the left, intermediate with ocean currents in the center, and slower with circulation of the solids on the right. Recall the principle of material hierarchy by which energy transformations also concentrate materials (fig. 4.12).

The circulation of water links air, ocean, and earth. The atmospheric winds drive ocean currents and cycles and accelerate arid land cycling with wind erosion. The ocean engines control the coastal parts of the land cycles. The water vapor from the sea is transformed into rain and snow over land, which sculpture watersheds and operate the sedimentary cycle. The pileup of snow forms glaciers that slide seaward, carving the earth. The earth engines help build the mountains, and these control atmospheric circulation. Maximum empower occurs when each form of energy reinforces the others.

All these engines are globally organized and coupled to sustain continents and mountains that have become hierarchical centers of the earth's surface. The planet's spatial hierarchy is visible in the astronauts' view from space (fig. 5.4a): The large blue ocean supports smaller areas of white-clouded storms, which converge on the 30% area of brown and green continents.

EARTH EMPOWER SIGNATURE

Figure 5.4b shows the annual energy flows (power) driving the earth system, and fig. 5.5b shows the emergy flows (empower). Figure 5.4a is a hierarchical overview of the whole earth. The main power sources are arranged in fig. 5.5a according to their transformities, indicating their position in the energy hierarchy. Figure 5.5b shows the empower of the four energy sources (solar energy, tidal energy, deep earth heat, and fuel reserves and minerals) supporting the assets of nature and society.[8]

The largest emergy flow of the earth system is the use of nonrenewable fuel and mineral reserves, which now exceeds all the others. Because fuels and minerals are not being replaced as fast as they are being used, the present system is not sustainable.

WHIRLCELLS OF THE EARTH

The earth forms a web of self-organizing units fueled by the heat engines (fig. 5.4) within the atmosphere, within the ocean, and within the land. The units are

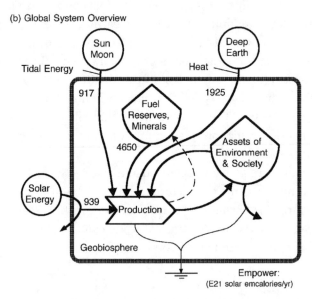

FIGURE 5.5 Overview of the geobiosphere and its energy sources. **(a)** Power and transformity of the main sources; **(b)** global systems overview and its empower basis.

arranged in diagrams from left to right in order of scale of territory, turnover time, and transformity. The empower flows and their transformities in figs. 5.5 to 5.8 show quantitatively how the air, sea, and land are hierarchically related. Energy flows (power) *decrease* from left to right as transformities and energy storage concentrations *increase* from left to the right.

ATMOSPHERE

The atmosphere contains abundant small circulation cells that converge and transform their energy into larger-scale storms that last longer and have a greater impact (fig. 5.6). The larger storms have less total energy flow but higher concentration,

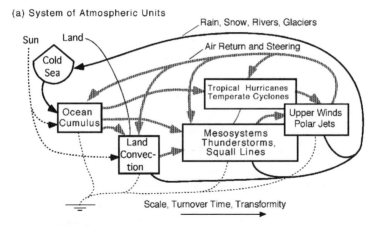

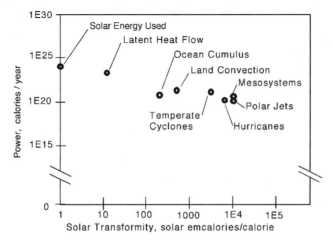

FIGURE 5.6   Energy transformation chain for the whirlcell units of the earth's atmosphere.

transformity, and control action. Storms govern what structures society can fit to a landscape. The largest whirlcell is the main hemispheric circulation of air that is most concentrated in jet streams.

Traditionally, meteorology has had a spatial emphasis that comes from the geographic pattern of weather data. Theory was developed about fields of fluid mass, force, momentum, vorticity, and energy. The larger-scale patterns were seen as an integration of the consequences of spatial fluid dynamics. Most weather forecasting models attempt to predict future geographic distributions in a map view by combining scattered data from the small scale.[9]

However, the model sought for the earth's fluid dynamics in this chapter is an alternative way of thinking. The visual image is of a food chain of self-organizing autocatalytic units (whirlcells), which draw in dilute available energy, transform, concentrate, feed back control actions, and recycle the air. The model represents a more aggregated, top-down view in which self-maintaining units move over the earth, dragging the fluid materials into a spatial pattern.

Figure 5.6 contains the principal classes of atmospheric whirlcells for planet earth, the main pathways of energy supply (left to right), and the high transformity control and water recycle pathways (right to left). The hierarchy of units is represented by plotting each unit on a graph of energy flow and transformity.[10]

Scale, territory, and turnover time also increase from left to right. The small breezes of local circulations are smallest on the left. Ocean cumulus and land convection activity, the territory of mesosystems, is next larger (squall lines and thunderstorms). The winter cyclonic storms and tropical hurricanes are in the center and may cover 1,000 miles, and the high atmospheric global circulation and its jet streams are farthest to the right, at the top of the atmospheric food chain. The territory of the polar westerlies 3 to 6 miles above the earth covers the polar half of the hemisphere. Control passes from right to left. By means of the forces of eddy fluid exchange, the jet streams control the large storms, and the large storms control the mesosystems, and these control what can develop on the small or local scales.

Atmospheric Energy Storage and Scale

The atmosphere stores available energy as both potential energy and kinetic energy. In agreement with the general hierarchy theory (chapter 4), storage increases with position in the energy hierarchy scale. For example, the kinetic energy stored in each eddy increases with the size of the eddy, although the energy flow is less.[11] The emergy stored over the earth increases with transformity (table 5.1).

Vertical Energy Hierarchy in the Atmosphere

Air high in the atmosphere is supported by air underneath. Like air in a stack of pillows, the air at the bottom is compressed by the weight above. The populations of molecular motions at the bottom support those higher up. Even for this static situation, more energy is involved at the bottom to support less at the top, as described in chapter 2.

Available energy also flows from the solar-heated earth surface upward. The kinetic energy of the whirlcells on the left in fig. 5.6 is mostly at the bottom of the atmosphere, whereas the circulations of the larger systems, such as the jet stream, on the right are mostly at a higher altitude.

Near the earth's surface, the air density is high, and the winds are the result of the general circulation plus the whirlcells of local convection. The kinetic energy available and dispersed in this layer is a cube of the velocity (Reiter 1969). Therefore, the wind energy available to windmills (if diverted from the ecosystems) is also proportional to the cube of velocity.[12]

Some of the water vapor carrying the heat of vaporization drawn from the sea surface is transferred upward as part of successive transformations in larger and

TABLE 5.1   Earth Emergy Storage Increases with Scale

Note	Item	Transformity (secal/cal)	Emergy Store (secal)
1	Meteorological structure	12–500	1.13 E23
2	Infrastructure of civilization	5.2 E6	3.8 E26
3	Glaciers	1.1 E6	7.6 E26
4	Soil	1.2 E6	1.87 E27
5	Oceanic structure	1.9 E7	3.8 E27
6	Learned information	7.7 E7	3.8 E28
7	Continental structure	8.9 E8	7.6 E33
9	Genetic information of life	3.8 E18	1.1 E34

NOTE: Storages based on global empower (sun, tide, and earth heat = 3.78 E24 solar emcalories per year) multiplied by assumed replacement times.
1. Meteorological structure replacement time = 0.03 yr. Transformity from Odum (2000b), table 2.
2. Infrastructure of civilization replacement time = 100 yr. Transformity from Brown (1980), fig. 3.25.
3. Glacier replacement time = 200 yr. Transformity from Odum (2000b), table 8.
4. Soil structure replacement time = 500 yr. Transformity calculated as follows:
   Area of continents = 1.5 E14 m^2;
   Soil depth = 0.45 m;
   % Organic matter = 1.5%;
   Energy content of organic matter = 5.4 kcal/g;
   Density of soil = 1.4 E6 g/cm^3;
   Energy in soil = (1.5 E14 m^3)(0.45 m)(1.4 E6 g/m^3)(5,400 cal/g)(0.015) = 1.58 E21 cal;
   Transformity = 1.87 E27 secal/1.58 E21 cal = 1.2 E6 secal/cal.
5. Oceanic structure replacement time = 1,000 yr. Transformity from Odum (2000b), table 6.
6. Learned information replacement time = 10,000 yr. Transformity from Odum (1996), table 12.4.
7. Continent structure replacement time = 2 E9 yr. Transformity calculated as follows:
   Volume of continents above sea level = 1.37 E17 m^2;
   Gibbs energy of rocks = 100 J/g (Odum 1996:302);
   Density of rock = 2.6 E6 g/cm^3;
   Energy in continents = (1.37 E17 m^3)(2.6 E6 g/m^3)(100 J/g)/(4.2 cal/J) = 8.5 E24 cal;
   Transformity = 7.6 E33 secal/8.5 E24 cal = 8.9 E8 secal/cal.
8. Genetic information replacement time = 3 E9. Transformity from Odum (1996:224).

larger whirlcells. The energy-containing water vapor is converted into rain at each stage as part of what is necessary to transfer a smaller amount to the next altitude. As a result, water vapor at high altitudes has higher transformity. Rains and snows that fall at high altitudes have higher real wealth value because of the high transformity of the air and that of the mountains. High mountain rains and snows have large geopotential energy that drives the rushing rivers and scraping glaciers. The great power and empower available in mountain waters generate hydroelectric power, irrigate valleys, and supply cities with water.

OCEANS

The circulation of the ocean is much slower, sluggish to initiate and slow to stop. The energies driving the sea develop water currents circulating on many scales, which are aided by the rotation of the earth basin underneath. These units of circulation are autocatalytic (fig. 5.2) and often are called *gyrals* (oceanic whirlcells). They have four main energy sources. The first is the driving forces of the winds. The winds generate waves, in which half of the energy resides in small vertical oscillations and the rest moves forward, eventually reaching the seashores. The second part of the energy comes from waters of different density (different salinity or temperature), which cause one layer to rise and flow above another. For example, freshwater rivers reaching the sea flow on top, driving major coastal gyrals. Third, the sun heats the sea surface, causing the water to expand above average sea level, especially with clear skies in the centers of the oceans. Elevated by the heating, these waters have potential energy to flow downhill. Fourth, tidal pull of the moon and sun drives water, causing elevated sea levels that progress around the world as a wave affected by the timing of the earth's rotation and the moon's orbit.

Hierarchy of Sea Circulation

Like the whirlcells of the atmosphere, the gyral units of the sea form an energy transformation chain connecting units of different size, turnover time, and transformity (fig. 5.7). The small gyrals are near the surface and on the interface with the land, where they are affected by the small, rapid processes at the earth's surface. Larger gyrals develop, analogous to the atmospheric storms, such as the vortices in the Gulf Stream. Large-scale gyrals fill each oceanic basin and are interconnected by the general circulation of the sea. In the southern hemisphere at 40° to 50° south latitude waters circulate rapidly around the whole earth, unblocked by continents. The ocean's central position in fig. 5.4 is appropriate because the components of the sea have higher transformities than the ordinary units of the atmosphere.[13]

The oceanic energy hierarchy is also organized vertically, with circulation scale increasing with depth. Deep circulations have larger dimensions and longer turnover times. The bioluminescent flashes of the animals increase in size, duration, and interval with depth (Clarke and Hubbard 1959). Deep waters circulate between oceans, forming a large-scale general circulation of the sea.

Oceanic Control of the Atmosphere

Whereas the atmosphere drives the ocean currents and often heats or cools the sea, water at a higher position in the energy hierarchy can control the atmosphere by arranging the storage of solar energy in seawater (figs. 5.6 and 5.7). Water is a thousand times denser than air and can store much more heat per unit volume.

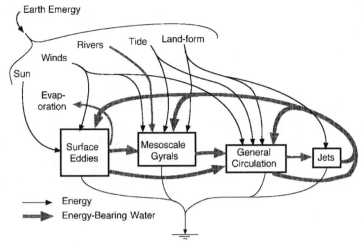

(a) System of Oceanic Circulation

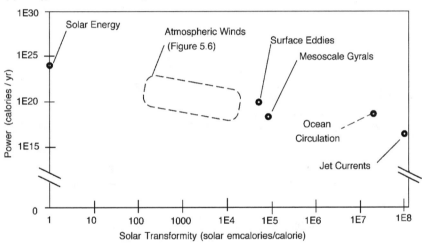

(b) Energy Flow and Transformity of Currents

FIGURE 5.7 Hierarchy of gyral (whirlcell) units in the ocean. **(a)** System of material and energy; **(b)** power and transformity.

The 3- to 10-year equatorial east–west El Niño and La Niña oscillations of weather involve different positions of heated tropical seawater along the equator. By its mutual coupling to the sea, the atmosphere gains a larger scale territory, turnover time, and transformity.

THE GEOLOGIC ENERGY SYSTEM

Over 5 billion years the earth's crust has developed a system of circulation of the solid but plastic earth that appears to have the same kind of "food chain" of self-organizing units found in ecosystems, the atmosphere, and oceans. In fig. 5.8a units are arranged according to their scale and position in the energy hierarchy, determined by their emergy per mass. Each is coupled to the others. The broad area of the seafloor is volcanic rock, spreading out from areas where molten magma emerges from the deeper earth. These basaltic rocks, and the marine sediments (oozes) that settle on them, converge and interact with the continental blocks and contribute to their mass and process. The oceans and atmosphere shown on the left generate rains and snows that fall on the land systems, causing erosion. The water and glaciers return to the sea, carrying eroded sediments (dotted pathways). The sediments deposit, accumulate, and form sedimentary rock, which contributes to the continental blocks. Some of the sediments under greater temperature

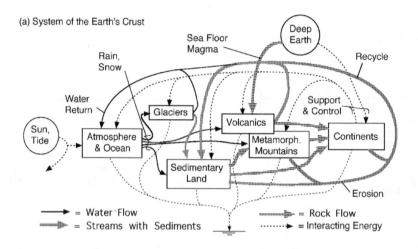

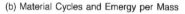

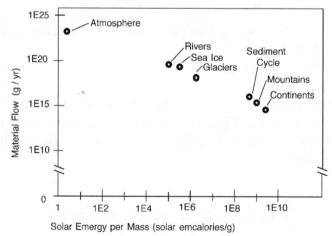

FIGURE 5.8 Self-organizing units in the geologic hierarchy. **(a)** System of materials and energy; **(b)** emergy per mass of circulating materials.

and pressure are slowly recrystallized to form granitic rocks of the continents. Rocks form and circulate to the right in fig. 5.8a as matter concentrates.[14] The cycle of the earth is closed with the recycle of eroded sediments as water returns to the sea. Some continental material returns to the deep earth. In each subsystem, accumulated storages of material and energy deliver their effects in pulses, including earthquakes, volcanic eruptions, floods, and iceberg formation.

The solid earth in its upper layers is brittle compared with earth under more heat and pressure below. Thus, the earth circulation causes the upper crust to break into plates as earth is moved along. Where the plates converge there are smaller-scale energy transformations.

Emergy Convergence in Mountains

Within continents, the self-organization of land generates mountains, which are hierarchical centers. Processes that converge mass uplift mountains and transform energy (fig. 5.9b). These processes also make mountain roots. The densities of the mountain mass are less than those of the deeper layers, so the mountains "float" on the pliable but denser earth mass below.

The higher the mountains, the faster their erosion (fig. 5.9a).[15] Wherever mountains persist, there is a balance between the mountain forming and erosion. Typically, the mountains are pulsed up and erode for a time before energies accumulate for another pulse. Thus mountains are centers of energy transformation, earth heat, and empower concentration. Mountains occur in areas of sedimentary rock, in volcanic areas, and where slow, granite-forming mechanisms upthrust old rocks. In each type of landscape system, mountains appear to be at the top of the energy hierarchy, with structural storages that are autocatalytic in reinforcing their own pattern. Higher mountains have higher emergy stores and emergy per mass.

Watershed Energy System

The rains falling on the land generate streams that carve the land as water flows downhill to the sea. In cold regions and at high altitudes, the snow falling on snow fields is compressed into glaciers, carving steep-sided valleys as they slip to the sea. Flowing waters carry the sediments that they have eroded toward the sea, interacting with the topography of the land to form stream networks. Like everything else, the watershed is organized as an energy hierarchy network.

The emergy of streams comes from rains and from elevated land that provides the platform and sediments. A common pattern is shown in fig. 5.10a,[16] where streams converge, eventually producing a large river. Available energy is transformed at each stage to produce larger-scale, higher-transformity flows downstream. Highest empower density and transformity are at the river mouth. Here is where civilizations often develop. For example, fig. 5.10b shows the energy flows of the Mississippi River as it uses the energy of elevated water (geopotential energy)

FIGURE 5.9 Hierarchy of mountains. **(a)** Erosion as a function of altitude (data from Chorley et al. 1984, based on 400-km^2 quadrats in the United States and Europe); **(b)** spatial concentration of uplift and erosion; **(c)** mountain circulation increases as mountain height increases.

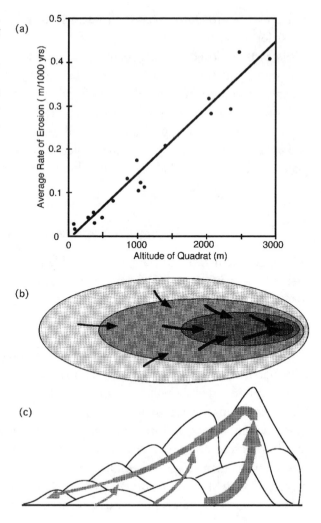

to organize the watershed and channels. Flowing downstream, power decreases, but the transformities of water increase.

Beyond its mouth, the river water spreads and diverges, either as a distributary or dispersing into estuaries (fig. 5.10a). These branching dispersals decrease in volume, transformity, and scale. Some watersheds, such as those from conical mountains are predominantly diverging. Some rivers disperse into deserts.

On an energy systems diagram, the converging stream stages are placed from left to right. Where the waters diverge again, these recirculation flows on energy diagrams are placed from right to left because they are losing concentration.

Streams may converge from the surrounding landscape into lakes that are centers of emergy concentration and transformity.[17] In some landscapes the rains percolating into the land collect and converge in underground passages to emerge as large springs, another example of high transformity and empower concentration.

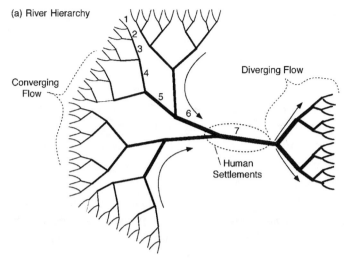

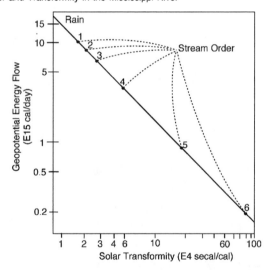

FIGURE 5.10 Energy hierarchy and watersheds. **(a)** Hierarchical convergence of stream branches; **(b)** geopotential energy flow as a function of transformity in the Mississippi River.

Work by Downstream Empower

Autocatalytic feedback is a general design characteristic of self-organization (chapter 3). It is an arrangement by transformed energy for performing useful work to reinforce the system. High-transformity waters have special capacity to increase system production with pulsing actions. The stream waters reaching the larger rivers downhill have the great power of large volumes arriving in floods. In natural organization, these floods spread into floodplains, distributaries, coastal marshes, and seashores, fertilizing with sediments, building land, causing currents, increasing aquatic production with organic matter and nutrients, and generating fertile fisheries, which are also high in the emergy scale of real wealth.

Many of the rivers of the world have been dammed for hydroelectric power and irrigation waters. In these, the emergy is diverted to cities or intensive agriculture, where the water's high emergy can be matched by combining with purchased fossil fuels as part of human enterprises. But the emergy was already being used in support of the landscape's economy before waters were diverted. In some dammed river systems, fish migrations up and down the rivers were important food sources. The question to be answered in considering whether the dams are desirable is, "Which alternative system, when matched with emergy from economic investment, generates the most real wealth?" Mark Brown and students compared the land productivity displaced by dams' reservoirs with the electric power generated (Brown and McClanahan 1996). Low dams with broad reservoirs (with small ratios of height to area) eliminated more emergy than they produced.

MINERAL WEALTH AND BIOGEOCHEMICAL CYCLES

Rich deposits of minerals historically were found scattered over the earth's crust as if a creator were scattering jewels. Each discovery produced a moment of economic bonanza, a boom town, and then a bust. Many publications have dealt with the typical pattern of mineral distribution and concentration of many deposits of low concentration and few deposits of high concentration. Economic geologists found that the distribution of mineral concentration was highly skewed, a pattern often fit by a lognormal distribution equation (fig. 5.11a).[18]

The distribution curves and the patchiness of mineral concentration seem to be explained by the principles of material hierarchy (chapter 4). Because materials are cycled by the flows of energy, and energy is hierarchically organized, distribution and circulation of materials become similarly arranged. Because available energy is transformed and used up to create products at the next level in the energy hierarchy, the amount of coupled materials that can be concentrated for each higher level has to decrease. The quantity of material that can be concentrated decreases in inverse relation to concentration. For the same flow of emergy, the quantity of minerals decreases with the transformations that increase concentrations (fig. 4.12). The spatial organization of material concentrations in centers is also explained by the energy transformation hierarchy, which maximizes power by concentrating its products.

Earth processes move trace materials along as part of the indiscriminate movements of mass (air, water, or earth) until each material reaches a large enough concentration for it to develop some autocatalytic process that uses its special properties to further concentrate it. For example, when copper dispersed as a trace metal in rocks reaches its critical concentration, it may cause an autocatalytic mineral formation. When phosphorus dispersed in waters reaches its critical

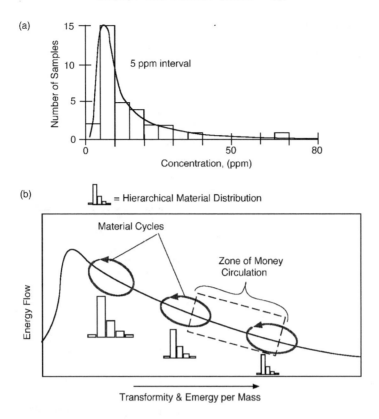

FIGURE 5.11 Material distribution and cycles in the energy hierarchy spectrum. **(a)** Typical distribution of minerals according to concentration (Ahrens 1954); **(b)** zones of transformity and mass emergy for different minerals.

concentration, it may cause organisms to collect, use, and concentrate it. Further concentration requires more energy transformations, so that the quantities concentrated decrease.

Energy for Mining

The cost and the amount of energy required to mine, concentrate, and process minerals also have a steep, hollow curve. Figure 5.12 is an example of the raw material and inputs of fuel and services required to concentrate minerals for commercial use. Low concentrations (on the left) take huge amounts of the inputs. The emergy required to concentrate minerals was calculated in fig. 5.12b so that services and fuels, in roughly equal amounts, could be added on a common basis.[19] There is a big difference between the commercially feasible deposits (higher concentrations on the right of fig. 5.12) and the concentrations in which the desired substance is dispersed among rocks (on the left). Deposits are not economical when a substance is below the concentration at which it can stimulate autocatalytic concentration.

FIGURE 5.12 Scarcity and the emergy required to concentrate minerals. **(a)** Tons required to produce copper (from Page and Creasy 1975); **(b)** emergy required.

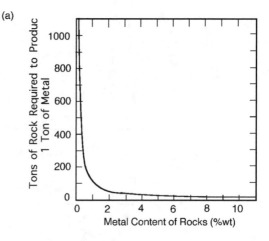

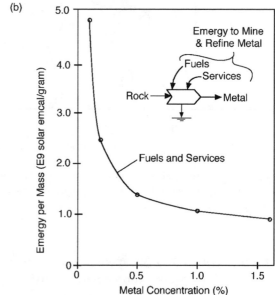

Transformity Range of Biogeochemical Cycles

Each chemical substance occupies a different range of transformities. Each of the small bar graphs in fig. 5.11b shows how the quantity of materials decreases as the hierarchical position and emergy per mass increase from left to right in a narrow zone of the hierarchy. For example, air circulates at the lower-transformity part of the earth system, water in the middle, and heavy metals further to the right. Part of the environmental problems of our time appears to result from displacement of chemical substances from their normal position in the energy hierarchy.

Interesting calculations have been made of the total material budget of nations (Wuppertal Institute, Germany; Adriansee et al. 1997). For example, a material budget of Germany includes not only all the things consumers use but also all the flows of natural materials through the landscape. Considering what a great difference there is in the emergy involved in flows of air and flows of steel, perhaps we should correct these budgets by multiplying each material by its unit emergy before making comparisons or calculating totals. An example was provided for Germany (Odum 2001) in which the mass of each material used in the economy was multiplied by its unit emergy, yielding a national emergy per mass index (1.6 E7 solar emergy calories [secal]/g). This value is intermediate between that of fluids (air and water) and the longer-lived continental structure (table 5.1).

Emergy in Materials

The real values of materials, in units of emergy per mass, are shown in fig. 5.13 and related to the quantity in the geobiosphere. The emergy per unit mass is inverse to the quantity. Materials of high value are scarce because more emergy is required to make them. The abundance of elements on Earth is the result of earth processes acting on the initial materials supplied to the earth in its formation. The original material endowment follows somewhat the abundance in the cosmos, which depends on the emergy used in hot stars in atomic element-making processes. The chemical elements with smaller atoms such as hydrogen (small atomic weight) are more abundant, and the large, heavy atoms such as uranium are scarce (fig. 4.18; Mason 1952; Ahrens 1954). This is another example of the energy hierarchy principle generating quantity–transformity graphs like fig. 4.5d. Earth materials are further organized by the energy transformations of self-organization on Earth. Materials range from 20

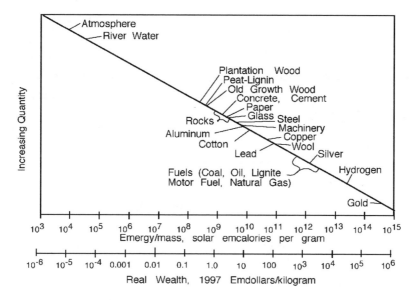

FIGURE 5.13 Emergy value of materials related to abundance. Emergy/mass data tabulated by Mark T. Brown with aid of student theses and dissertations (Haukoos 1995; Buranakarn 1998; McGrane 1998).

secal/g for air to 6 E11 secal/g for gold. Chapter 9 explains the emdollar scale in fig. 5.13, which provides economic equivalents of emergy values.

People have incorporated the concept of regarding scarce things as valuable into their culture. This concept is understandable not only because it takes more energy to concentrate scarce materials but also because scarce materials can reinforce their support systems more effectively. This is a result of their unique properties that potentially have more opportunity to amplify inflows.

A PULSING EARTH

As explained with fig. 4.9, the frequency of pulsing in the geobiosphere follows the transformities: small and frequent in the atmosphere, slower in the ocean, and long, accumulating, and catastrophic in the volcanoes and earthquakes of the more solid earth (Alexander 1978). Gutenberg and Richter (1949) found that larger earthquakes have longer accumulation times and more energy delivered per time. These data are a downsloping straight line on the pulse frequency graph in fig. 4.9.

The Geologic Scale of Civilization

Civilization has become a geologic force, affecting the whole earth. The surface of the earth has been redeveloped with dammed rivers, open pit mines, urban sprawl, highway networks, agriculture, and coastal developments. The human economy now is in the upper controlling part of the earth hierarchy (to the right in fig. 5.4b). It has devoured some of the older pathways of energy use and material cycling.[20]

Civilization is itself a high-transformity, unsustainable pulse (chapter 7), consuming the fuel storages of the earth cycles faster than they are replaced. The increased carbon dioxide from this consumption pulse threatens the biochemistry of coral reefs, increases intensity of atmospheric storms, extends droughts, and accelerates soil erosion.

Information Management of Earth

The geobiosphere has long been carpeted and controlled by life, most recently predominantly with the activities of 6 billion humans. Even before the emergence of people, earth processes were controlled by the ecosystems. The living cover controlled the weathering of rocks, the formation of soils, the organization of sediments, the building of reefs, the channels of streams, and the distribution of rains. But the organisms of ecosystems, and now humans, are programmed by their inherited genetic information, modified by learned information.

Many organisms and the activities of plants, animals, and microorganisms operate in hours, days, or a few years. But the information of life is shared across the planet and sustained through the ages. Although species have evolved in fantastic variety, the main part of the genetic information of life is in common, shared by the millions of organisms. Genetics of life culminating in human genes took a billion years to develop, a very long turnover time with a large transformity. The main essence of life has a large global territory. It was evolved as part of earth crust development along with the development of the continents. The memory of what worked was selected by the interactions of ecosystems and geologic operations on the landscape. The programs for earth operation are in the genetics of life. Genetic inheritance is a geologic property.

Gaia

The term *Gaia* was introduced by James Lovelock (1979) for the idea that life organizes the earth for the continuation of life. This author studying microcosms (Odum and Lugo 1970) suggested how ecosystems could control their climate. The Gaia concept focuses the window of our attention on that part of the universal energy hierarchy with the environmental–informational interface. It is a dramatic way of looking from the larger scale down. The shared information of life is the highest unit value (highest transformity) on Earth. But no level can operate without the levels above and below. A corollary of the energy hierarchy principle requires upscale–downscale interaction (chapter 4). Apparently all of the scales are self-organizing at the same time with the same principles. Transformities accompany other measures of scale (fig. 4.19), from molecules to stars.[21]

In the past, the self-organizational processes of the larger scale carried the information of life forward after each pulse of progress. As the pulse of the fuel-civilization of our time recedes and the earth cycles to a lower-energy state, we may hope to carry the essentials of global genetic and learned information forward for pulses yet to come. We will consider how this may be possible in chapter 13.

Summary

The flows of energy organize the planet Earth, its atmosphere, oceans, lands, and life with spatial centers, concentrations, and pulses according to the hierarchy of energy transformations. Energy systems models give us a way to overview the fantastic complexity of phenomena on planet Earth. Joined together as a natural cogeneration system, heat engines supporting the geobiosphere are coupled up and down and side to side, with excess energy in one place bypassing and reinforcing elsewhere. The heat engines drive chains of whirlcell units analogous to the food chains of ecosystems, with power flows converging upscale and energy control pathways passing downscale. With a common pattern, the similar spatial designs of self-organization show up in the pattern of atmospheric storms, the networks of rivers, the distribution of mountains and plains, the concentrations of valuable minerals, the distribution of animals, and the centers of civilization. Transformities indicate the place of all phenomena in the global energy system, where genetic information of life is the highest. Over time, the nested pulsing of the smaller scales provides the skewed noise within the rhythm of infrequent pulses of the larger scale, one of which is the fuel-driven civilization of our millennium.

BIBLIOGRAPHY

Adriansee, A. S., A. Bringezu, Y. Hammond, E. Moriguchi, D. Rodenburg, D. Rogich, and H. Schutz. 1997. *Resource Flows: The Material Basis of Industrial Economies.* Washington, DC: World Resources Institute.

Ahrens, L. H. 1954. The lognormal distribution of the elements (Part 1). *Geochimica et Cosmochimica Acta,* 5: 49–73; (Part 2) 6: 21–131.

Alexander, J. F. 1978. *Energy basis of disasters and the cycles of order and disorder.* Ph.D. dissertation, University of Florida, Gainesville.

Brandt-Williams, S. 1999. *Evaluation of watershed control of two central Florida Lakes: Newnans Lake and Lake Weir.* Ph.D. dissertation, University of Florida, Gainesville.

Brown, M. T. 1980. *Energy basis for hierarchies in urban and regional systems.* Ph.D. dissertation, University of Florida.

Brown, M. T. and T. McClanahan. 1996. Emergy analysis perspectives for Thailand and Mekong River dam proposals. *Ecological Modeling,* 91: 105–130.

Brown, M. T. and S. Ulgiati. 1999. Emergy evaluation of the biosphere and natural capital. *Ambio,* 28(6): 486–492.

Buranakarn, V. 1998. *Evaluation of recycling and reuse of building materials using the emergy analysis method.* Ph.D. dissertation, University of Florida, Gainesville.

Chapman, P. R. 1974. The energy costs of producing copper and aluminum from primary sources. *Metals and Materials,* 8: 107–111.

Chorley, R. J., S. A. Schumm, and D. E. Sudggen. 1984. *Geomorphology.* London: Methuen.

Clarke, G. L. and C. J. Hubbard. 1959. Quantitative records of luminescent flashing of oceanic animals at great depth. *Limnology and Oceanography,* 4: 163–179.

Diamond, C. 1984. *Energy basis for the regional organization of the Mississippi River Basin.* M.S. thesis, University of Florida, Gainesville.

Gilliland, M. W. 1975. Energy analysis and public policy. *Science,* 189(4208): 1051–1056.

Goldberg, E. D. 1965. Minor elements in seawater. In J. P. Riley and G. Skirrow (eds.), *Chemical Oceanography,* 163–194. London: Academic Press.

Gutenberg, B. and C. F. Richter. 1949. *Seismicity of the Earth and Associated Phenomena.* Princeton, NJ: Princeton University Press.

Haukoos, D. S. 1995. *Sustainable architecture and its relationship to industrialized building.* M.S. thesis, University of Florida, Gainesville.

Lewis, J. M. 1998. Clarifying the dynamics of the general circulation: Phillips 1956 experiment. *Bulletin of the American Meteorological Society,* 79(1): 39–60.

Lovelock, J. E. 1979. *Gaia: A New Look at Life on Earth*. Oxford, UK: Oxford University Press.

Mason, B. 1952. *Principles of Geochemistry*. New York: Wiley.

McGrane, G. 1998. *Simulating whole-earth cycles using hierarchies and other general systems concepts*. Ph.D. dissertation, University of Florida, Gainesville.

Miller, R. L. and E. D. Goldberg. 1955. The normal distribution in geochemistry. *Geochimica et Cosmochimica Acta*, 8: 53–62.

Odum, H. T. 1982. Pulsing, power, and hierarchy. In W. J. Mitsch, R. K. Ragade, R. W. Bosserman, and J. A. Dillon Jr., eds., *Energetics and Systems*, 33–60. Ann Arbor, MI: Ann Arbor Science.

——. 1996. *Environmental Accounting, Emergy and Decision Making*. New York: Wiley.

——. 2000a. Emergy evaluation of an OTEC electrical power system. *Energy*, 25: 389–393.

——. 2000b. *Emergy of Global Processes, Folio #2, Handbook of Emergy Evaluation*. Gainesville: Center for Environmental Policy, University of Florida.

——. 2001. Material circulation, energy hierarchy, and building construction. In C. Kibert, ed., *Construction Ecology Materials as a Basis for Green Building*, 37–71. London: Spon.

Odum, H. T., M. T. Brown, and S. L. Brandt-Williams. 2000. *Introduction and Global Budget, Folio #1, Handbook of Emergy Evaluation*. Gainesville: Center for Environmental Policy, University of Florida.

Odum, H. T. and A. Lugo. 1970. Metabolism of forest floor microcosms. In H. T. Odum and R. F. Pigeon, eds., *A Tropical Rain Forest*, 35–54. Washington, DC: Division of Technical Information, U.S. Atomic Energy Commission.

Page, N. J. and S. C. Creasy. 1975. Ore grade, metal production and energy. *Journal of Research, U.S. Geological Survey*, 3(1): 9–13.

Palmerini, C. G. 1993. Geothermal energy. In T. B. Johansson, H. Kelly, A. K. N. Reddy, and R. H. Williams, eds., *Renewable Energy*, 550–591. Washington, DC: Island Press.

Partridge, G. W. 1975. Global dynamics and climate: A system of minimum entropy exchange. *Quarterly Journal of Research of the Meteorological Society*, 101: 475–485.

——. 1978. The steady state format of global climate. *Quarterly Journal of Research of the Meteorological Society*, 104: 929–945.

——. 1981. Thermodynamic dissipation of the global climate system. *Quarterly Journal of Research of the Meteorological Society*, 107: 531–547.

Reiter, E. R. 1969. *Atmospheric Transport Processes, Part I, Energy Transfers and Transformations*. Oak Ridge, TN: AEC Critical Review Series, U.S. Atomic Energy Commission, Division of Technical Information.

Romitelli, M. S. 1997. *Energy analysis of watersheds*. Ph.D. dissertation, University of Florida, Gainesville.

Swenson, R. 1989. Emergent attractors and the laws of maximum entropy production. *Systems Research*, 6(3): 187–197.

Tilley, D. R. 1999. *Energy basis of forest systems*. Ph.D. dissertation, University of Florida, Gainesville.

Veizer, J. 1988. The earth and its life: Systems perspective. *Origins of Life and Evolution of the Biosphere*, 18: 13–39.

Wallace, J. M. and P. V. Hobbs. 1977. *Atmospheric Science*. New York: Academic Press.

Wetzel, K. 1984. An attempt of an interpretation of the lognormal distribution of chemical elements in rocks. *Chemie der Erde-Geochemistry,* 43: 161–170.

Wiin-Nielsen, A. and T. Chen. 1993. *Fundamentals of Atmospheric Energetics.* New York: Oxford University Press.

NOTES

1. Named after Sadi Carnot (1796–1832), who showed how to calculate the total heat received and mechanical work done when a fluid expands and contracts in a cyclic process.

2. This approach analyzes the physical earth as populations of nonliving autocatalytic units. Traditionally, the *circulation* of fluids has been analyzed by writing equations for the interacting forces on each block of fluid volume (visualized in a grid on a map of the earth's surface). The equation of motion for each fluid parcel has force terms for pressure differences, the Coriolis acceleration by the earth's rotation of the landscape, the centrifugal force where fluids are spinning, the friction of one layer affecting another, and gravity. Then these equations are used to calculate the patterns of fluid circulation in space and time for thousands of blocks of air, water, or land, often using supercomputers. The processes of the small scale are used to predict patterns on the large scale.

3. In other words, they recognized the maximum power principle but apparently without knowledge of Lotka's papers. Partridge (1975, 1978) showed that nonlinear (autocatalytic) dynamics was required to maximize heat and latent heat transfer from the sea, adjusting the earth to an optimum temperature.

4. The author's simulation model of Earth's photosynthetic production as a function of the cloud–rain balance generates maximum production with intermediate cloudiness (Odum 1982). Too much cloud and precipitation increases albedo, reducing photosynthesis. Too little cloud and rain increases surface water stratification, limiting photosynthesis (Partridge 1981).

5. The freezing point of seawater, with an average salinity of 3.5%, is –1.87°C. Unlike freshwater bodies, in which the densest water is 4°C, seawater is densest at freezing. The densest seawater is the coldest, high-salinity water, which is formed in polar regions when water on the sea surface freezes. Freezing leaves some of the salt behind in the cold water, which sinks. When water descends to the great pressures of the deepest parts of the sea, it is heated by compression a few tenths of a degree.

6. The author made an emergy evaluation of an OTEC operation for the shores of Taiwan for which economic calculations had been made (Odum 2000a). The ratio of emergy in the yield to that in the necessary economic inputs was 1.5 to 1, which is less than that of wood (chapter 7). The costs were excessive.

7. Martha Gilliland (1975) found net emergy contribution from a geothermal energy operation in California in a volcanic region where there is a strong temperature gradient. See also my revised calculation (Odum 1996).

8. Current emergy evaluation of the earth's main sources was provided in folio #1 of the emergy handbook (Odum et al. 2000) based on global emergy evaluation by Brown and Ulgiati (1999).

9. John Lewis (1998) reviewed the way early simulation models helped show the energetic exchanges between the day-to-day weather phenomena of observers and forecasters (storms and fronts) and the larger-scale patterns of the whole hemisphere (general circulation that includes the average flows of air at the ground and the strong upper-level jets).

10. Calculations based on folio #2 of the *Handbook of Emergy Evaluation* (Odum 2000b).

11. Wiin-Nielsen and Chen (1993) plotted graphs of atmospheric wind energy in relation to scale. More kinetic energy was stored in larger eddies, increasing with replacement time (decreasing wave number).

12. The kinetic energy stored in moving air is proportional to the square of the wind velocity. At a fixed place such as a windmill on the ground, the kinetic energy is flowing by with the same wind velocity. The energy available is the product, which makes wind energy a cube of velocity.

13. The turnover time for the larger volumes of water mass in the deep ocean to be replaced by circulation is 500–1,000 years. The water itself is replaced by volcanic emissions in about 10 million years. The replacement time of chemical substances by their biogeochemical cycles ranges from 2.6 E6 years for Na to 100 years for Al (Goldberg 1965).

14. Emergy per mass calculations for the earth are from folios #1 and #2 of the *Handbook of Emergy Evaluation,* which used the erosion rates from fig. 5.9 (Odum et al. 2000; Odum 2000b). Atmosphere flow was estimated as mass recirculated in 10 days:

$(5.14 \text{ E}14 \text{ m}^2)(1,014 \text{ g/cm}^2)(1 \text{ E}4 \text{ cm}^2/\text{m}^2) = 5.21 \text{ E}21 \text{ g/earth};$

$(5.21 \text{ E}21 \text{ g/earth})(365/10 \text{ turnovers per year}) = 1.9 \text{ E}23 \text{ g/yr};$

$15.83 \text{ E}24 \text{ sej/yr}/(1.9 \text{ E}23 \text{ g/yr}) = 83.3 \text{ sej/g};$

$(83.3 \text{ sej/g})/(4.186 \text{ J/calorie}) = 19.9 \text{ secal/g}.$

15. Erosion assists uplift from below by reducing the weight on top, analogous to removing weights from a floating boat. The deposition of the eroded sediments in surrounding areas also assists uplift by increasing the pressure in surrounding areas, in essence pushing the mountains up, a process of adjustment called isostasy.

16. From Environmental Accounting (Odum 1996), based on calculations by Diamond (1984). Recent emergy evaluations of other watersheds include hierarchical distribution of chemical potential energy relative to seawater based on increasing transformity of rains and geologic contributions at higher altitudes (Romitelli 1997; Tilley 1999).

17. Sherry Brandt-Williams (1999) evaluated the emergy of lakes as hierarchical centers of watersheds.

18. There are many papers on the skewed distribution of materials with concentration, which have included many theories (Ahrens 1954; Miller and Goldberg 1955; Wetzel 1984). Kriging is a statistical method of representing patchy distribution of mineral concentrations.

19. Figure 5.12b (Odum 2001) was calculated from fig. 5.12a (Page and Creasy 1975) by adding the three main emergy inputs necessary to concentrate the copper metal from dilute ore. For each point in the graph, the mass of rock used was multiplied by the average emergy per mass of the earth cycle. Chapman (1974) provided a graph of fuel required to refine copper from ores. Fuels used were multiplied by the transformity,

and dollar cost data were multiplied by the 1975 emergy/money ratio to get the service emergy.

20. As they moved up the scale of dominance of the earth, humans often displaced the controlling occupants of hierarchical centers with whom they were competing. For example, humans displaced grizzly bears, wolves, and other large animals. With the impacts of global scale, civilization displaced many major rivers and deltas.

21. The increase of mass accumulations with turnover times of geologic processes was shown by Veizer (1988). See note 8, chapter 4.

CHAPTER 6

ENERGY AND ECOSYSTEMS

FORESTS, STREAMS, seas, and living reefs are ecological systems, called ecosystems for short. They cover the earth with life and control the geobiosphere. They process energy, materials, and information to support an amazing diversity of species. They are the life support system for humanity. In this chapter, we consider how ecosystems illustrate the principles of self-organization: energy hierarchy, metabolism, spatial concentration, material cycling, and pulsing from chapters 1–4. Understanding mechanisms at the level of ecosystems helps us understand the larger scale of society in later chapters.[1]

SELECTING BOUNDARIES

The whole biosphere, its metabolism and circulating materials, was viewed as a single system in fig. 1.3c. Zooming in for a closer view, we see forests, deserts, and seas, which only rarely have distinct boundaries. Looking even closer, we find organisms and nonliving components organized to work together. The biosphere has subsystems within subsystems. What we consider depends on the scale of view. To simplify our study of structure and function, let us define systems by imposing arbitrary boundaries, as in fig. 6.1. Humanity, its machines, its networks of communication, and money circulation, emphasized in later chapters are part of the largest ecosystems and belong in their systems diagrams.

Whereas ecosystems are divided into smaller ecosystems geometrically in fig. 6.1, subsystems may also be separated for study in other ways. Each species' population can be isolated as a subsystem, which is what happens when human society is considered without nature. Similarly, as illustrated in figs. 2.1 and 2.9, mineral cycles (e.g., carbon, nitrogen, phosphorus, oxygen) can be separated for diagramming and study.

Smaller ecosystems are linked with other ecosystems by the flows of air and water between them and by larger species of animals with territories that include two or more ecosystems. For example, hawks, tigers, and sharks are supported by, and affect, a mosaic of different ecosystems that are adjacent. Seasonally migrating birds

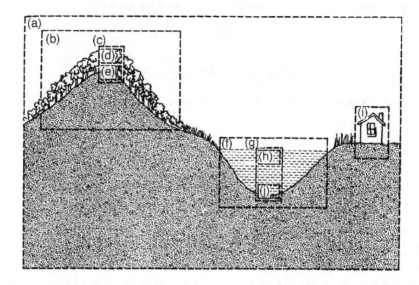

FIGURE 6.1   Forested mountain, fertile lake, and house with various ecosystems arbitrarily defined according to the purpose of study. **(a)** An inclusive system with the total landscape included; **(b)** a forested mountain; **(c)** a single prism of forest for intensive study; **(d)** a photosynthetic zone; **(e)** includes regenerative soil, litter, and animal zones; **(f)** a lake as a whole; **(g)** a vertical prism for representative calculations of some of the main parts of the lake; **(h)** the euphotic (lighted photosynthetic) zone subsystem; **(i)** the house, further elaborated in fig. 2.11; and **(j)** mud and a hypolimnetic subsystem (colder section of a density-stratified aquatic ecosystem).

nest in one ecosystem and winter in another thousands of miles away. Each species couples its life cycle to each area when there is a pulse of energy support there. Examples are sandpipers on beaches, salmon in estuaries, and whales in the sea.

Silver Springs Ecosystem

Let's introduce ecological systems with the river ecosystem in Silver Springs, Florida (Knight 1980; Odum 1955) (fig. 6.2). In large springs where waters flow out of the ground with constant properties, the union of sunlight with clear water containing optimal quantities of nutrients produces fertile beds of waving green bottom plants, covered with diatoms and supporting a food chain of many animals and microorganisms.[2]

The simplifications we make to understand complex ecosystems such as Silver Springs are called *minimodels*. When describing systems, nearly everyone explains the organization of parts and processes with words. Sometimes we draw boundaries in the vertical cross-sectional side view, as in fig. 6.1. Sometimes we simplify our view by making a map of main patterns as viewed from above (fig. 6.2a). Sometimes we represent ecosystems with artistic simplification (as in the Elizabeth A. McMahan drawing in fig. 6.2b). Sometimes we tabulate inventories of species.

Energy Systems Model

Using broad pathways for visual effect, fig. 6.2c is a systems view of the main energy flows passing through blocks, each of which represents an important part of the ecosystem. In each block the energy is transformed to higher quality, and much of its availability is used up. The first block contains the plant producers, which in turn contain a photosynthetic section and a block representing the consumer parts of plants. Consumer populations are grouped in one of four stages of consumption. The first (box H) has herbivores, the second stage (box C) has carnivores, and the third (box TC) has top carnivores. There is an organic pool, called detritus (box D), contributed to and used by many species. The organic detritus is important to all the food pathways. This is typical of many ecosystems. Detritus is a mixture of dead organic matter and living bacteria, which transform substances into long-lasting components.

Because species are grouped, such diagrams present a simplified picture of the energy transformations in each stage that upgrade some energy and disperse degraded energy as heat in the various work processes of the populations. The model in fig. 6.2c was historically important in the period when ecologists were learning to compartmentalize (aggregate in boxes) and build minimodels. These six compartments were called trophic levels. Each was further removed from the source of energy, and each step had a drop of about 10% in potential energy in the transformed flow to the next level (downstream in the energy transformation series). Most species usually receive energy contributions from more than one kind of food, including organic pools, animals, plants, and microbes, rather than from just one of these categories. It is possible to classify a species by the food category that is largest in its diet, or one can even assign a part of a species to different boxes.

Emergy Evaluation and Transformities

The protein content and nutritional quality of organic matter increase with the concentrating process down the energy stream. There is chemical diversification in the body structures of these more complex higher animals, and they draw energy from a larger territory and exert greater control within the ecosystem. Quantitative insight related to their quality comes from evaluation of empower and transformities (chapter 4). Figure 6.2d further simplifies the model of the ecosystem retaining the familiar categories of herbivore, carnivore, top carnivore, and detritus. Transformities increase along the energy transformation chain, indicating the higher values and implying greater influence of the top components (Odum 1996).

Production and Respiratory Metabolism, P and R

In fig. 6.2 the Silver Springs ecosystem is further simplified into two sources, two storages, and the flows of energy (fig. 6.2e) and carbon (fig. 6.2f) in the processes

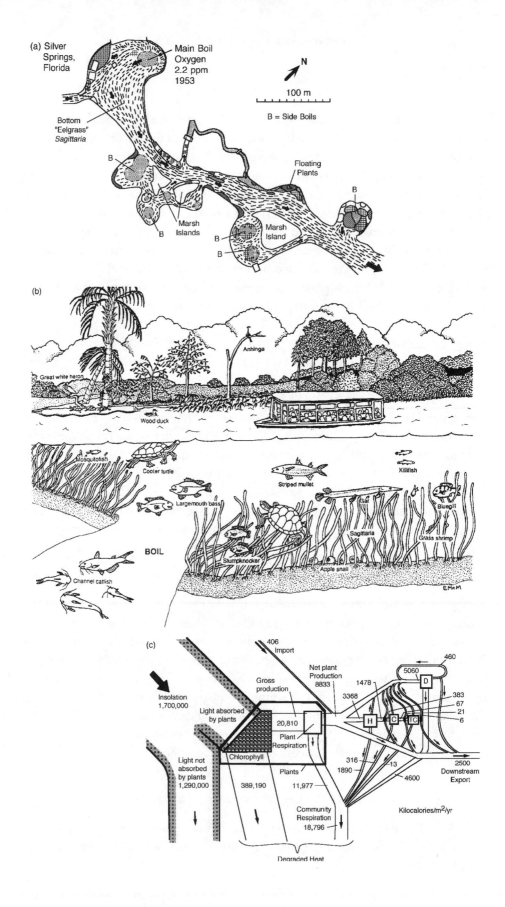

(a) Silver Springs, Florida

Main Boil
Oxygen
2.2 ppm
1953

N

100 m

B = Side Boils

Bottom "Eelgrass" Sagittaria

B

B

B

B

B

Marsh Islands

Marsh Island

Floating Plants

B

(b)

Great white heron

Anhinga

Wood duck

Mosquitofish

Cooter turtle

Killifish

Striped mullet

Largemouth bass

Gar

Bluegill

Sagittaria

Grass shrimp

BOIL

Slumpknocker

Apple snail

Channel catfish

E McM

(c)

406 Import

Net plant Production 8833

460

5060

Insolation 1,700,000

Light absorbed by plants

Gross production

20,810

1478

3368

D

383
67
21
6

Light not absorbed by plants 1,290,000

Chlorophyll

Plant Respiration

H

C

TC

Plants

389,190

11,977

Community Respiration 18,796

316
1890

13

4600

2500 Downstream Export

Kilocalories/m^2/yr

Degraded Heat

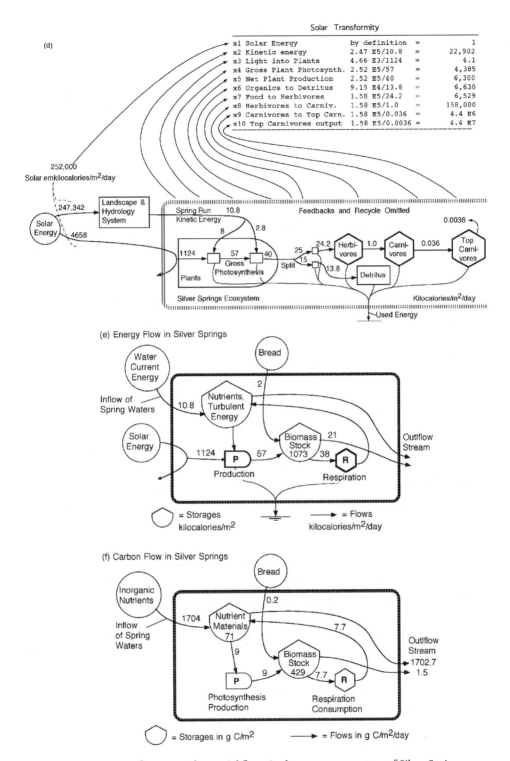

(d)

Solar Transformity

x1	Solar Energy	by definition	=	1
x2	Kinetic energy	2.47 E5/10.8	=	22,902
x3	Light into Plants	4.66 E3/1124	=	4.1
x4	Gross Plant Photosynth.	2.52 E5/57	=	4,385
x5	Net Plant Production	2.52 E5/40	=	6,300
x6	Organics to Detritus	9.15 E4/13.8	=	6,630
x7	Food to Herbivores	1.58 E5/24.2	=	6,529
x8	Herbivores to Carniv.	1.58 E5/1.0	=	158,000
x9	Carnivores to Top Carn.	1.58 E5/0.036	=	4.4 E6
x10	Top Carnivores output	1.58 E5/0.0036	=	4.4 E7

(e) Energy Flow in Silver Springs

(f) Carbon Flow in Silver Springs

FIGURE 6.2 Summary of energy and material flows in the stream ecosystem of Silver Springs, near Ocala, Florida (Odum 1955). **(a)** Sketch of headwaters of Silver River; **(b)** underwater view drawn by Elizabeth A. McMahan (Odum et al. 1998); **(c)** summary of energy flows (Odum 1955); **(d)** abbreviated energy diagram of the ecosystem at Silver Springs, Florida, showing the increase of transformities from left to right down the food chain (Collins and Odum 2000); **(e)** system aggregated into compartments of production and consumption with recycle; **(f)** aggregate view showing flows of carbon, typical of the nutrient cycles. C, carnivores; D, detritus; H, herbivores; TC, top carnivores.

of production and respiration, according to the basic plan of ecosystems given in fig. 2.9. Silver Springs and many forest ecosystems, on average, produce more than they consume, exporting some organic matter in runoff downstream. However, most streams have more organic input and consumption than they produce, exporting the inorganic nutrient byproducts downstream.

Ecosystem Comparisons

Although ecological systems differ from each other, most have enough features in common to justify describing some basic patterns that are common throughout the world. Although the fishes and plankton of the sea seem very different from the birds and tree leaves of the land system, the energetic properties in the stage diagrams are quite similar. Comparing plankton in the sea with the leaves of the rainforest was a famous thesis of Bates (1960) and Steeman-Nielsen (1957).

The Vertical Design of Ecosystems

If we choose boundaries to include both production (P) and respiration (R), we find the functions stratified in vertical layers as a response to the sunlight entering from above and the action of gravity in separating gas, liquid, and solid phases (fig. 6.3). The photosynthetic production and its chlorophyll-bearing structures become localized on top in the light, and the consumption and larger consumers are concentrated below. Photosynthesis tends to deplete raw materials on top, releasing oxygen and accumulating new organic matter. The top zone receives the sharp diurnal and seasonal pulse of solar insolation and serves as a roof. The organic matter is carried downward by gravity or by the tube systems in the inner bark of tree trunks. Removal of the organics to a lower zone leaves the upper production zone oxidized. The concentration of respiratory action below accumulates carbon dioxide, makes conditions acid, and releases again the mineralized critical nutrients needed by the plants for photosynthesis. This stratum therefore is called the regeneration zone. Oxygen tends to diminish because respiratory metabolism dominates, and in some systems the oxygen reaches zero as other chemically reduced substances form, such as hydrogen, hydrogen sulfide, and ammonia.

Ecosystems have various means for returning regenerated nutrients back up to the photosynthetic production zone. This can occur through root transpiration in trees, through the vertical movements of the swimming or flying consumers who release wastes along the way, by release of gas bubbles that carry absorbed materials to the surface, and through other physical currents to which a particular system is adapted.

In fig. 6.4 ten systems are contrasted, with P and R zones stratified according to the plan in fig. 6.3. The vertical scale ranges from a mile or more in the sea to a few centimeters in mats and encrusting communities. In the thin aquatic systems, the

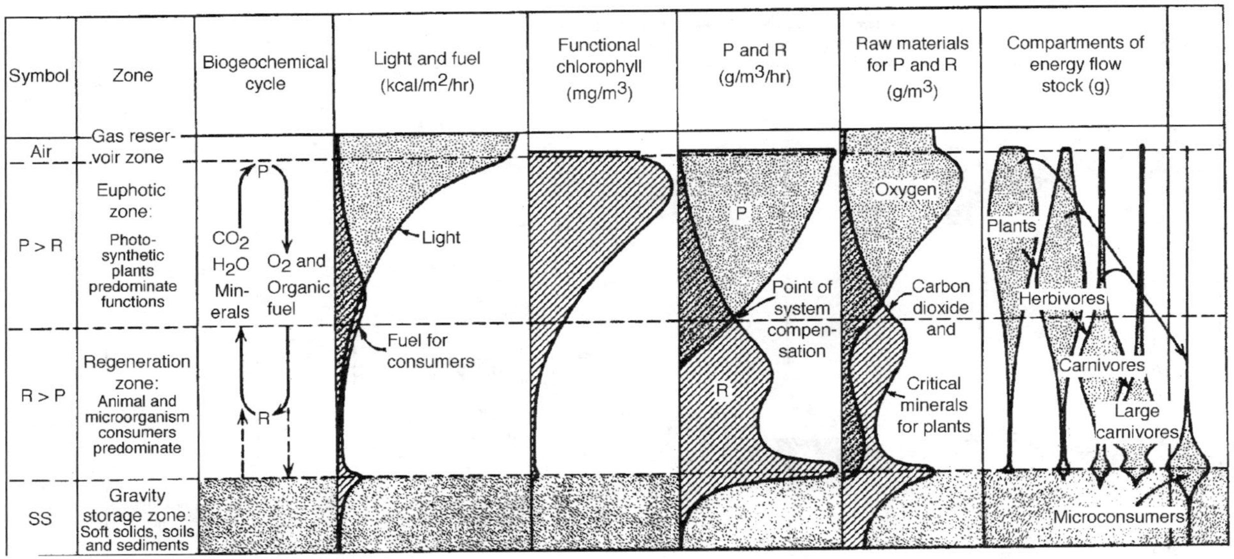

FIGURE 6.3 Principal vertical zones of many ecological systems on land and in water.

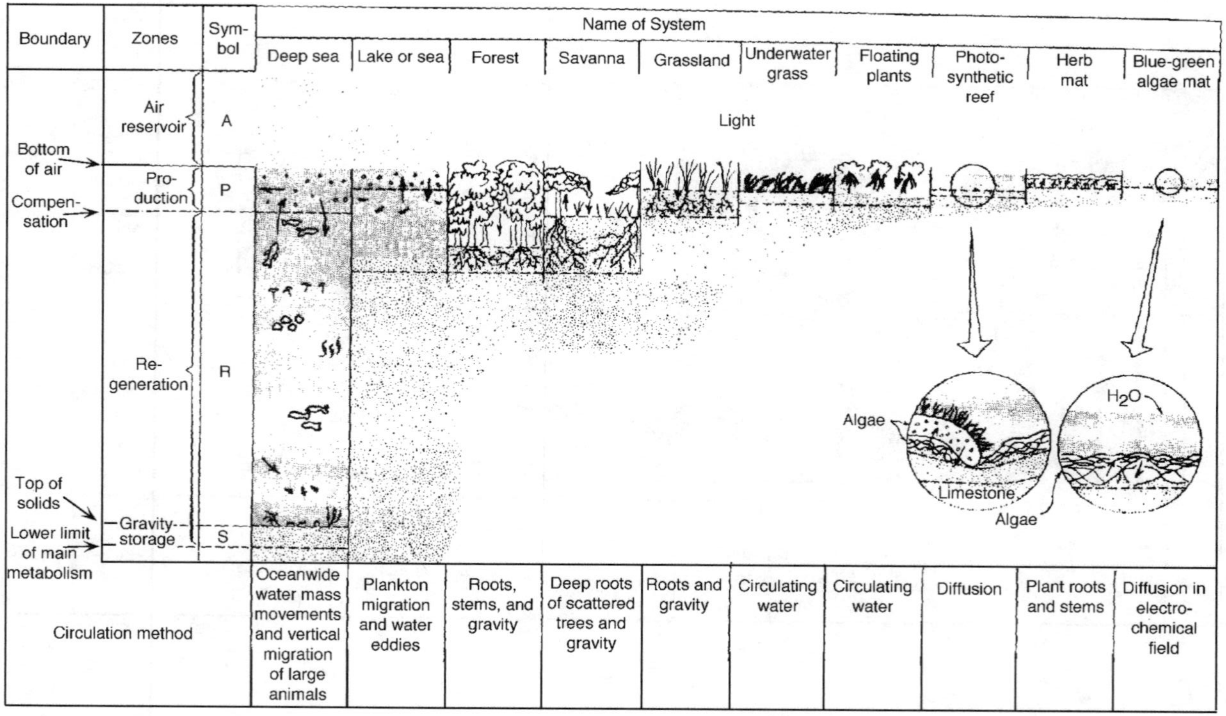

FIGURE 6.4 Vertical zones in 10 contrasting ecological systems defined in fig. 6.3.

P and *R* zones are mainly within porous solids (e.g., reef). In systems such as the deep sea, where the solid surface is distant, little organic matter reaches it and only after many years through indirect routes of the deep ocean currents and through successive handling by vertically migrating populations. When the mud surfaces are near the productive euphotic zone, they serve as centers of fuel accumulation and consumption. In general, in aquatic systems more of the circulatory work is done by fluid movements, whereas in land systems specialized plants carry on nongaseous circulation.

Spatial Organization

Ecosystems are spatially organized into centers according to the energy hierarchy (chapter 4, fig. 4.8). Trees are organized about their trunks and animals about their nesting centers. The large size of many animals tends to concentrate the respiratory function, with each organism collecting food over a wide area. Many consumers, such as ants, antelopes, and people, have clustered populations because their complex functions are favored by grouping.

Because incoming solar radiation is spread evenly on an area, photosynthesis and light-absorbing chlorophyll tend to be spread evenly. As we pass from forest to field to swamp and aquatic medium, or to the agricultural systems controlled by humans, there tends to be an even cover of green despite great differences in the kind of plants providing the support. When light intensity is less, plants adapt by increasing the chlorophyll. For instance, cloud-covered Ireland is very green.

Figure 6.5 shows the hierarchical pattern of uniformly spread-out production and clustered consumption. This pattern is observed in diatoms and tube feeders on glass slides that have been submerged in fertile streams for six weeks. It is also observed from an airplane flying at a height of 10,000 feet over the wheat fields (production) and farm villages (consumption) of Kansas. The centralization of tree functions in a forest also follows the pattern, with limbs and roots spreading out from the trunk.

The Functional Biomasses of Nature

Biomass includes the bodies of living organisms and the nonliving organic matter in peat, wood, litter, and sediments. Biomass graphs (fig. 6.6) are pictorial inventory reports that can be used with energy systems diagrams to help describe networks and functions. Obtaining data on biomass representative of the large systems of nature such as forests and seas is a very difficult task that has occupied ecological research for 50 years with thousands of methods. When small spots are studied or small samples taken, the data are not representative because of the spatial organization that is characteristic of most ecosystems. Sampling large sections of systems is laborious and expensive. Figure 6.7 shows a helicopter-borne net, 50 feet in diameter, which the author used in shallow Texas bays to estimate the bio-

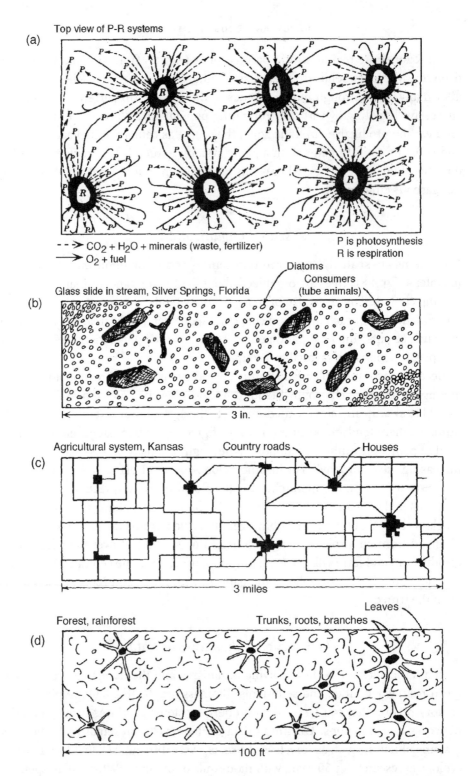

FIGURE 6.5 Patterns of spatial hierarchy in ecological systems viewed from above. **(a)** Metabolism showing productive photosynthesis (*P*) dispersed evenly over the surface and respiration (*R*) clustered in centers and linked to production through convergence of circulating pathways; **(b)** periphyton in Silver Springs; **(c)** agrarian landscape from aerial view; **(d)** canopy and trunks in a tropical rainforest.

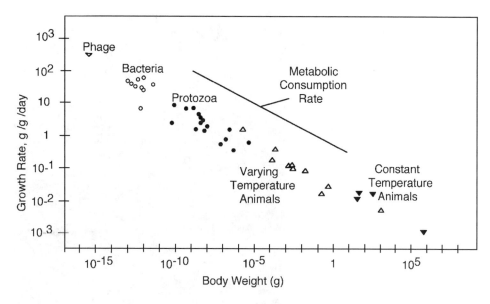

FIGURE 6.6 Growth rate, metabolism, and weight of animals and microorganisms (Straskraba 1966).

mass of fish (Jones et al. 1963). Consistent with the ideal stated as a macroscopic view (chapter 1, fig. 1.2), sampling was done as if by a giant, far above the system, oblivious of the fine-scale variation. Unfortunately, we had no giant.

Ramon Margalef (1957) and Sven Jorgensen (1997) proposed that maximum biomass was a "goal" of the self-organization of ecosystems in their adaptation and succession. This concept follows from the maximum empower principle because building biomass to a large optimum size is a necessary part of maximizing the productivity of autocatalytic processes (chapter 3, fig. 3.6).

Biomass and Scale

Chapter 4 explained why storage concentrations increase with scale. The biomass-es of old-growth trees, elephants, and whales are examples. The larger the territory of support and influence, the larger the stored quantity in the center. For animals, the body biomass is the center of its territory. The larger the unit, the slower the depreciation and turnover time. Energetically, larger concentrations have lower dispersals because of the smaller surface per unit volume. Such units meet the system requirement that higher-quality (higher-transformity) units store longer be-fore sending pulsing feedback control actions out to the surrounding territory.

At first glance the biomass seems to vary widely. A person strolling in a forest sees a great mass of plant matter and a few small animals. From a fishing boat in the sea, usually no plants are visible at all, but large fish appear from time to time. With such different proportions of foliage and creatures, we might conclude that the fundamental bases of these living patterns are different.

FIGURE 6.7 Helicopter-borne drop net ready for release of electromagnetic supports over Corpus Christi Bay, Texas (Jones et al. 1963).

However, the basic photosynthesis in the sea and in the forest is done by tiny cells with rapid turnover times. In lakes and the sea, the water provides the structural support, whereas in the forest, the ecosystem supports the photosynthetic cells with wood structure (leaves, limbs, roots, trunks, and peaty soil components).

The woody storages above ground and soil organics below ground are high in the energy transformation chain, with less energy, slower turnover time, and longer times between pulses and restarts. Perhaps the greatest ecosystem biomass is to be found in the giant trees of virgin rainforest (fig. 6.8) and the great trees of the redwood and sequoia forests of the western United States. These contain huge emergy content because of the hundreds of years of environmental work needed to produce them. In the sea comparable structure is the one or two parts per million organic matter in the huge seawater mass, with similar turnover time to the organics in the forest.

Figure 4.6d and fig. 6.6 are examples of the many published graphs showing the declining unit growth rate and metabolism with scale (Calder 1984; Peters 1983; Zeuthen 1953). This is a property of the universal energy hierarchy explained in chapter 4. At higher hierarchical levels less energy is processed, and there are smaller depreciation rates and greater storage concentrations. Pulsing renewal has greater impact (e.g., when a great tree falls).

FIGURE 6.8 Large biomass of old-growth rainforest. **(a)** Virgin forest of Dominica; **(b)** old growth from Costa Rica.

Photosynthetic Production

As explained previously (fig. 2.1, fig. 3.10), photosynthetic production combines sunlight with materials to produce organic matter. Wherever ecosystems are adapted, photosynthesis is greater when there is more sunlight and the necessary materials are unlimited.

Gross production is the first step in transforming solar energy into chemical potential energy. Measuring gross production directly is difficult because products made by photosynthesis are being used up by the plant's respiration processes at the same time.[3] *Net production,* the difference between the photosynthesis and the concurrent consumption, is usually measured. An index of gross production often is calculated by adding the rate at which organic matter or oxygen is used up in the dark to the rate of net production in the light. Or the rate of carbon dioxide generation at night is added to the carbon dioxide fixed during the day. The efficiency of gross production (energy produced/energy absorbed) in daylight is about 5%. The efficiency is inverse to the light intensity (chapter 7, appendix fig. A4).[4]

Production in Underwater Ecosystems

For underwater ecosystems, the main inputs are the sun, nutrients such as phosphorus, nitrogen, and potassium, and the physical stirring of the water (caused by wind, downhill flows of water, tides, differences in water density from heating, evaporation and rivers). The sunlight provides the most energy, but when each input is multiplied by its respective transformity and expressed on a common empower basis, solar, physical, and chemical sources are similar.

Models like that shown in fig. 2.4 or intricate physiological methods are required to estimate processes when P and R are nearly balanced and the plant cells small, as in the tropical sea. In tiny cells, more photosynthesis causes an instantaneous increase in respiration in the light (fig. 2.3) and an almost instantaneous drop in respiration after dark. These areas may have high gross production that does not show up in depositions or gaseous exchanges outside the cells. Because of the difficulty measuring gross production, there are still many doubts about the distribution of the basic photosynthetic process over the earth. Most maps of the world's productivity are based on net production. These show the areas where there are net foods available to other organisms.

Production and Transpiration

In typical woody ecosystems on land, photosynthesis depends on the process of *transpiration* by which freshwater from rain reaching soils is drawn up the trunks of trees by the evaporation of water through the microscopic stomate holes in leaves. On land, terrestrial plants use the flow of dry air to evaporate water from leaf pores.

This creates pressure differences (25 atmospheres or more) between leaves and roots that draw soil water and many nutrients from the ground to the production processes. The sun's infrared rays and the photons of visible wavelengths used by photosynthesis heat the leaves, increasing the pressure of the leaf's water vapor and causing it to escape into the air. The winds also renew concentrations of carbon dioxide at the leaf pores, where it diffuses in for photosynthetic production of organic matter. In other words, terrestrial plants use atmospheric inputs including the sun, turbulent energy, dry air, carbon dioxide, and rainfall, all coupled together as shown in the systems model in fig. 6.9a. Because transpiration integrates all these inputs, it is proportional to net production of terrestrial ecosystems, as shown by Rosenzweig in fig. 6.9b.

Biomes and Productivity

In the grand view of the biosphere, the environment has many spatial subdivisions: seas, tundras, deserts, and great rivers. Several generations of observers have thought that division boundary lines could be drawn to group together systems with similar structure and function called *biomes*. Thus the areas of coniferous forest of the colder regions have been grouped together, even though the conditions and actual species of plants and animals constituting the system on one continent are mostly different from those on another continent.

Because there are many input factors and sequences to which ecosystems adapt, each with specialized species, there are many kinds of ecosystems on Earth. In order to have a simplified global overview, classifications of ecosystem types (biomes) often have been made with only two factors, ignoring temporal sequences. Figure 6.10a shows a popular classification of major terrestrial biomes on a graph of rainfall and temperature on logarithmic coordinates.[5] This system was developed to show the major role of transpiration in productivity, where energy for transpiration was in rough proportion to temperature on a logarithmic scale. In other words, fig. 6.10a represents many biomes because rainfall and energy for transpiration (indicated by temperature) are the main emergy sources to which ecosystems organize their functions.

The climatic belts of the earth determine the location of rainfall and the conditions for transpiration (warm dry air). The satellite view of Africa (fig. 6.10b) shows the location of some of the biomes and their climatic conditions. Rainforest is at the equator, where converging winds cause air to rise, form clouds, and generate gentle rainstorms. Seasonal tropical forest (moist forest) is found to the north and south on either side. Next comes the savanna (seasonal grassland with scattered trees). Deserts with little vegetation are around 35° latitude, where the high-pressure zone of the atmospheric circulation causes air to descend, heat by compression, and evaporate clouds.

Much of the energy of the sun, tide, and earth heat goes through the oceans, generating zones of aquatic climate to which ecosystems adapt, developing different

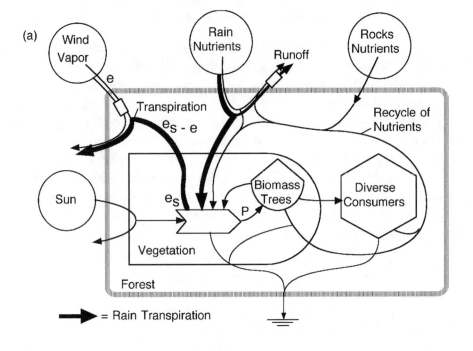

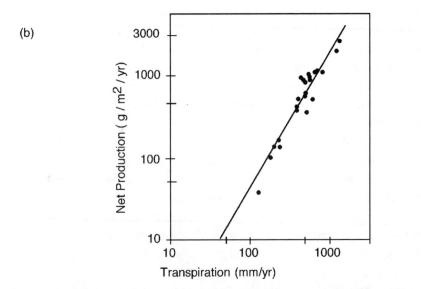

FIGURE 6.9 Transpiration role in terrestrial production. **(a)** Systems diagram of the role of transpiration in integrating energy sources contributing to photosynthetic production; **(b)** net production as a function of transpiration in several biomes (Rosenzweig 1968).

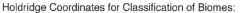

(a)

Holdridge Coordinates for Classification of Biomes:

Altitude:	Nival	Alpine	Subalpine	Montane	Lower Montane	Lowland
Latitude:	Polar	Subpolar	Boreal	Cool Temperate	Warm Temp Subtropical	Tropical

Rainfall, (mm/yr)

8000						Rain F.
					Rain Forest	Wet Forest
4000						
				Rain Forest	Wet Forest	Moist Forest
2000						
			Rain Paramo	Wet Forest	Moist Forest	Dry Forest
1000						
		Rain Tundra	Wet Paramo	Moist Forest	Dry Forest	Very Dry Forest
500						
	Polar	Wet Tundra	Moist Puna	Steppe	Thorn Steppe	Thorn Woods
250						
	Polar	Moist Tundra	Dry Bush	Desert Bush	Desert Bush	Desert Bush
125						
	Polar	Dry Tundra	Desert	Desert	Desert	Desert

Temperature (°C) 1.5   3.0   6   12   24   35

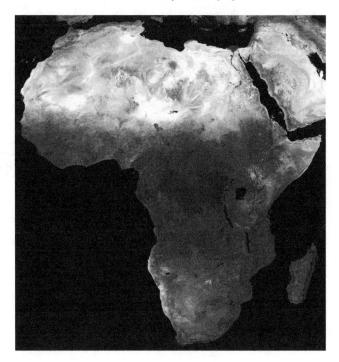

FIGURE 6.10 Holdridge classification of biomes by rainfall and temperature on logarithmic scales. Temperature also represents transpiration, an exponential function of temperature. **(b)** Satellite view of Africa showing the distribution of main biomes (Tucker et al. 1985).

kinds of aquatic ecosystems (aquatic biomes), each with distinctive forms of life. The circulation pattern of the sea has gyres and loops that allow characteristic biological communities to remain in about the same place, even though the water is in continuous motion and ultimately intermixes. For example, the blue-water plankton ecosystem in the center of the subtropical oceans occurs in a low-nutrient, clear-water mass that is gently circulating. In many tropical seas where trade winds blow from east to west, plankton migrate up for part of the day, riding with the current to the west. Then at night they descend to lower layers, with undercurrents to the east riding in a reverse direction.

Underwater biomes include oligohaline estuaries, estuarine oyster reefs, tropical coral reefs, kelp beds, intertidal encrusting communities, hypersaline bays, blue-water plankton systems, blue-green algal mats, seasonally stratified oligotrophic lakes, under-ice marine ecosystems, the chemosynthetic life of undersea volcanic vents, the life in the oceanic deeps, and many others.

Ecosystems in regions with sharp seasons (e.g., arctic seas, monsoon areas of Asia) tend to store production during their growing periods to last during their unproductive periods. These systems demonstrate their productivity by their storages. Systems in steadier surroundings tend to use their production as it is made, and net accumulations are small.

Many of the most fascinating subsystems are uncommon and involve little of the overall budget of the earth. Although not listed as biomes, these subsystems may represent systems that dominated in past ages, systems that will cover the earth in the future, systems that help us visualize possibilities on other planets, systems that can be engineered for the benefit of the economy, and systems that have great value for future scientific progress (see chapter 12). For example, the ancient conical structures found in the geologic record and once thought to be from extinct algal biota were found alive in a coastal environment of modern Australia.

Ultimately, zones occupied by the major subsystems may correspond to climate zones, which determine the amount of available energy and the energetic requirements for adapting to desert, temperate, tropical, and arctic conditions. In naming subsystems of the earth it is informative to choose some physical feature or population that is most prominent and may be said to dominate the emergy flows. The set of input flows expressed on a common empower basis is the ecosystem's *emergy signature*.

Comparison of Ecosystem Productivity

Some of the highest and lowest values of production are summarized in table 6.1. The energy flows vary daily and seasonally because the input energies vary in a similar manner. Generally speaking, the seasons of bright light are shorter in the temperate latitudes than in the tropical areas, so the energy input and the power flows are about twice as high in tropical areas (4,000–5,000 kcal/m^2/day) as in temperate areas (1,500–4,000 kcal/m^2/day). In polar areas the energy input

TABLE 6.1 Magnitudes of Primary Production[a]

System	Rate of Production of Organic Matter (kcal/m²/day)	Efficiency (%)[b]
Systems with little measurable production (most photosynthesis is masked by internal cycling, lags, and storages or is nonexistent)		
Subtropical blue water[c]	2.9	0.09
Deserts[d]	0.4	0.05
Arctic tundra[d]	1.8	0.08
Fertilized systems with high production that is stored in seasonal growth, organic storage, or outflowing yield		
Algal culture in pilot-plant scale[e]	72	3.0
Sugar cane[f]	74	1.8
Water hyacinths[g]	20–40	1.5
Tropical forest plantation, Cadam[h]	28	0.7
Systems with measurably high production that is used mainly the following night (minerals being recycled, P and R similar)		
Sewage pond on 7-day turnover[i]	144	2.8
Coral reefs[j]	39–151	2.4
Tropical marine meadows[k]	20–144	2.0
Tropical rainforest[h]	131	3.5
Waste-receiving marine bay, Galveston, Texas[i]	188–232	2.5
Clear spring stream with vegetation-covered bottom, Silver Springs, Florida[g]	70	2.7

[a] Day net production plus night respiration. This procedure omits much immediately used photosynthesis. For more detailed data, see general reviews by Pearsall (1954), Ryther (1963), and Westlake (1963). Data that were originally given in grams of oxygen were converted to kilocalories using 4 kcal/g. Data originally given in grams of carbon were converted using 8 kcal/g.

[b] Percentage of total sunlight received. About half of sunlight is directly involved in photosynthesis, but the heat radiation received may also contribute to such auxiliary processes as evapotranspiration, photorespiration, and rates of biochemical reactions.

[c] Ryther (1963).

[d] Rosenzweig (1968).

[e] Tamiya (1957).

[f] Odum and E. P. Odum (1959, 1985).

[g] Odum et al. (1998)..

[h] Odum and Pigeon (1970).

[i] Ludwig et al. (1951).

[j] Halfrich and Townsley (1965).

[k] Odum (1967).

and power flows are smaller (2,000 kcal/m²/day), although they are very intense in the short summer.

Listed first in table 6.1 are low values characteristic of both land and water in much of the subtropics, where shortages of critical raw materials cause recycling mechanisms to develop. Listed next are net producers cultivated for maximum photosynthesis by inflows of fertilizing nutrients. These values are 100 times higher than those of the limited areas. These systems provide storages and yields for people. Finally, in the third list are some natural ecosystems well organized for effective recycling of minerals by regenerative respiration. They have photosynthetic rates as high as those of the fertilized-yield systems. Only with very intensive fertilization does net production by plants exceed the gross production of diverse, well-organized recycling (example, chapter 12, fig. 12.11).

Diversity

Some ecosystems are simple, with a few species, and others are complex, with many species. In an analogous way, some human societies are simple, with a few occupations, and others are complex, with many occupations. On both scales, more variety means more complexity in the network of energy and material flow. A useful index of diversity is the number of different kinds found in counting 1,000 individuals.[6] Table 6.2 contains examples. Whether simple or complex systems develop depends on which generates the maximum metabolism for the conditions (maximum empower principle from chapter 3). Two principles involving productivity explain most of the conditions of diversity observed in nature and microcosms.[7]

Species Diversity and Physiological Adaptation

The first principle is the complementary allocation of productive energy to physiological function and interspecies organization. When environmental conditions are extreme, the necessary physiological adaptations take more of the energy of the ecosystem's production, and therefore there are fewer species. For example, brine shrimp have many special enzymes, and estuarine oysters have specially adapted kidneys. Without energy for interspecies organization, competition excludes species. The first examples in table 6.2 are severe environments with few species or occupations.

When environments provide optimal conditions for life, then the species present need not have special physiological abilities. Energy can go into structures and behavior for adapting species to each other. There is more energy to support complex network specializations and more division of labor. The last entries in table 6.2 have more kinds of units and more specialization.

Nature achieves its greatest manifestations of beauty, intricacy, and mystery in the very complex systems: the tropical coral reef, the tropical rainforest, the benthos-dominated marine systems on the west coasts of continents in temperate

TABLE 6.2   Species and Occupational Diversity

System	Roles Found When 1,000 Individuals Are Counted[a]
Systems with few species in stressed environments	
Mississippi Delta, low-salinity zone	1
Brines[b]	6
Hot springs	1
Polluted harbor, Corpus Christi, Texas	2
Humans on a military frontier[c]	10
Complex systems in stable environments	
Stable stream, Silver Springs, Florida[d]	35
Rainforest, El Verde, P.R.	75
Ocean bottom[d]	75
Tropical sea[d]	90
Human city	111

[a] When fewer than 1,000 are counted, statistical agreement between duplicate counts is not good; when larger numbers are counted, work is unduly tedious, and discontinuities between different systems tend to be crossed. All units in the same size realm are counted.
[b] Nixon (1969).
[c] From Infantry Table of Organization as summarized by Stanton (1991).
[d] Yount (1956).
[d] Sanders (1968).

zones, the bottom of the deep sea, and some ancient lakes of Africa. These places are environmentally stable, and the species networks that develop there have great specialization and division of labor in the same way that the city of New York has enormous diversity of occupations. Complex networks must maintain controls for stability, but there is energy for control because the environments are initially stable, necessitating no great energy expenditures for physiological adaptation.

The large and concentrated flows of energy into an industrialized city support extreme specialization of occupational functions not possible for the same number of people in an Eskimo village with climatic stress. Similarly, the rainforest has more circuits than the brine ecosystem. Complexity and diversity on the scale of society thus are related to the flows of energy and the budgeting of these flows between various functions. This principle explains why a decrease in diversity accompanies pollution stress.

In many other environments of Earth the conditions for life vary sharply by day, season, and year. The estuaries of Texas are salty in years of land drought and then are suddenly flooded with freshwater when the rains come. The Mississippi River fluctuates in its pathways to the sea, leaving bays saline at times and fresh in times of high runoff. Sharp seasonal changes occur in the polar tundra, deserts, and waters that receive the fluctuating wastes from human sewers. In these environments energies go into sustaining life against stress and into the constant replacement of decimated stocks. Little energy finds its way to diversification and speciation. The few species that occupy these realms must be generalists.

When energy is channeled into only a few species, there may be large net yields of organic matter instead of the work spent on specialization. The patterns of color, visual cues, and beauty are lacking, but nature is exciting in another way, for the few dominant stocks of creatures are characterized by great blooms, pulses, crops, and migrations. Society can secure food yields from this kind of system, but only irregularly. The annual pulse in the high-salinity Laguna Madre of Texas and Mexico is an example of high production but low diversity.

Species Diversity and Unused Resources

The second principle of diversity is the inverse relationship between excess resources for net production and diversity. Where conditions are not extreme, ecosystems have two main diversity regimes for maximizing empower.

• If there are *excessive inflows of available resources, light, or nutrients,* then there is overgrowth by competitive but poor-quality producers that yield *high net production.* They do not use much of the products to sustain high-quality, long-lasting structures. The diversity is low. They may grow in steady state, depositing organic matter made with the excess inputs. People may not like soupy green swimming waters, rafts of water hyacinths, or cattail beds because of their low diversity and low-quality structures, but they are nature's way of dealing with excess nutrients, converting them into high-quality organic storages in sediments. In a recent survey of sites with high nutrients in Puerto Rico, 85 species were found in the role of net producing *low-diversity overgrowth* (Kent et al. 2000).

• When available *resources are not in excess,* a high-diversity, cooperative ecosystem develops with efficient recycling that eliminates any limiting factors other than the sun. Structures are longer lasting, with most of the production going into roles that increase gross production and maintenance. High rates of consumption and recycle favor high rates of gross photosynthesis, but there is little net production.

Energy and the Number of Connections

The number of possible connections that require emergy increases rapidly with the number of species (fig. 6.11a). To coexist, either some emergy-using arrangements must be made to prevent interspecies stress and destructive competition, or

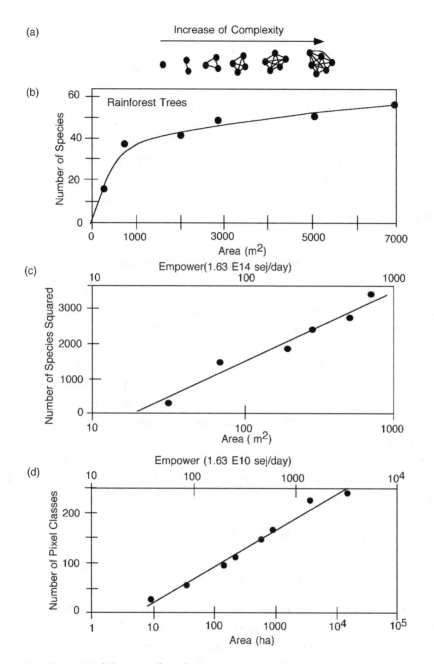

FIGURE 6.11 Emergy and diversity of rainforest. **(a)** Increase of connection possibilities with numbers of species; **(b)** plant species found as a function of empower area (Smith 1970); **(c)** number of species squared; **(d)** types of rainforest pixels counted from satellite as a function of area and empower (Keitt 1991).

systems emergy is lost by the interactions. Figures 6.11b and 6.11c show the plant species found with increasing area in the rainforest in Puerto Rico by counting on the ground. In fig. 6.11d, Keitt (1991) shows the increase in recognizable pixels of satellite images with increasing area.

Species–area graphs indicate the area required for a number of species. In many ecosystems the empower (e.g., energy in sun, rain, wind, land) supporting the ecosystem is in proportion to the area. Therefore these graphs can be used to determine how much empower is required to support species diversity (Odum 1999) in the short run. Chapter 8 explains what is required to sustain species information in the long run, and chapter 9 explains the economic valuation of diversity using emdollars.

Biomass and Diversity Minimodels

Principles relating diversity to energy flow are combined in the energy systems diagram in fig. 6.12a. Production is shown generating biomass storage, which then provides an interaction with incoming species to generate a storage tank of diversity (number of species). The extra diversity draws on the biomass storage in proportion to the square of the species number, which limits the diversity. The gross production is increased by the diversity and the efficiency brought by its division of labor. Computer simulation generates the kind of curves of biomass growth and diversity often observed as the succession goes through a stage of simple net production replaced in part with complexity.

Figures 6.12b and 6.12c summarize the net metabolism. The early growth stage has high net production and small respiration, when use of resources is being developed. The later stage has greater gross production and respiration and very little net production. Figure 6.12d is a typical simulation of the minimodel (fig. 6.12a). The rate of net production peaks first, then biomass, and finally diversity.

Oil and Low Diversity of Fossils

When there are only a few poorly organized and unspecialized species, organic matter is not efficiently used, and large quantities go into storage in the sediment and fossil deposits. Thus oil and coal deposits may be related to conditions of severity or fluctuation (e.g., sediments originally formed in hypersaline areas, arctic estuaries, or river mouths). Many of the sedimentary areas of the world carry no oil because they had stable conditions and living networks that were efficient. Recognition of this simple principle can save money spent on oil exploration. Count the species diversity from the little animal skeletons in the drill cuttings of exploratory mining. Possibilities for finding oil are predicted when the species diversity is found to be less than five species per 1,000 animal skeletons counted. Such systems have been observed with high levels of organic matter and residues in their sedimentary depositions.[8]

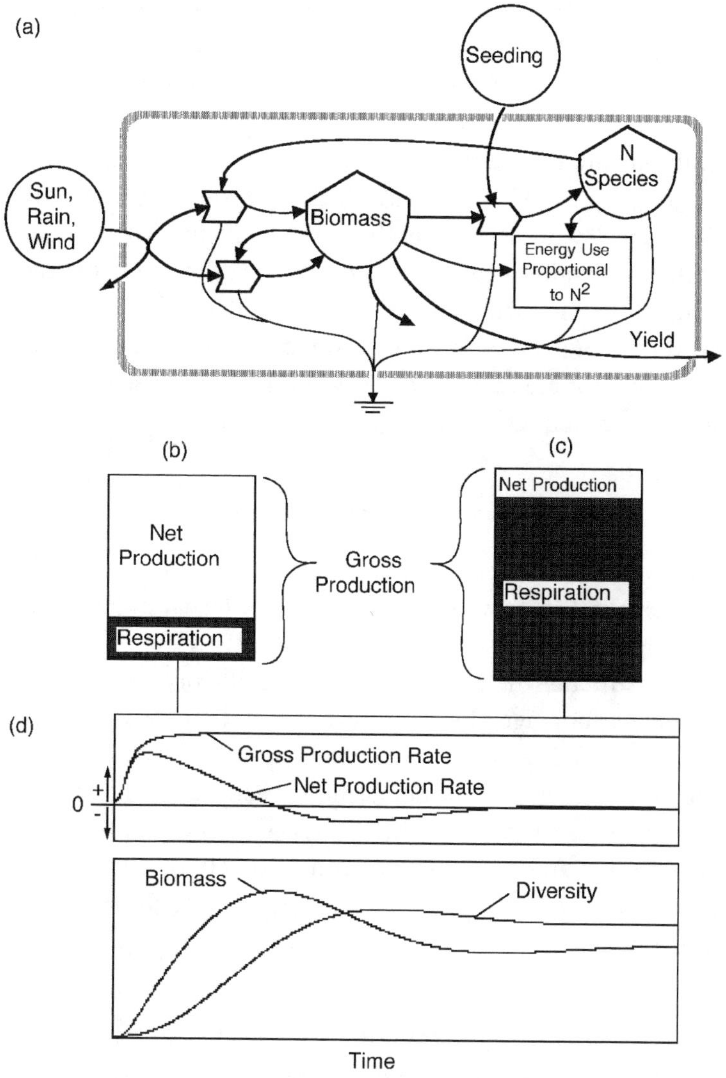

FIGURE 6.12 Minimodel called CLIMAX (Odum and Odum 2000) simulating succession in which excess sunlight and nutrients are drawn into use. Appendix fig. A11 has the equations. **(a)** Energy systems diagram; **(b)** production and respiration during overgrowth colonization; **(c)** production and respiration at climax; **(d)** typical simulation showing the sequence of biomass and diversity.

PATTERNS OVER TIME

Ecosystems, like other systems, pulse with cycles of growth, descent, and restart (chapter 3, fig. 3.13). Metabolism fluctuates in ecosystems even when the conditions of light and temperature are held constant, as in the aquarium ecosystems in experimental growth chambers (fig. 2.6). The small-scale microbial ecosystems with small components rise and restart frequently, in a few hours. The cycles of large forests take centuries. Genetic and ecological changes required for modern coral reefs involve geologic time scales. Within the larger systems with longer cycles there are small-scale components with short-term pulses.

Depending on our scale of view, we call these variations with time oscillations, succession, or evolution. For the same energy, there is more productivity in the long run with pulsing (chapter 3). Some of the timing for pulsing is supplied by the outside energy sources. The inputs of the sun's energy and other resources rise and fall each day, in seasons each year, and in multiyear cycles. Naturally, photosynthetic production goes up and down with these outside cadences. The ecosystem may adjust its parameters so that the outside frequency maximizes its energy transformations. Or, where the frequency provided by the outside energy sources is not appropriate (wrong scale) for maximum empower, the ecosystems may develop their own internal mechanisms of oscillation.

Prey–Predator Pacemakers

A prey–predator oscillation can serve in ecosystems as a pacemaker (analogous to a heart pacemaker) to provide an appropriate internal pulse.[9] Arctic animals, jackrabbits, lynx, and snowy owls are famous for their cycles of abundance and scarcity. Systems models help us understand oscillatory mechanisms. The Lotka–Volterra theory long ago showed how a lag in the eating and being eaten of a prey–predator sequence could create undulations in the stocks (Lotka 1925). Many demonstrations of this principle have been made in laboratory populations in simplified situations in which there were no additional controls. Examples are given by Drake et al. (1966) and Utida (1957) (fig. 6.13). Simulation studies show that minor changes in timing factors can shift a steady system to an oscillating one.[10]

Successional Sequences

The sequences of succession depend on the starting conditions, outside sources (materials, energy, and species), and scale of time and space. The self-organizational process for microscopic ecosystems can develop its potential in a few days; the largest systems, such as rainforests, and ecosystems that generate land form structures, such as reefs, take centuries. Systems that developed complex steady states were called climaxes in the early studies of ecology. Although mature ecosystems are not permanent (chapter 3, fig. 3.12), the name *climax* is still appropriate.

Pulsing of Production and Consumption

Ecosystem production and consumption can operate an oscillation by accumulating organics from production and using them with the frenzied pulse of high-transformity consumers described with fig. 3.13. The model in fig. 6.14a (equa-

FIGURE 6.13 Example of Lotka–Volterra prey–predator oscillation in laboratory culture. **(a)** Protozoans eating bacteria (Drake et al. 1966); **(b)** grain beetles (solid line) and wasps (dashed line) (Utida 1957 from Browning 1963:78).

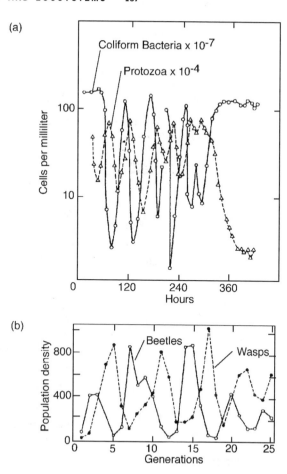

tions in appendix fig. A9) alternates production and consumption and generates the sequence of accumulated organic products and pulsed consumer assets. On a small scale, these pulses are pacemakers within an ecosystem for maximum performance of components. On a larger scale, the rhythm is ecological succession. Total photosynthetic production accumulates living and dead biomass, which is consumed in a frenzied pulse, either from internal mechanisms (episodic fire) or by external destruction from the pulses of the larger-scale system (earthquakes, volcanoes, floods).

Figure 6.14b is the result of simulating the minimodel (fig. 6.14a). Net production generates biomass and structure, followed by diversity and efficiency in the typical pattern of succession (fig. 3.12). The stage of low-diversity overgrowth and net production is followed by the climax stage with high-transformity animals, net consumption, and high diversity.

C. S. Holling (1986) provided a figure-eight graph that has been popular for describing ecosystem sequences (fig. 6.14d). The four stages are numbered 1–4, which are comparable with numbers 1–4 in fig. 6.14c.

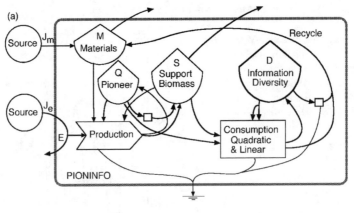

(a)

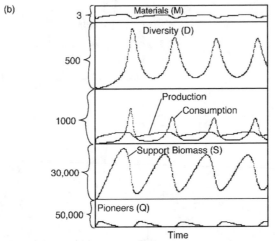

(b)

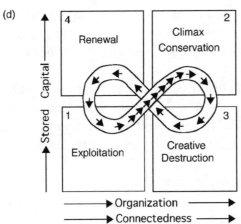

(c) PIONINFO Simulation

Support Biomass (S)

(d)

FIGURE 6.14 Pulsing patterns of ecosystem succession, with stage numbers to aid comparisons. **(a)** Model of production, pulsing, and diversity, called PIONINFO (Odum 1999). See appendix fig. A9 for equations. **(b)** Typical computer simulation. **(c)** Points on the graph of biomass that correspond to C. S. Holling's 4 stages of the adaptive cycle. **(d)** C. S. Holling's figure-8 diagram (Holling 1986; Odum and Odum 1959).

Diurnal Change

Each day ecosystems generate their food and use most of it the next night. This quiet metabolism shows up in the rise and fall of oxygen and carbon dioxide described in chapter 2. For example, the oxygen downstream in Silver Springs (fig. 6.15a) is high in the day and lower at night. By the time the water has gone a kilometer, its properties have been changed by the ecosystem metabolism. The records of metabolism inferred from the oxygen changes and the flow of water at Silver Springs were used to calculate the collective metabolism of the ecosystem. Although the water temperature was constant throughout the year, the production varied with the sunlight, smaller in winter and larger in summer. Although the details are different, diurnal measurements of oxygen under water and carbon dioxide in terrestrial ecosystems are principal ways of calculating their rates of production and consumption.

Figure 6.15b shows the photosynthesis of a coral reef obtained by measuring the change in dissolved oxygen in the waters while flowing across the reef's complex cover of corals, calcareous algae, and hundreds of animal species.

Figure 6.15c shows the metabolism in a forest that was enclosed in a giant cylinder, with air drawn in at the top and flushed out the bottom with a fan 2 meters in diameter. The sun increased the flow of water from roots up through the trees and out through the microscopic pores in the leaves, the process of transpiration. The photosynthesis and respiration were indicated by the uptake and release of carbon dioxide. Most of the forest respiration was in the soil and lower part of the forest and was accurately measured. The photosynthesis was underestimated because there was some eddy exchange of air out of the cylinder at the top.

The best measurements of land metabolism were made by evaluating the carbon dioxide exchange between forest and atmosphere. In the tropical forest of Samoa the gross primary production was 9.65 g carbon/m^2/day (Odum 1995). (See fig. 3.5.)

Seasonal Cycles

The solar energy reaching ecosystems of the earth varies seasonally. Even the equator has a 25% variation in insolation. Most areas have seasonal changes in clouds, winds, and rains. In response, metabolism varies with the seasons. Over much of the world, production and consumption tend to stay in phase, each stimulating the other (fig. 6.16). For example, in the Texas bays (fig. 6.16b) microbial metabolism increases with temperature, and animals migrate in from the sea, using organic matter stored from the previous summer. As a result, respiration increases at the same time as the photosynthesis.

Ecosystems in regimes with sharp seasons (e.g., arctic seas, monsoon areas of Asia) tend to store production during their growing periods to last during their

(a) Oxygen Increases Down Stream

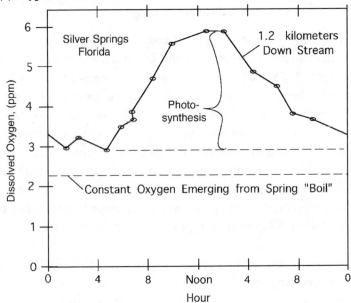

(b) Upstream-downstream Change in Oxygen

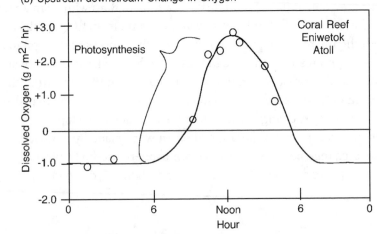

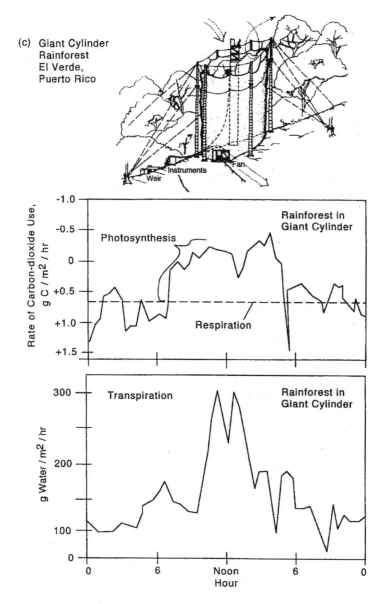

FIGURE 6.15 Diurnal patterns of metabolism in ecosystems. (a) Silver Springs, May 24–25, 1954 (Knight 1980; Odum 1955); (b) interisland coral reef, Japan, Eniwetok, July 1954 (Odum and Odum 1955); (c) metabolism and transpiration in a rainforest enclosed in a giant cylinder, El Verde, Puerto Rico, June 22, 1966 (Odum and Jordan 1970).

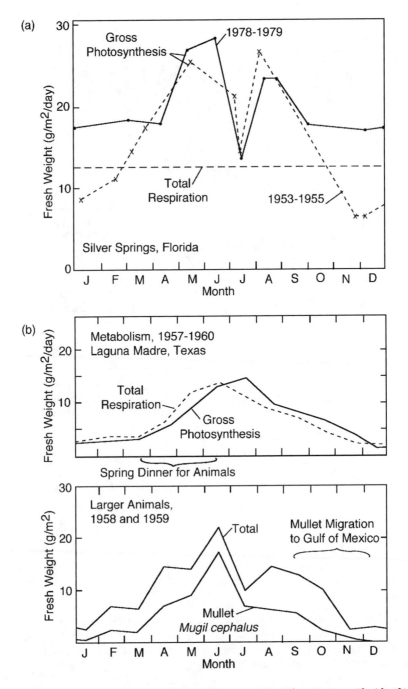

FIGURE 6.16 Seasonal patterns of metabolism and biomass **(a)** in Silver Springs, Florida; **(b)** in the Laguna Madre, a high-salinity Texas estuary south of Corpus Christi. Population data were supplied by Thomas Hellier (1962), Arlington College, Texas.

unproductive periods. They show how systems can adapt to the pulsing of the larger-scale surroundings. These systems demonstrate their productivity by their storages. Systems in steadier surroundings tend to use their production as it is made, and net accumulations are small. However, in plankton ecosystems of temperate and arctic waters, the long winter accumulates excess nutrients that set off a bloom of competitive overgrowth in the spring, followed by diversification and a more complex ecosystem later in the summer (fig. 6.17b).

ORGANIZATION OF ECOSYSTEMS

An ecological system is a network of food and mineral flows in which the major pathways are populations of animals, plants, and microorganisms, with each specialized to participate in a different way. Some species of plants and microbes develop different biochemical compounds so that they can be consumed only by consumers that have adapted specialized enzymes. Each species population is isolated by its behavior or chemistry so that its food and mineral flows are not entangled with those of other species. The separation of species keeps the genes separate also. Like wires in a radio, the ecosystem pathways are insulated against short-circuiting. Self-organization with many species allows division of labor that increases efficiency. A system with more species and more organization may be compared to a radio that has more parts. An ecological network of species may involve several hundred different populations.

Power Circuits Versus Controls

Like a radio, an ecosystem has power circuits and control circuits. The abundant species at the base of food chains are the power circuits. Power flows from left to right in systems diagrams.

The higher-transformity consumers at the end of the food web process information and supply control. Higher animals regulate populations, pollinate, plant seeds, place their wastes, and sometimes defend against destructive influences. Squirrels do work of planting acorns in a forest, which is a small part of the forest energy flow, but their work controls the growth of new oaks. On the systems diagrams, such control pathways are drawn from right to left. The principles and diagrams are similar for ecosystems and for people controlling urban power lines (chapter 8).

Color Coding

In a complex ecosystem, as in a complex radio, pathways are color coded with bright colors, which require biochemical energy for pigment synthesis. These colors aid

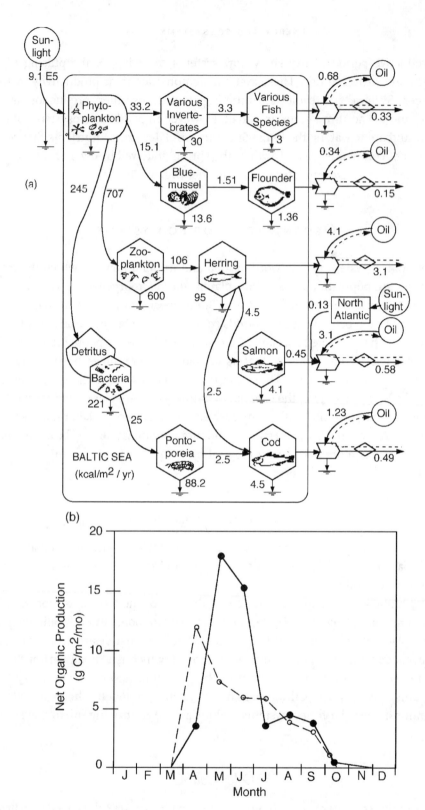

FIGURE 6.17 Ecosystem in the Baltic Sea, from Jansson and Zucchetto (Jansson and Zucchetto 1978; Zucchetto and Jansson 1985). **(a)** Systems diagram with energy flows; **(b)** seasonal record of metabolism in the Baltic Sea with photosynthetic spring bloom (solid line) after winter period of respiratory consumption (dashed line) (Stigebrandt 1991).

the programs of behavior of organisms with simple nervous systems. People in cities also use color in dress and uniforms, which helps social organization.

The many channels with special chemical and physical structures require much metabolic energy for their maintenance. Thus, complex ecosystems have high respiratory metabolism. The power budgets of the whole network are large, but no one product or species receives much energy. Energy goes into the work for the system without piling up offspring, wood, coal, or any other products. Complex systems are very efficient in using all the organic matter they have, leaving little to the fossil record.

Behavior and System Programming

From the ancient lore of natural history, people learn about animals and plants and their life histories, when they breed, how they live and interact, and the meaning of little aspects of their behavior. The sciences of behavior document the facts in a systematic way and describe the various structures and operations. Thus, a male bird with feather movement 1 sets off response 2 in the female, which sets off action 3 in the male, and so on, continuing until reproduction is performed. The process is like a treasure hunt for which we must first fit together the pieces of a map. The treasure is a complex function such as reproduction or nurturing. The sequence of steps forms the same kind of flow chart that a computer programmer draws in explaining the successive steps of a program.

Sometimes we study the interesting patterns of animal behavior as quaint characteristics of nature without recognizing their contributions. We often study single species without asking what the behavior does for the rest of the ecosystem. According to the systems hypothesis of loop reinforcement, the power demands of behavior develop and survive because complex behaviors feed back control services. Complex migrations of crab, shrimp, mullet, and other fishes from the Atlantic coastal bays and estuaries go out to sea in the winter and come back in during the spring, rapidly reinjecting a consumer system as the power input of the sun increases with the warmer weather. Without the migration there would be a lag in development, with much input power unused. The seasonal pulse of production, respiration, and migrating stocks in the shallow Laguna Madre of Texas is shown in fig. 6.16b (Odum 1967).

The eccentric behaviors of birds, bats, and blooms are analogous to preprogrammed computer units that control the timing of computer networks. The huge variety of known species[11] is like a bin of electronic parts, tubes, transistors, resistors, and relays. Just as the properties of electronic parts might be listed in a catalog, we need tabulations of the system performances of the earth's species, including information on such things as storage capacity, power demand, cycle frequencies, and number of input connectors. The design of existing ecological systems will then be better understood, and planning for future designs can begin. In a similar way, the various occupations and operational groups of society constitute a bin of parts for the design of social systems, and the same method may apply.

Up close, observers sometimes see so much beauty and magnificence in details that they can't see what the species accomplish for the whole. In past cultures, people thought about species as a repatriation from Noah's ark, God's handiwork, a service to humanity, or an accidental result of genetic processes operating among islands. But to prevail, species have to contribute somewhere within their system in order to be reinforced by the self-organizing process. The first question for good natural history is, "What does your species do for its system?"; for good anthropology it is, "What does your occupation do for society?"

Understanding Complex Ecosystems

Simplification is required for the human mind to understand such complex systems as rainforests and coral reefs. Some abandon the systems view to study some of the parts or select a chemical substance to follow. Others select simpler ecosystems with fewer species in order to study principles. It is easier to diagram and evaluate the few larger components of the cold Baltic Sea (fig. 6.17) than the more complex systems of tropical seas. Some see ecosystems as an anarchy of species in a competitive struggle for existence without much control. They don't believe there is any organization to study.

Some study the species and their interactions, but in complex ecosystems this generates more detail than can be collectively understood at the system level. Study of rainforests on islands such as Puerto Rico involves a smaller pool of species than in mainland rainforests. Another approach aggregates species into groups according to main functions, as in the trophic level model in fig. 6.2d. Diagrams such as fig. 6.18 of the montane rainforest in eastern Puerto Rico summarize structure and functions of the rainforest. Even if the same number of units fit on one page, each person makes a different diagram because their different knowledge leads them to select different components to emphasize. However, such diagrams make clear to another person the intended emphasis. Words are too ambiguous and inefficient for representing networks. Having a number of such diagrams helps bring out different patterns and principles.

Others count the species and relate diversity counts to energy support and other factors (fig. 6.11). Some develop large, complex computer models, looking for synthesis to emerge from the combinations. Others keep their computer simulation models simple, like controlled experiments, examining a few features while aggregating the rest (figs. 6.12 and 6.14a).

Zooming the human mind to a larger scale bypasses ecosystem complexity. The ecosystems become units to be described with holistic measures such as total number of species, total biomass, total information, total DNA, total photosynthesis, total respiration, total chlorophyll, overall efficiency of light conversion, energy balance, or chemical cycles. However, all scales are important and interesting, although most scientists concentrate on one scale. By definition, complex ecosystems cannot be understood by study at one scale. Energy systems diagrams are helpful because they show connections between small scales on the left and larger scales to the right. Ecosystem diagrams represent the energy hierarchy.

ORGANIZATION IN THE COMPLEX RAINFOREST

Tracing pathways on the systems diagram of a rainforest (fig. 6.18) helps guide discussion to interesting properties. One can start by pointing to the outside sources from the climate and the earth. Complex rainforests are found where the climate does not have a sharp pulse in drought, temperature, or other factors that force a system to start over each year. Photosynthesis (gross production) is about 32 g dry matter/m^2/day. The plant photosynthesis is divided between many different species, each of which has a similar photosynthetic function but is made distinctive by special chemical insulating agents. The cycle for discarding old leaves and recycling minerals is provided by fungi and bacteria, with species adapted to deal with the special chemistry. Various arrangements of roots and microorganisms at the ground level further organize closed-loop mineral cycles, which are mutually stimulating.

Both the producers and the microorganisms are regulated by specialized consumers. Leaf-eating insects are adjusted in their actions to take about 7% of the leaf matter, as long as each plant species is evenly dispersed. Similarly, a great many species of small fungus-eating flies and other small consumers may keep the microorganisms from becoming too numerous. Both the leaf-grazers and the microbe eaters are specialized to deal with the chemical insulating compounds of plants. As long as the various plants are dispersed in small numbers for each species, the plant and microbe eaters cannot find enough continuous specialized food to become epidemic. If any species of plant starts to accelerate its relative growth and become dominant, it will be counterattacked by accelerating actions of the consumers, whose small size gives them a rapid response time. Thus diversity and a check-and-balance system prevent epidemics and maintain stability, a theory supported by much evidence (Voute 1963).

Higher-level consumers, such as the lizards and frogs, and secondary-level microorganisms and eaters of detritus, such as the earthworms, need less chemical specialization because they are not eating the organisms with special chemical insulators. Instead, they are eating the small animal consumers that are not so different in chemical composition. Thus food lines converge, with the higher consumers shifting from one species to another if any become numerous. The convergence also provides a variety of chemical molecules, supplying more complex nutrition, which, in turn, allows the more complex functions and tight packaging required for the rapid movements, elaborate behavioral adaptations, and diversity

FIGURE 6.18 Ways of thinking about rainforest diversity. **(a)** Energy systems diagram of guilds and pathways at El Verde, Puerto Rico (Odum 1995); **(b)** location of some systems functions. Invert., invertebrates; Micro., microbes.

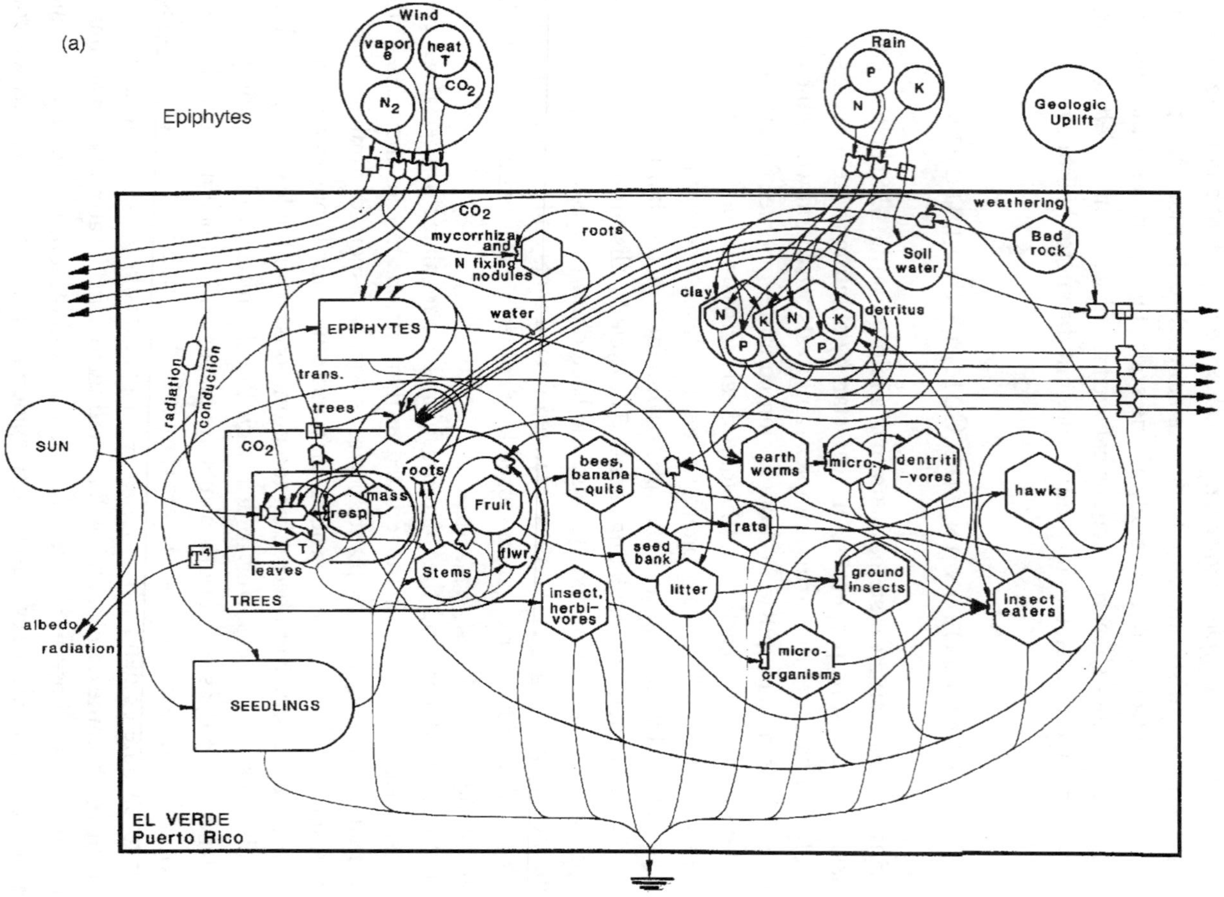

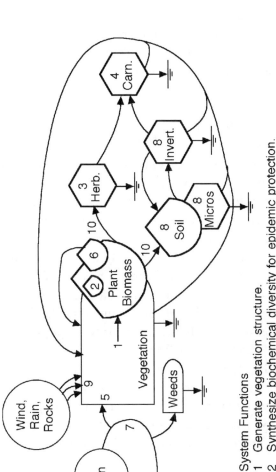

(b)

System Functions

1 Generate vegetation structure.
2 Synthesize biochemical diversity for epidemic protection.
3 Specialize plant consumers to limit plant populations.
4 Generalize animal consumers to limit animal overgrowth.
5 Cover with adapted photovoltaic chlorophyll.
6 Copy information, fruit, and reseed.
7 Shade to control weeds.
8 Build soil structure.
9 Combine sources in interactive energy matching.
10 Balance functions with parallel energy chains.

of the higher organisms. The specialization in higher organisms permits complex patterns of system control, through pollination and seeding.

Ultimately, food chain convergence leads to larger birds, mammals, and humans, providing nutritive diversity. The energies of photosynthesis, through dozens of different species, support the work processes of the system, such as light use, duplication, mineral cycling, epidemic stabilization, structural organization, and soil repair. Some of the functions of the various parts of ecosystems are shown in fig. 6.18b. Purposeful mechanisms are self-organized into a decentralized program of ecosystem control. Thus the ancient ecological systems continue on quietly, efficiently, cleanly self-controlled year after year despite impacts and extinctions.

SUMMARY

In this chapter we examined the structure and functions of ecosystems and their energy basis. After a billion years of evolution, our biosphere has many kinds of ecological systems, each adapted to different environmental conditions and each with adaptive mechanisms. The common characteristics of ecosystems were introduced using the stream ecosystem of Silver Springs as an example. Then similarities between ecosystems were found in vertical stratification, spatial organization, temporal sequences, and energy transformation series. Biomass distribution and its metabolism were related to the energy hierarchy. Diversity was found to depend on the allocation of production according to availability of excess resources for net production overgrowth and the priority for physiological adaptation. Patterns over time adapt to external rhythms and internal pacemakers as needed to pulse for maximizing empower. Pulsing cycles are called oscillation, succession, and evolution, depending on the scale. Ways of understanding complex systems were reviewed using the rainforest as an example and network diagrams to clarify the connections between scales. Many similarities exist between complex ecosystems and complex societies, whose energy basis is considered next.

BIBLIOGRAPHY

Note: An asterisk denotes additional reading.

Bates, M. 1960. *The Forest and the Sea.* New York: Random House.

Beyers, R. J. and H. T. Odum. 1993. *Ecological Microcosms.* New York: Springer-Verlag.

Browning, T. P. 1963. *Animal Populations.* New York: Harper & Row.

Calder, W. A. III. 1984. *Size, Function, and Life History.* Cambridge, MA: Harvard University Press.

Campbell, D. 1984. *Energy filter properties of ecosystems.* Ph.D. dissertation, University of Florida, Gainesville.

*Clarke, G. L. 1946. Dynamics of production in a marine area. *Ecological Monographs,* 16: 321–335.

Collins, D and H. T. Odum. 2000. Calculating transformities with an eigenvector method. In M. T. Brown, ed., *Emergy Synthesis: Theory and Applications of the Emergy Methodology, Proceedings of the First Biennial Emergy Analysis Research Conference Sept. 1999.* Gainesville, FL: Center for Environmental Policy.

Drake, J. F., J. L. Jost, A. G. Frederickson, and J. M. Tsuchiya. 1966. The food chain. In J. F. Saunders, ed., *Bioregenerative Systems,* 87–95. Washington, DC: National Aeronautics and Space Administration (NASA) SP-165.

Forbes, S. A. 1887. The lake as a microcosm. *Bulletin of Scientific Association Peoria, Australia,* 77–87, reprinted *Illinois Natural History Survey,* 15: 537–550.

Hall, C. A. S., M. R. Taylor, and E. Everham. 1992. A geographically based ecosystem model and its application to the carbon balance of the Luquillo Forest, Puerto Rico. *Water, Air, and Soil Pollution,* 64: 385–404.

Helfrich, P. and S. J. Townsley. 1965. Influence of the sea. In F. R. Fosberg, ed., *Man's Place in the Island Ecosystem, 10th Pacific Science Congress,* 39–56. Honolulu, HI: Bishop Museum Press.

Hellier, T. 1962. *Fish production of the Laguna Madre.* Ph.D. dissertation, University of Texas, Austin.

Holdridge, L. R. 1959. *Mapa Ecologico de Guatemala con la Clave de Clasificacion de Vegetales del Mundo.* San Jose, Costa Rica: Instituto Interamericano de Ciencias Agricolas.

Holling, C. S. 1986. The resilience of terrestrial ecosystems: Local surprise and global change. In R. E. Munn, ed., *Sustainable Development of the Biosphere,* 292–317. Cambridge, UK: Cambridge University Press.

Jansson, A. M. and J. Zucchetto. 1978. *Energy, Economic and Ecological Relationships for Gotland, Sweden: A Regional Systems Study.* Ecological Bulletin No. 28. Stockholm: Swedish Natural Science Research Council.

Jones, R. S., W. B. Ogletree, J. H. Thompson, and W. Flenniken. 1963. Helicopter borne purse net for population sampling of shallow marine bays. *Publication of the Institute of Marine Science, University of Texas,* 9: 2–6.

Jorgensen, E. S. 1997. *Integration of Ecosystem Theories: A Pattern,* 2nd ed. Dordrecht, The Netherlands: Kluwer.

Kang, D. 1998. *Pulsing and self organization.* Ph.D. dissertation, University of Florida, Gainesville.

Keitt, T. H. 1991. *Hierarchical organization of energy and information in a tropical rainforest ecosystem.* Master's thesis, University of Florida, Gainesville.

Kent, R., H. T. Odum, and F. N. Scatena. 2000. Eutrophic overgrowth in the self-organization of tropical wetlands illustrated with a study of swine wastes in rainforest plots. *Ecological Engineering,* 15: 255–269.

Knight, R. 1980. *Energy basis of control in aquatic ecosystems.* Ph.D. dissertation, University of Florida, Gainesville.

Lotka, A. J. 1925. *Elements of Physical Biology.* Baltimore, MD: Williams & Wilkins.

Ludwig, H. F., W. J. Oswald, H. B. Gotaas, and V. Lynch. 1951. Algae symbiosis in oxidation ponds: Growth characteristics of *Euglena gracilis* culture in sewage. *Sewage and Industrial Wastes*, 23: 1337–1355.

Margalef, R. 1957. La teoria de la informacion en ecologia. *Memorias de la Real Academia de Ciencias y Artes de Barcelona*, 32(13): 373–449. Translation in *Society of General Systems Yearbook*, 3: 360–371.

Nixon, S. 1969. *Characteristics of some hypersaline ecosystems*. Ph.D. dissertation, University of North Carolina, Chapel Hill.

Odum, E. C. and H. T. Odum. 1985. System of ethanol production from sugarcane in Brazil. *Ciencia e Cultura*, 37(11): 1849–1856.

Odum, H. T. 1955. Trophic structure and productivity of Silver Springs, Florida. *Ecological Monographs*, 27: 56112.

———. 1967. Biological circuits and the marine systems of Texas. In T. A. Olson and F. J. Burgess, eds., *Pollution and Marine Ecology, Proceedings of the Conference on the Status of Knowledge, Critical Research Needs, and Potential Research Facilities Relating to Ecology and Pollution Problems in the Marine Environment*, 99–157. New York: Interscience.

*———. 1985. *Self Organization of Ecosystems in Marine Ponds Receiving Treated Sewage*. Chapel Hill: University of North Carolina Sea Grant SG-85-04.

———. 1995. Tropical forest systems and the human economy. In A. E. Lugo and C. Lowe, eds., *Tropical Forest Management and Ecology*, 343–393. New York: Springer-Verlag.

———. 1996. *Environmental Accounting, Emergy and Decision Making*. New York: Wiley.

———. 1999. Limits of information and biodiversity. In H. Loeffler and E. W. Streissler, eds., *Sozialpolitik und Okologieprobleme der Zukunft*, 229–269. Vienna: Austrian Academy of Science.

Odum, H. T., R. Cuzon, R. J. Beyers, and C. Allbaugh. 1963. Diurnal metabolism, total phosphorus, Ohle anomaly, and zooplankton diversity of abnormal marine ecosystems of Texas. *Publication of the Institute of Marine Science, University of Texas*, 9: 404–453.

*Odum, H. T. and C. M. Hoskin. 1957. Comparative studies on the metabolism of marine waters. *Publication of the Institute of Marine Science, University of Texas*, 5: 16–46.

Odum, H. T. and C. F. Jordan. 1970. Metabolism and evapotranspiration of the lower forest in a giant plastic cylinder. In H. T. Odum and R. F. Pigeon, eds., *A Tropical Rain Forest*, I-165–I-189, Oak Ridge, TN: AEC Division of Technical Information.

Odum, H. T. and E. C. Odum. 2000. *Modeling for All Scales*. San Diego, CA: Academic Press.

Odum, H. T., E. C. Odum, and M. T. Brown with illustrations by E. A. McMahan. 1998. *Environment and Society in Florida*. Boca Raton, FL: Lewis–St. Lucie Press.

Odum, H. T. and E. P. Odum. 1955. Trophic structure and productivity of a windward coral reef at Eniwetok Atoll, Marshall Islands. *Ecological Monographs*, 25: 291–320.

Odum, H. T. and E. P. Odum. 1959. Energy in ecological systems. In E. P. Odum, *Fundamentals of Ecology*, chapter 3. Philadelphia: Saunders.

Odum, H. T. and R. Pigeon, eds. 1970. *A Tropical Rain Forest*. Oak Ridge, TN: AEC Division of Technical Information.

Patten, B. C. 1965. Community organization and energy relationships in plankton. *Oak Ridge National Laboratory Reports in Biology and Medicine* ORNL-3634. Oak Ridge, TN: Oak Ridge National Laboratory.

Pearsall, W. H. 1954. Growth and production. *British Association for the Advancement of Science,* 11: 232–241.

Peters, H. P. 1983. *The Ecological Implications of Body Size.* New York: Cambridge University Press.

Rosenzweig, M. L. 1968. Net primary productivity of terrestrial communities prediction from climatological data. *American Naturalist,* 102: 67–74.

*Ryan, S. 1990. Diurnal CO_2 exchange and photosynthesis of the Samoa Tropical Forest. *Global Biogeochemical Cycles,* 5(1): 69–84.

Ryther, J. H. 1963. Geographic variations in productivity. In M. M. Hill, *The Sea,* Vol. 2, 347–380. New York: Interscience.

Sanders, H. L. 1968. Marine benthic diversity. *American Naturalist,* 102: 243–282.

Smith, R. F. 1970. The vegetation structure of a Puerto Rican rain forest before and after short-term gamma irradiation. In H. T. Odum and R. Pigeon, eds., *A Tropical Rain Forest,* D103–D140, Oak Ridge, TN: AEC Division of Technical Information.

Steeman-Nielsen, E. S. 1957. The chlorophyll content and the light utilization in communities of plankton algae and terrestrial higher plants. *Physiologia Plantarum,* 10: 1009–1021.

Stanton, S. L. 1991. *World War II Order of Battle.* New York: Galahad.

Stigebrandt, A. 1991. Computations of oxygen fluxes through the sea surface and the net production of organic matter with application to the Baltic and adjacent seas. *Limnology and Oceanography,* 36(3): 444–454.

Straskraba, M. 1966. Taxonomical studies on Czechoslovak Conchostraca III. Family Leptestheriidae, with some remarks on the variability and distribution of Conchostraca and a key to the Middle-European species. *Hydrobiologia,* 27: 571–589.

Tamiya, H. 1957. Mass culture of algae. *Annual Review of Plant Physiology,* 8: 309–334.

Tucker, C. J., J. R. G. Townshend, and T. E. Goff. 1985. African land-cover classification using satellite data. *Science,* 227(4685): 369–376 and front cover.

Utida, S. 1957. Cyclic fluctuations of population density intrinsic to the host parasite system. *Ecology,* 38: 442–448.

Voute, A. D. 1963. Harmonious control of forest insects. In *International Review of Forestry Research,* Vol. 1, 326–383. New York: Academic Press.

Westlake, D. F. 1963. Comparisons of plant productivity. *Biological Reviews,* 38: 386–425.

Yount, J. L. 1956. Factors that control species numbers in Silver Springs, Florida. *Limnology and Oceanography,* 1: 286–295.

Zeuthen, K. E. 1953. Oxygen uptake and body size in organisms. *Quarterly Review of Biology,* 28: 1–12.

Zucchetto, J. and A.-M. Jansson. 1985. *Resources and Society Ecological Studies #56.* New York: Springer-Verlag.

Zwick, P. D. 1985. *Energy systems and inertial oscillation.* Ph.D. dissertation, University of Florida, Gainesville.

Notes

1. Forbes's article (1887) on the lake as a microcosm was important to the science of limnology in attracting people to a whole systems view. See the review of microcosms in Beyers and Odum (1993).

2. Measurements since 1920 have shown the waters emerging in the springs of Florida nearly constant in temperature and chemical composition. Waters percolating through natural ecosystems and limestones were low in organic matter and moderately low in phosphorus and nitrogen salts. Unfortunately, spring waters have begun to change as population growth and economic development of the watersheds are causing lower oxygen levels and higher nutrient levels. The ecosystem in Silver Springs in 2001 had more blue-green algae and fewer fish. The Rodman dam blocked most of the mullet that require migration to the sea for reproduction. The large catfish also have disappeared.

3. Production can be measured as oxygen is released, carbon taken up, weight increased, and so on. In the chain of biochemical steps, oxygen is the first readily measurable byproduct, which is released by the plant cells after light energy is incorporated as chemical potential energy. Other substances that are taken in (such as carbon dioxide, nitrogen, and phosphorus) are involved further downstream in the reaction chain after energies are dispersed in work and after flows are distributed in branching routes and time-delaying storages. Thus oxygen gives higher measurement estimates of photosynthetic production. Most systems have concurrent respiration in the day in the same cells, using much oxygen that would have emerged. Chlorophyll activated by light emits back radiation (fluorescence). The respiration at night is not a good measure for small cells because sometimes it is only a small fraction of that during the day.

4. Charles Hall and students used a unit model of this type to simulate and map the spatial distribution of transpiration and productivity over the steep topography of the Luquillo Mountains in Puerto Rico (Hall et al. 1992).

5. The plot on two coordinates, rainfall and temperature in fig. 6.10, was simplified from Holdridge's (1959) original classification, which used triangular coordinates to represent conditions for biomes. Triangular coordinates can be used to represent three variables when the sum is a constant. However, Holdridge's third coordinate was not an independent third variable but the ratio of rainfall Y to transpiration based on temperature X. All scales were logarithmic, and one was plotted negatively. Thus, $\log X - \log Y = \log X/Y$. Figure 6.10 omits the ratios but retains the Holdridge names and definitions in terms of rainfall and temperature. Holdridge used logarithm of temperature as a surrogate for transpiration, but the relationship is very approximate. Transpiration is proportional to wind velocity and the vapor pressure difference between leaf pores (stomata) and the air. The vapor pressure in the air is an energetic property of the air mass. The vapor pressure of the plant depends on the leaf temperatures, and these depend on the sun's insolation and the air temperature.

6. Margalef (1957) provided several measures of diversity. One of these is the slope of the graph of species found and logarithm of the individuals counted. Other measures used information theory. See chapter 8.

7. For review and examples, see chapter 5 of *Ecological Microcosms* (Beyers and Odum 1993).

8. See the curve of species diversity inverse to organic matter (Odum 1967).

9. In systems engineering, a chain of two storing units (integrators) with a loop-back is a second-order system because the differential equation for the whole system has a second derivative. Such systems may oscillate regularly, and when the input is variable, undesirably wide fluctuations may result. The oscillator may get into phase with the external variation of input and go into resonance with variations in storages and flow that are larger than those in the input. Campbell (1984) showed how ecosystems could capture more pulsing energy by adjusting their storages to oscillate with matching frequencies. Electronic sine generators send out an undulating, up-and-down voltage that follows the position of a shadow projected by a nail rotating on a wheel. Sine generators can represent the energy from the sun as it rises and falls during the day or by season because of the trigonometry of the angle at which light rays strike the earth's surface when the sky is clear. Patten (1965) suggested that the sine wave generator made with two coupled electronic integrators was a model for the oscillations of two animal populations in a chain, such as a prey and predator relationship. Any population is an integrator summing its food incomes and energy drains.

10. Optimum frequencies for maximum empower of ecosystems were studied by Dan Campbell (1984), Paul Zwick (1985), and Daeseok Kang (1998).

11. The possible use of rare species as special parts justifies great expense in preserving them from extinction because their original development costs were large and spread over evolutionary time. Emergy values are very high for endangered species (Odum 1996).

CHAPTER 7

EMPOWER BASIS FOR SOCIETY

THE EMERGENCE of human societies from minor components of nature to the dominant modern technological civilization is a story of shifting empower. Progress has become so rapid and dramatic in the last two centuries that many people believe anything is possible. This chapter first considers the power basis for earlier societies, the limits to solar energy, and some of the ways food, clothing fiber, shelter, and heat were supplied to human society. Next urban civilization is explained. Then emergy is used to evaluate alternative energy sources and the carrying capacity of the earth for people in the future.

PHOTOSYNTHETIC EFFICIENCY

To understand the energy basis for past societies, we have to recognize the way photosynthesis is limited by the dilute nature of solar energy reaching Earth. There is much confusion about the efficiency with which sunlight is converted to organic products. With the help of fig. 3.3, consider how plants have evolved to maximize the production that is possible. Humanity developed cultures and customs in amazing variety within these limits.

Photosynthetic production of the organic products (food, clothing fiber, shelter, heat) that support society is limited by the dilute concentration of sunlight reaching the earth. In the first step of photosynthesis photons separate electrons from organic structure, which gains a positive charge as a result. Shown in fig. 3.3a, the positive charges of most plants interact with water to make oxygen, and the negative charges drive the load of organic production biochemistry. In other words, the green pigmented structures of plants are natural photovoltaic cells.

To understand the thermodynamic relationships of sunlight and efficiency, refer again to fig. 3.2, which shows how power output depends on load. In that figure the loading backforce of the output process is varied and maximum power is at an intermediate load, with an intermediate optimum efficiency. Isolated chloroplasts follow the same law (fig. 3.3b). If the load is constant but the input energy is varied, fig. 3.3c results, which shows that the efficiency is inverse to the increasing input

power. Equations are given in appendix fig. A4. In other words, there is a thermodynamic optimum efficiency for maximum production at intermediate light.[1] As
an example, recall the graph in fig. 3.5 showing that there is a thermodynamic
optimum for a whole ecosystem.

Given a longer time, plants adapt to light levels by adding or decreasing chlorophyll so as to maximize energy conversion. Ireland and Oregon are dark green
in winter because oceanic clouds shade the light, so more chlorophyll is needed to
catch the scarce light.

The Earth Fit of Solar Societies

Before the time of recorded history, the biological and social evolution of the human species was nurtured by photosynthesis of the ecological systems to which
tribes were adapted. When there were no special energy sources for societies except the input of solar energy, tides, and deep heat of the earth, humans were a
small part of the overall scheme but were protected by the earth's great stability,
complexity, rhythms, and staying power of the natural system. Because humans
had little energy available to direct at the mother ecosystems, they had little ability
to overpopulate, and population densities often were limited by resources. When
Environment, Power, and Society was published there were still remnants of ecosys-
tem-based human cultures for anthropologists to study, but now most have been
relegated to memory. Figures 7.1 through 7.5 summarize the energy networks of
societies evolved and adapted to the renewable energies of their environment.[2]

Hunting and Gathering Systems

Human societies evolving in ecosystems were at the end of complex food chains
based on photosynthesis. In the central African rainforests the Pygmy population
in the complex tropical forest used only a small part of the many channels of fruit
and animal products available (Turnbull 1916, 1963). Figure 7.1 displays human
population at the top of the energy hierarchy with diverse converging energy transfers, each of small magnitude. The forest concentrates energy and upgrades the
biochemical diversity needed for human nutrition. Similar small populations were
found in the Amazon basin and other places where the vegetation, soil structures,
and geochemical cycles were self-regulated and steady. Virgin forests have huge
trees (figs. 6.8 and 7.1c), root structures, and a lacework of dead biomass in logs
and branches on the forest floor. Human population densities were about one person per square mile.[3] Some gardens were developed in clearings, but extensive
agriculture in the complex forest was prevented by the high diversity of insects and
weeds, which helps to limit any one plant species.

Even in stable forest regimes, the culture of human populations can provide
pulsing rhythms, described as universal in chapter 3. Rappaport (1971) explained

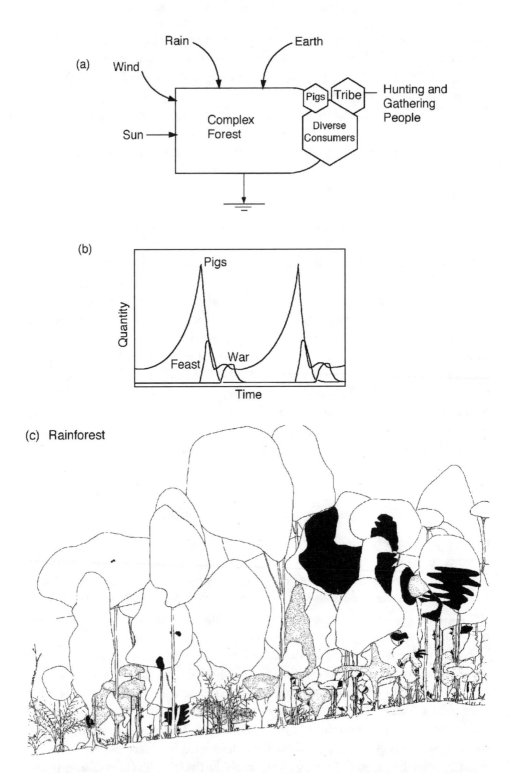

FIGURE 7.1 Hunting, gathering, and gardening society in low density within the bosom of complex ecosystems. **(a)** Energy system; **(b)** computer simulation of a model of cyclic tribal life in forests of New Guinea by Christine Paddoch (1972), representing the studies by Rappaport (1971); **(c)** cross-section view of rainforest (Smith 1970). Most of the trees shown are different species.

how the forest people of New Guinea accumulated semidomestic pigs, followed by a tribal feast that initiated an incidence of intertribal war that was terminated when the energies were no longer in excess. Figure 7.1b shows a computer simulation of the pulsing flip-flop cycle as controlled by a cultural pacemaker.

Society on Coral Atolls

The island peoples on atolls of the open Pacific were part of a system of stable hydro-climate and complex coral reefs. An atoll system of humanity and nature was described by Alkire (1965) and is shown as an energy systems diagram in fig. 7.2. The same kind of complex organization of diversity found in the structure, culture, and functions of human society is found in complex tropical ecosystems (chapter 6). With stable climatic regimes the atoll system developed a complex network of tree products, root crops, and colorful marine inhabitants. Many channels of food for the human population were present, although no great energy flow was available in any one. Staple root crops, coconuts, and breadfruit from land supported a population of one per acre, but the protein source was derived from a much larger area of reef and lagoon. Calculated using the whole atoll as the basis for support, human density was thirteen people per square mile. The percentage of the marine energy budget for humans was small, but the small footprint of the society within the ecosystem was also its protection. The energy resources available to humans were insufficient for the human population to damage its life support. Diversity of the ecosystem was a protection against epidemic decimation of their food supply network.

Surviving cultures had to fit into their ecosystem, developing a great knowledge of species properties, the seasonal cycles, and the energy network. Medicinal herbs, poisons, building materials, and animal products were available in great biochemical diversity but in small quantities, so much energy was required for gathering. Adapting a culture to such diverse permutations of input may have been instrumental in the development of the flexibility of human intelligence. Adaptation of societies was guided by the same kind of natural selection of cultural designs as has been suggested for populations and ecosystem designs (maximum empower principle, chapter 3).

Support from Agriculture Versus Complex Ecosystem

Every schoolchild learns that agriculture began with domestication of plants and farm animals. The idea is to substitute domesticated plants and animals for the more complex or transient networks of nature. As diagrammed in fig. 7.3, channeling environmental energy away from the ecosystem to a simpler energy transformation chain theoretically provides more energy to society and ultimately more activity and carrying capacity for the culture. But the simplified system of fewer species has fewer means for rapid spread of plant cover, less biochemical diversity, fewer self-controls, less protection against epidemics, fewer means to limit weeds, and less soil-building capacity.

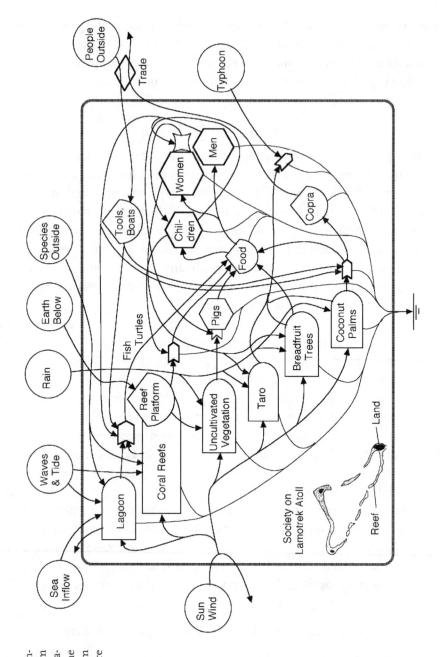

FIGURE 7.2 Example of low-density society in a stable system of several ecosystems, the Pacific coral atoll at Lamotrek. The diagram summarizes in system form the data given by Alkire (1965).

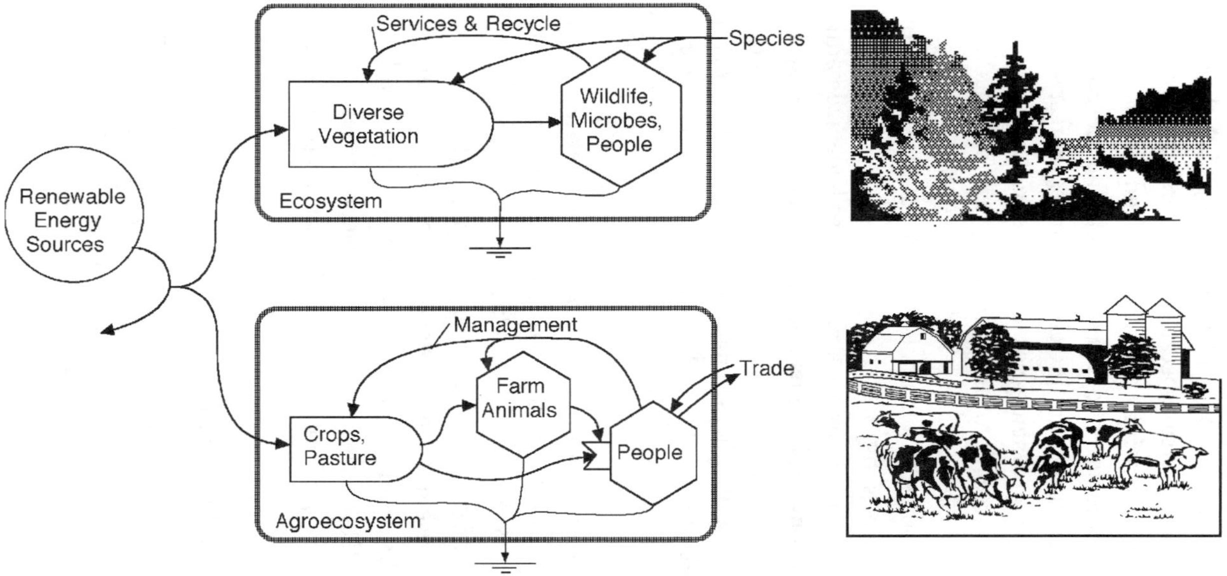

FIGURE 7.3 Substitution of agroecosystems with crops, farm animals, and higher population densities of people for the hunting and gathering ecosystem, both operating on renewable energy of sun, wind, and earth.

When inputs of climate and land are stable (fig. 7.2), the diverse ecosystem can displace the simplified version because the diverse one is more efficient. Diverse ecological systems successfully compete with and displace simplified agroecosystems when environmental conditions are stable and there are no periods of severe cold, drought, or other life-limiting circumstances. Forest soil stores seeds, which explode in germination when exposed to the sun. Unless special energies are directed to prevent succession, the complex system with its division of labor and services draws in more resources and displaces crops whose energies are going only into weedy storage. If left alone, succession covers the ground and displaces agricultural crop plants (fig. 3.12).

Where water and nutrients are scarce, the well-adapted natural ecosystem overcomes these limits by applying its budget of available energy. For example, desert plants have water conservation adaptations such as water-storing structures, deep roots, and means for opening stomata only at night. Nutrients are tightly recycled, aided by the work of animals. Under natural conditions, the gross photosynthesis of the complex system probably cannot be exceeded, for most of its yield assists production.

Even though natural complexes export little, they do have high gross photosynthesis (table 6.1). The natural systems found in each area may have already overcome limits, except the inescapable one of available light. Self-management by complex consumers is diagrammed at the top of fig. 7.3. The potential yield of food is small because much of the energy goes into necessary work. In the bottom diagram of fig. 7.3 people and their farm animals replace wild species.

Table 7.1 lists data on production and energetic efficiency of agricultural systems converting solar energy. Compare these with data on production by the natural sys-

TABLE 7.1 Agricultural Food Production

Crop	Harvest of Organic Matter (kcal/m^2/da)	Efficiency in Use of Sunlight (%)[a]
Not subsidized by fossil fuels		
Farms, United States, 1880[b]	1.28	0.03
Grain, Africa, 1936[c]	0.72	0.02
Industrialized agriculture		
Rice, United States, 1964[d]	10	0.25
Grain average, North America, 1960[b]	5	0.12

[a] Calculated with 4,000 kcal/m^2/day sunlight.
[b] Parker (1963).
[c] Brown (1965).
[d] President's Science Advisory Committee on World Food Production (1967).

tems developed through millions of years (table 6.1). The gross photosynthesis (P) of the natural systems in each belt of the earth indicates the upper limit of the sun's conversion for any alternative system using the same inputs. Usually the efficiency of light conversion by agroecosystems is much less because weeding and harvest leave the ground bare or poorly covered with chlorophyll for part of the year.

Agriculture in Pulsing Regimes

The story for the zones of the earth where there are sharp variations and severe fluctuations in climate and hydroclimate is quite different. Where drought and heavy rains alternate or where sea salinities vary drastically, a system either diverts great energies into special physiological adaptations to the fluctuations or, when conditions are too severe, continually establishes and reestablishes its system after unfavorable periods. Special adaptations are developed for recolonization or seasonal cycles. The systems of nature that developed for the fluctuating environments have little energy to evolve many specialized species for a division of labor. Most of the energy goes to a few species that program timely seed dispersal, germination, growth, energy storage, flowering, fruiting, and arranging stages.

In other words, the systems of nature that have evolved in the zones of fluctuation on the desert borders, at the mouths of rivers, and in the polar regions are systems of a few species of pulsating populations, temporary large storages, and channeled energies. In such regions the natural system produces a few species in surges not unlike agriculture. In fact, primitive agriculture was really a capture of weedy net production systems in which the storages were concentrated in humans and their animals. Thus, systems of solar energy–based agriculture are competitive in areas of fluctuating environment but are subject to its variations, incurring frequent famine and losses to epidemic insects.

One of the ways humans adapted to pulsing climatic regimes was to become nomads, using energy storages to move with the fluctuating zones of favorable conditions. In so doing, people were one of the means for rapid recolonization and planting. Pulses of locusts, caribou, bison, fish, and shrimp migrate in these zones of the earth. Migrating at times and places of animal migrations helped tribes accumulate energy for travel. Vaneeta Hoon (1996) evaluated energy supporting the annual movement of buckwheat-based people from valleys to Himalayan highlands.

Pulsating agriculture based completely on solar energy takes advantage of seasonal pulses of rainfall to draw a high percentage of energy into a culture. The disadvantages of short duration and fluctuations limit primary production, but fluctuations, whether from natural causes or arranged by the farmer, are necessary to a harvest. Unfavorable seasons keep out more complex associations of animals and plants that would divert energies from the harvest. Pulses apparently favor humans because of their planning and social skills. A program of seasonal planting competes well with the competition of annual weeds. Some very effective systems of agricultural society evolved in the monsoon, prairie, and savanna belts of the earth, which have growing periods alternating with dry conditions.

Seasonal Rice and Cattle

Rice is one of the most successful ancient systems for supporting dense populations (one person per acre) based wholly on natural energy in monsoon climates. Farm animals are used as energy transformers to do the necessary work of planting, fertilizer spreading, high-protein synthesis, energy storage, and weeding. Main energy pathways for the rice–cattle system in India are shown in fig. 7.4. People and their animals have the energy storage to inject a fast start for the agricultural crop, putting it ahead of possible competitors when the cool, moist winds and rains from the Indian Ocean return in June. At this time, agriculture and the ecosystems start over after the parching spring temperatures of up to 120°F (49°C).

Religion as an Energy Program: Sacred Cows

Figure 7.4 shows energy flows in the system of rice and sacred cows that was typical in India. The work of humans and bullocks accomplished necessary functions, which included nutrient cycling and processing of dung for fuel and fertilizer. The network runs well on sunlight energy, with control loops and features of stability in

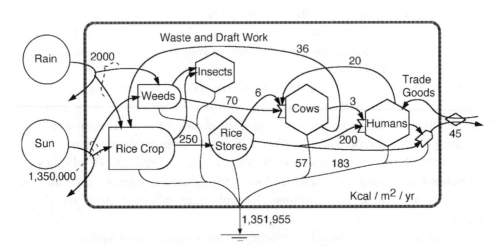

FIGURE 7.4 Sacred cow agroecosystem of India, operating with the sharp seasonal pulse of monsoon wet and dry seasons. Relationships were suggested by Harris (1965). Data are for tropical dense populations with 640 people per square mile, 0.1 animals per person; Indian grain yields 250 kg/acre/yr (Brown 1965). One farm animal is shown for every 10 people. One-third of the food calories of cattle remain in feces. Work and fecal fertilization are taken as half of animal metabolism. The animal protein intake for India was about 6 g/person (Brown 1965); 2% of the food crop was fed to animals. Animal metabolism was 8,000 kcal/day (Kleiber 1961). Farm work was taken as 10% of total worker hours of population.

the otherwise unstable environment of the monsoon. In India the cows are sacred to part of the population. Some people have advocated eliminating the sacred cows in order to skip the extra mouth and shorten the food chain. But Marvin Harris explained how the cows glean nutrition from different food chains, produce the bullocks that are necessary for the agricultural pulse, recycle the minerals, and provide some critical proteins through milk (Harris 1965). He quoted Gandhi as saying that the cows are sacred because they have practical value. Figure 7.4 shows their useful roles. Perhaps one of the tasks of religion is to provide appropriate programs of energy control. See chapter 11.

Artificial Pulse and Shifting Agriculture

Whereas grain agriculture in prairie and savanna climates was much like the seasonal restart of grassy natural systems, agriculture was quite different when complex forests had to be displaced. Despite competition from nature, agriculture was successfully introduced into stable forest regimes through a pulse generated with human effort. Forest plots were cut and the biomass energy of the logs and brush burned to open the ground to sunlight, release nutrients, and kill off insects and competing plants. The potential energy of the elevated crowns against gravity, amassed over many years, helped the work when trees fall. Thus, the energy accumulated in succession was used to undo its own work. Crops were grown for several years before the nutrients were reduced or the surrounding forest could reinvade. Then the land was allowed to go back into succession (fallow land) to restore the soils and energy to support another agricultural pulse later. A family could move from one plot to the next, not returning to the original plot for 20 to 50 years. This was shifting agriculture, and it worked as long as the population density was small. Both ecosystems and human cultures were supported.

A network diagram for shifting agriculture in fig. 7.5 shows an alternation between two main land uses: forest growth and subsistence gardens. Humans act as a system switch, controlling the start of agriculture through a small expenditure of cutting energy. The timing and nutrient storages permit an effective fall burn and subsequent crop growth. Systems engineers will recognize the start and abandonment as an automatic flip-flop. Families can receive continuous support by moving their settlements in rotation after a plot loses its fertility. The farmer's plot, if it were not abandoned, would not hold the soil structure, provide enough energy to control the successional invaders (weeds and insects), or hold stabilized chemical cycles. Carter (1969) details the labor and social structure that accompany a workable shifting agriculture system in Guatemala.

The cycle of regrowth and soil restoration is rapid as long as the plots are surrounded by complex forest that is old enough to supply seeding. The natural ecosystem has many species for quick reforestation of openings. The diversity of the successional forest species and insects also works against any crop plantings. Many crops grown in high-diversity situations, such as cassava, yucca, and tapioca,

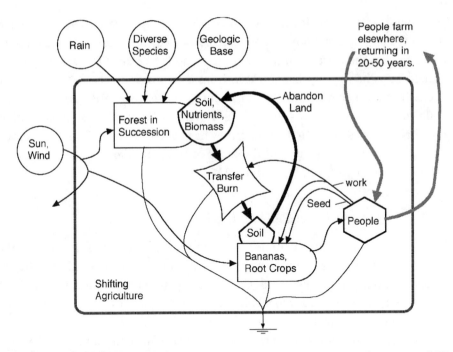

FIGURE 7.5 Energy diagram for shifting agriculture. Humans serve as a switching timer, shifting the stored forest emergy into food crops for several years before being forced to move on by nutrient shortages, loss of soil structure, insects, and weeds. Heavy lines indicate circular pathways of soil and people.

have edible roots, which have some natural poisons against insects that have to be washed out before consumption by people.

Agroecosystems Using Ecosystem Structure

In the moist tropics, many tree crops, such as cacao, are successful. Some crops such as coffee and tea are planted under the partial shade of trees, whose crowns are trimmed to let some daylight reach the crop. The crown trees provide a stable microclimate, roots to stabilize soil structure, transpiration to draw out excess soil water, nodules that fix nitrogen, and diverse habitat for animals that control insects. Yields are less than those of crops in open fields, but they are less vulnerable to disturbance. Under the canopy the ecosystem provides protection and support functions to the crop plants.

An example from the Congo uses corridors of rotation and is based on an old system of shifting cultivation developed by the Bantu peoples (Turnbull 1963). The short cultivation phase ends with a shade crop, which helps reestablish the complex successional vegetation quickly, thus restoring the mineral cycle, soil structure, and the canopy that will permit a new cycle after later cutting.

Before the Industrial Revolution in agriculture, the world was fed with thousands of varieties of plants adapted to yield food, clothing fiber, and housing. These agricultural systems used only renewable energies and knowledgeable management by rural people. There were agricultural varieties for all the special climate conditions over the earth. With the decline of fossil fuels, these varieties will be needed again. There is an urgent need to keep these varieties from going extinct during the current period of intensive agriculture based on nonrenewable energy, which is discussed next.

NONRENEWABLE, FUEL-BASED CIVILIZATION

Beginning two centuries ago, many countries began to develop an entirely new basis for power using coal, oil, virgin forest storages, atomic fuels for nuclear power, and other nonrenewable resources to supplement renewable energy. Concentrations of available energy of moderately high transformity had accumulated from the environmental work by the earth's emergy for millions of years. Although knowledge was required to develop machines to couple these power inputs to human society, the real basis for progress was the entirely new energy sources that were an order of magnitude greater than previously available to people. Excess energies were available for further research to obtain more fossil fuels and to apply them in supporting people. The subsequent progress was like a flash explosion. The energy added to agriculture increased yields, as shown in fig. 7.6b. In chapter 6 we learned that most of the energy of the natural ecological cover was used for the many functions required for survival: competing, recycling, and holding structure. When these functions were carried out using the new energy sources, more of the gross photosynthesis became available for storage and yield to human consumers. The work that used to be done by the plants, animals, and rural farmers using environmental energy was done directly and indirectly by fossil fuels.

Fossil fuels were substituted for feedback work processes of the older systems, interacting with but not replacing the photosynthetic use of solar energy. There is no evidence that 100 years of research has increased the efficiency of photosynthetic conversion of solar energy. Apparently a billion years of evolution had already achieved the thermodynamic limit described with figs. 3.3 and 3.5.

As more and more calories from outside were substituted for the necessary work, more of the gross production could go into net food yield. By substituting for system work, fuel-based inputs released more yield. The fossil fuel energies were applied to the system at the point where there was feedback work and often in a magnitude greater than could be derived by concentrating solar energy. More energy was applied to defeat invasions of alternative natural ecosystems. Fossil fuel energy was directed to supply any necessary materials. The added energy was amplified by increasing flows of limiting factors. Figure 7.6b shows the system after substitution, and table 7.1 summarizes the yields attained. Lester Brown (1965) called the jump in agricultural yields "take off."

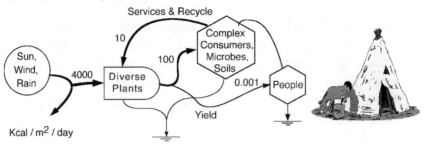

(a) Complex Native Agrosystem

Services & Recycle

10

Sun, Wind, Rain

4000

Diverse Plants

100

Complex Consumers, Microbes, Soils

0.001

People

Yield

Kcal / m^2 / day

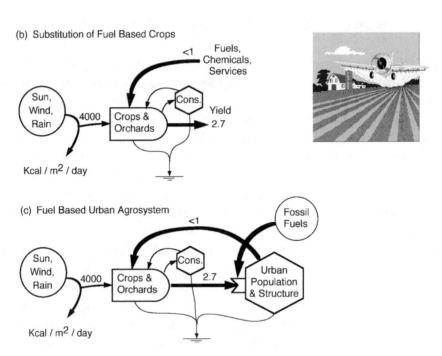

(b) Substitution of Fuel Based Crops

<1

Fuels, Chemicals, Services

Sun, Wind, Rain

4000

Crops & Orchards

Cons.

Yield
2.7

Kcal / m^2 / day

(c) Fuel Based Urban Agrosystem

Fossil Fuels

<1

Sun, Wind, Rain

4000

Crops & Orchards

Cons.

2.7

Urban Population & Structure

Kcal / m^2 / day

FIGURE 7.6 Comparison of yields to humans in hunting–gathering–gardening system with those of transformed agriculture based on fossil fuels and specialized crop varieties. **(a)** Small net yield to low-density society from self-sufficient varieties in complex ecosystems; **(b)** high yields after fossil fuels have substituted industrial services for the functions of those natural species, releasing more of the basic production to yield; **(c)** transformed system of intensive agroecosystems and urban society. Work flows aided by the urban system include mechanized and commercial preparation of seeding and planting, replacing the natural dispersion system; fertilizer excesses, which replace the mineral recycling system; chemical and power weeding, replacing the woody maintenance of a shading system; soil preparation and treatment to replace the forest soil-building processes; in-secticides, which replace the system of chemical diversity and carnivores for preventing epidemic grazing and disease; and development of varieties that are capable of passing on the savings in work to net food storages. New varieties are developed as diseases appear, thus providing the genetic selection formerly arranged by the forest ecosystem. The different kinds of available energy flows are not measures of equivalent work until expressed in emergy units. In fig. 7.6a people living at a density of 1 per square mile (2.6 km^2) and daily food consumption of 2,500 kcal consume only about 0.001 kcal from each square meter of their territory. In the intensified agriculture of fig. 7.6b and 7.6c, 170 people per square mile supported 32 times this number in cities when the level of grain production in the United States was about 1,000 kcal/m^2/yr (Brown 1965). Costs of services from the fuel-based economy were about \$54 (1965 \$) per acre, and the emergy/money ratio was 2.5 E12 solar emcalories per dollar. The empower feedback from society (counting only services) was the product, 1.35 E14 solar emcal/acre/yr (8.9 E4 solar emkcal/m^2/day), 22 times the emergy of the sunlight.

Abundant food rolled out from huge fields that were developed with the new industrialized agriculture. The fields were sown by machinery, tilled by tractors, and weeded and poisoned with chemicals. Epidemic diseases were kept in check by great teams of scientists in distant experiment stations developing new and changing varieties to stay ahead of the evolution and adaptation of disease. An example is the intensive cornfield agroecosystem of Illinois (fig. 7.7). Soon a few people were supporting many, and most of the rural population left the little farms to fill new industrial cities, where many of the jobs were manufacturing the inputs to the new agriculture (fig. 7.6c).

Potatoes Made Partly of Oil

The high yields from industrial agriculture generated a very cruel illusion because the citizens, the teachers, and the leaders did not understand the energetics in-

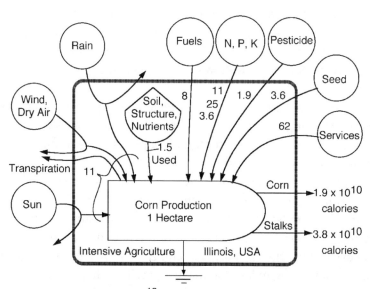

Inputs x 10^{13} solar emcalories per year

Solar Transformities:

Industrialized Corn:
$$\frac{(11+1.5+8+62+11+25+3.6+1.9+3.6) \times 10^{13}}{1.9 \times 10^{10}} = 6.7 \times 10^4 \text{ secal / cal}$$

Corn Stalks:
$$\frac{(11+1.5+8+62+11+25+3.6+1.9+3.6) \times 10^{13}}{3.8 \times 10^{10}} = 3.4 \times 10^4 \text{ secal / cal}$$

FIGURE 7.7 Emergy evaluation of intensive corn agriculture in Illinois (after Pimentel and Hall 1984).

volved and the various means by which the energies entering a complex system are fed back as subsidies indirectly into all parts of the network. With conceit, the enriched people imagined that progress in agricultural yields resulted from new know-how in the use of the sun. A whole generation of citizens thought that higher efficiencies in using the energy of the sun had arrived. This was a sad hoax, for people of the developed world no longer eat potatoes made from solar energy. In fig. 7.6b the food reaching the humans is produced mostly by the energy subsidies in support of all the human services required. People are really eating potatoes made partly of oil.

Much of the power flow that supports intensive agriculture is not used on the farm but is spent in the cities to manufacture chemicals, build tractors, develop varieties, make fertilizers, and provide input and output marketing systems, which in turn maintain mobs of administrators and clerks who hold the system together. As we stand on the edge of the vast fields of grain, with tractors and production as far as the eye can see, we are tempted to think human brilliance has mastered nature. However, the plain truth is that fuels are being substituted for plant and animal functions, releasing more of the photosynthetic product to go to consumers. Wherever the flow of industrially organized fossil fuel is missing or the work from a fossil fuel culture is eliminated, the agriculture possible is only what it once was or even worse, for the know-how of self-sufficient agroecosystems and varieties is disappearing.

People from an industrial–agricultural region who go to a low-energy country to advise on improving agriculture can help only if there is a cheap fuel supply for another zone of fossil fuel agriculture. As fuel prices rise and fuel use decreases, the advice will come in the opposite direction.

Citizens in industrialized countries think they can look down on the old system of humans, animals, and subsistence agriculture that provided a living for a few people from an acre or two in India when the monsoon rains were favorable. Yet if fossil and nuclear fuels were cut off, we would have to find people still farming in older ways to show the currently affluent citizens how to survive on the land while the population was being reduced to make it possible (chapter 13).

Figure 1.4 (chapter 1) compares an island economy before and after the new fuel input, summarizing the changes from environmental base to fossil fuel–based agriculture in macroscopic view.

Rotation Needed for Intensive Agriculture

Many kinds of intensive agriculture are not sustainable because they are based on initially rich soils that are not sustained by low-diversity agroecosystems. Rotation of the lands through a high-diversity successional restoration is required, just as in the earlier shifting agriculture. Such rotation includes seeding from a complex ecosystem with diverse plants, animals, and microbes for soil restoration. Thus sustainable intensive agriculture requires that ecosystem areas be distributed among the intensively farmed areas. Figure 7.8 shows the optimal ecosystem seeding area for maximum economic output. Figure 7.8b was generated from a computer simulation.

Diminishing Returns of the Energy Subsidy to Agriculture

Every kind of energy flow must interact with and mutually amplify another flow of higher or lower transformity (chapter 4). To increase production, fossil fuel inputs must amplify conversions of environmental and marine resources. However, the increasing inputs of machinery, fertilizer, and pesticides in agriculture have a diminishing return as other factors, such as water and sunlight, become limiting. World agriculture is reaching its ceiling because most of the fossil fuel–based

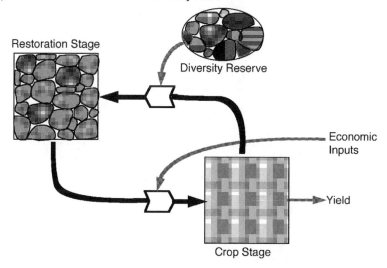

(a) Land Rotation Seeded from Diversity Reserves

Restoration Stage

Diversity Reserve

Economic Inputs

Yield

Crop Stage

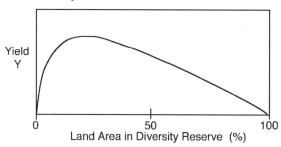

(b) Effect of Diversity Reserve on Yield

Yield
Y

0   50   100
Land Area in Diversity Reserve (%)

FIGURE 7.8 Role of high-diversity reserve areas in accelerating the restoration stage in land rotation. **(a)** Three kinds of areas required for sustainable land rotation; **(b)** result of computer simulation showing yield as a function of the area reserved for high-diversity ecosystems to supply seeding (Odum 1962).

inputs are in excess, and there is little fertile land available to expand production and little water for irrigation. In fig. 7.9 the curves of productivity as a function of increased irrigation energy, fertilizer, pesticides, and equipment follow the classic limiting factor curve explained with fig. 3.10b.[4]

The various kinds of agricultural systems form a series, from the primitive ecological one with people as a minor component to the modern regime with human food from petroleum subsidy of intensified agriculture (fig. 7.9). Yield increases with added emergy use but with diminishing returns.

As the fuel use increased, the number of agricultural species and genetic varieties decreased. At the same time, however, the variety of urban occupations involved increased. By this theory the number of occupations of people and the number of species in the system are complementary; an increase in one entails decreasing the other.

Algal Culture and Fallacious Dreams

Another cruel illusion was proffered by laboratory scientists and writers who proposed to feed the world on algae, which they implied were productive on a different

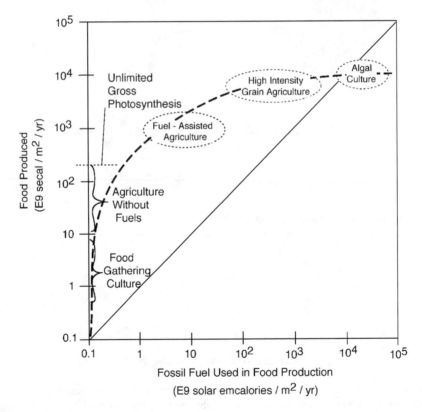

FIGURE 7.9 Food yields as a function of the fossil fuel emergy used directly and indirectly, both expressed in empower units.

order of magnitude from agriculture. The carrying capacity for a person (sunlight area required) was said to be only half of a square meter. Two main fallacies were involved, but before they could be proved, those with a little knowledge were attracted by the subject and popularized it with articles published in leading magazines. The ocean was said to be able to support expanding populations indefinitely with algae. When practical attempts at farming algae were made, the yields came out to be the same as the maximum yields from agriculture and the products not as edible.

One of the fallacies was the assumption that the high photosynthetic efficiency of chloroplasts (green units in plant cells) at low light intensity (20–40% at 1/100 daylight) is the same at bright light. This is a thermodynamic error. There is an inherent inverse relationship of light intensity and efficiency in energy conversions (fig. 3.3).

The second fallacy lay in consideration of the algae in pure cultures, free from their supporting ecological system. In ecosystems, photosynthetic producers require mineral cycles, the diversion of much energy into various networks for stability, and the maintenance of extensive structure and controls for survival, as we have already discussed. In the laboratory tests, all these elements were supplied by the fossil fuel culture through thousands of dollars spent annually on laboratory equipment and services to keep a small number of algae in net yields. On the scale of pilot plant tests, costs representing human service inputs were huge. Until all these dollar costs and human services were considered, the advocates imagined that they had higher efficiencies of net yield. The erroneous omissions are illustrated in fig. 7.10. Gross yields were the same as high yields in other systems, and net yields were higher because fossil fuels had provided the work necessary for system operation, in the same way that fossil fuels now subsidize potatoes.

Electricity from Algae

As mentioned in chapter 3, Beyers and Odum (1993) studied the production of electricity from photosynthesizing blue-green algal mats (fig. 3.4). The solar transformity of this low, spatially dispersed voltage was (2,000 solar emcalories)/(0.32 calories of electricity) = 6,250 solar emcal/cal. To provide 110-volt power, which has a solar transformity of 170,000 secal/cal, a further concentration of 27 times is required. These data are helpful in showing how dispersed solar energy conversions must be concentrated after capture to support the intensive needs of an information society.

Fuel-Rich Farm Animals

One of the conceits of our fossil fuel–based agriculture is pride in new varieties of beef cattle, sheep, chickens, and other domestic strains of animals that yield much more protein food than the older varieties used by primitive societies. The citizen

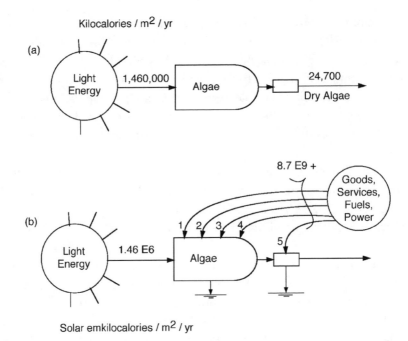

FIGURE 7.10 Energy and emergy flows in a food-producing algal pilot plant operated by Arthur D. Little company (Fischer 1961). A yield of 20 tons/acre was obtained. A year of insolation provided 1,460,000 kcal/m^2 input (400 langleys/day), with gross production 10%. A high rate of flushing maintained cells in a rapidly growing state, with a high ratio of photosynthesis to respiration. Potential energy was calculated from algal weight as 5 kcal/g dry weight. Costs of the installations were spread over an estimated 10-year lifetime. The dollar costs of $2.8/m^2/yr were converted to emergy using 10,000 kcal of fuel emergy per dollar spent. **(a)** Diagram showing solar energy conversion (kilocalories) with all other inputs omitted; **(b)** diagram including all inputs expressed in emergy units (solar emkilocalories): **(1)** fertilization, **(2)** stirring, **(3)** containing and distributing, **(4)** controlling growth, and **(5)** concentrating for harvest.

assumes that this is a permanent advance that secures for agriculture a high yield and that we will never have to go back to self-sufficient animals.

This belief is based on energetic fallacies. Genetic breeding and biotechnology have carefully eliminated many of the special energy-using abilities of these animals so that their food can go into maximum net growth. The farmers supply all the old functions with fossil fuel subsidies. Figure 7.11 compares animal husbandry with and without the work of fossil fuel.

We now have chickens that are little more than standing egg machines, cows that are mainly udders on four stalks, and plants with so few protective and survival mechanisms that they are immediately eliminated when power-rich management is withdrawn. Such varieties are complementary to high-emergy agriculture and cannot be used without it. Because animals are concentrated on the landscape and genetically similar, they are vulnerable to epidemic diseases such as the international outbreaks of hoof and mouth disease in 2001.

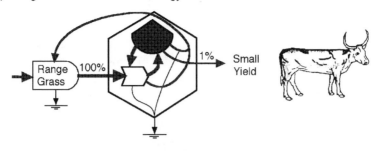

(a) Range Cattle on Renewable Energy

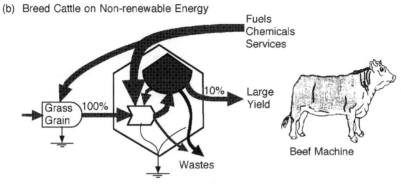

(b) Breed Cattle on Non-renewable Energy

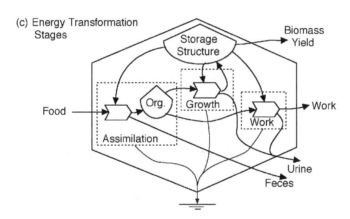

(c) Energy Transformation Stages

FIGURE 7.11 Comparison of yield of self-sufficient animals harvested from renewable, environmental energy range with the transformed herds and many functions replaced with fossil fuel inputs. **(a)** Energy systems diagram of cattle on environment-based range; **(b)** diagram of high-yielding cattle with intensive use of inputs based on fossil fuels; **(c)** diagram of energy transformations in consumers evaluated in table 7.2. Org., organic matter.

Referring to table 7.1, we find the efficiency of gross plant production to be about 2% and that of transfer to the animal consumers (herbivores) to range from 0.5% to 49% (table 7.2). The high efficiencies of growth by the steer under fattening diets occur when the animals are not running about doing their own support. Some of the food substances may be digested, assimilated, and deposited without much metabolic processing. Even higher yields are obtained if protein foods are used. This may violate the principle from chapter 4 that energy is wasted when energy is used of higher transformity than necessary or is used for a function that

TABLE 7.2   Efficiencies of Food Chain Transfer in Animals[a]

System	Ingestion into Assimilation (%)	Assimilation into Growth (%)[b]	Overall Ingestion into Growth (%)[b]
Steers under fattening diet[b]	66	74	49
Marine zooplankton in rich laboratory conditions[c]	59	57	34
Elephant on natural range[d]	32	1.5	0.48

[a] If population is exporting at the same rate as new growth occurs, the population is at steady state. Animals include populations of stomach microorganisms that eat and process food, yielding nutritive substances to host.

[b] Brody (1945:82).

[c] Means of medians of 10 groups of data from literature and 12 other experimental studies (Conover 1964).

[d] Petrides and Swank (1964).

works with a lower-transformity input. Mad cow disease developed from feeding cows meat packing wastes (agrocannibalism).

To operate agriculture without energy subsidy, we must have cows that can fend for themselves in reproduction, protect themselves from weather and disease, move with the food supply, and develop their own patterns of group behavior. Competitive plant species are required that can also provide reasonable growth in unfertilized soils, despite losses from insects, and so forth. Naturally, the yields of net food have to be less, and no amount of genetic breeding will help.

Overfishing and Natural Capital

Boats, fuels, fish-locating sonar, refrigeration, and trained labor for fishing became cheap because of fossil fuels. A huge onslaught on fish populations and the global fisheries began with development of high-powered ships and giant nets. They stripped the lakes and seas of one fish stock after another, trading sardines for herring, then salmon, cod, and finally haddock. As each fish species became scarce, its price increased, helping the markets to pull down the remaining stocks until there was little reproduction. Fish stocks are necessary to sustain fish production. Stocks necessary for autocatalytic growth and reproduction are called natural capital. Efforts to manage each fish population to seek the optimum stock for maximum yield failed because the models did not include the whole ecosystem. Fish stocks reduced by fishing were displaced by competitive populations of little food value. For example, sometimes jellyfish replaced edibles.

In other words, fossil fuel fisheries mined the seas. To be sustainable, a system organization has each unit contributing work back to its system so as to reinforce itself in a loop (chapter 4). Even intensive agriculture works to sustain the productivity of the land, but fossil fuel fishing fed nothing back to reinforce the fishery food chain. Whereas fisheries had the highest net yield of protein food for the effort before industrial ships, soon the fish were mostly gone, and the fish prices were higher than those of luxury beef.

Fuel-Subsidized Aquaculture

Some commercial fishing from public waters was replaced by industrialized aquaculture of freshwater and saltwater ponds. Directly and indirectly, fossil fuels changed aquaculture in ways parallel to terrestrial agriculture. In fig. 7.12, representative data on yields for three energy situations are shown by the energy systems diagrams. The first group, with only sun, yields up to 4 g dry weight/m²/year (40 kg/hectare/yr), depending on natural fertilizer conditions. With the energy of inorganic fertilizer added, yields are higher, and in the third group, with organic foods added that feed the fish more directly, yields go to 200 g dry weight/m²/year (2,000 kg/hectare/yr). Both inorganic and organic inputs became cheap and easy to add, as fossil fuels enriched nearly everything.

All over the tropical world, intensive aquaculture of shrimp was developed in coastal ponds using fuels, electric pumps, fertilizers, chemicals, and human services, all based directly and indirectly on fossil fuels. Thousands of ponds were constructed in coastal Ecuador, displacing the mangroves and that source of liveli-

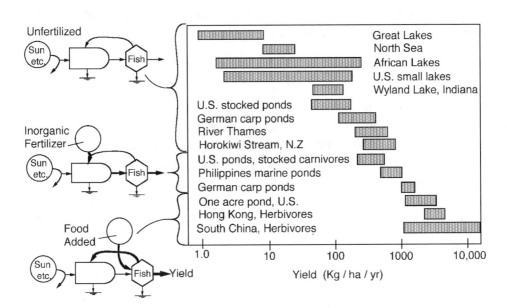

FIGURE 7.12 Yield rates from fisheries and aquaculture, redrawn from Mann (1965) with energy systems diagram to show the energy sources. Inputs increase from top to bottom: unfertilized waters above, waters receiving inorganic fertilizer nutrients next, and ponds receiving organic foods below.

hood for local people. The ponds were seeded with larval stages from declining natural shrimp populations and from artificial laboratory culture. Shrimp raised in ponds for several months were harvested and sent to developed countries at prices competitive with those of shrimp from coastal trawlers. The prices were too high for most people in Ecuador, and much of the emergy was sent out of the country, with much less emergy in the buying power of the money received in exchange (see chapter 9). A few people made money, but the displaced local people drifted to the cities seeking jobs, the same displacement that occurred when fossil fuels intensified agriculture. The shrimp ponds and their exports enriched developed economies that imported them at the expense of the undeveloped countries (see explanation of emergy imbalance in trade in chapter 10). Over time, yields began to diminish. Genetic variety was lost; the ponds developed diseases and competitors. Intensive aquaculture, like intensive agriculture, may require rotation and a reservoir of natural diversity.

In cold temperate waters, intensive, fuel-subsidized salmon culture was developed. Floating pens (cages with water passing through screened walls to the young salmon kept inside) were placed in estuaries and seeded with juvenile salmon from fish hatcheries. Many of the fish hatcheries were subsidized by fossil fuels. They were based on salmon tricked into returning from the sea to the hatchery because salmon follow the smells from waters in which they were first released. The pens require that the estuarine waters around them be clean and low in nutrients. The density of pens is limited when the wastes from the pens make the water unsuitable because of the algal and microbial blooms caused by the excess nutrients (Odum 2001). Emergy evaluation found pen-raised salmon with similar transformity to the natural migratory stocks: 1.7 E7 secal/cal (Odum 2001).

Direct Conversion of Fossil Fuel to Food

Among many ideas for using fossil fuels to support agriculture is their direct conversion into food. Humans cannot eat fossil fuels directly because the fuel molecules are not digestible. But in World War II, under the stress of failing food supplies, oxidized petroleum was used to make an edible fat in Germany. Many microorganisms readily digest hydrocarbons, and the microbial products can support a food chain. Various microbial systems were tested. McPherson (1964) reported yields from a microbial, methane-based system tested by the Shell Development Company. The efficiency of producing organic food that is half protein was around 10%, not unlike that of natural food chains.[5] On an emergy basis, the gas input was 2.2 E12 secal/kg, and the service input was 1.44 E12 secal/kg (total input = 3.64 E12 secal/kg). If the kilogram of food contained 4.6 E6 gram calories, the transformity of food from natural gas conversion was 8.1 E5, a value in between corn and beef (quotient of input emergy/product energy).

Converting the fossil fuel to food was less efficient than using the fossil fuel to amplify solar production by supplying limiting materials. Recall the energy matching principle from chapter 4 in which each quality of energy has maximum effect when it interacts with energy of higher or lower transformity.

Empower and Civilization

Electric Empower

Over the last century, the emerging civilization stepped up to electric power. Fuels could be converted into electric power, a more flexible form of energy of higher transformity, one capable of supporting people in concentrated cities, extending the time of work through night lights, and processing information. Fuels were consumed in boilers, and the difference in temperature between the steam and the environment was used to operate heat engines arranged to turn electric generators.

Electric power was also generated from river dams that diverted the geopotential energy of elevated water away from its normal work developing river basins, sediments, wetlands, floodplains, valley productivity, and coastal ocean currents.[6] Many of the rivers of the world were dammed, taking energy away from these life support processes. As we explained in chapter 5, precipitation in mountains has a high empower content, organizing much of the surface geology of the earth. The mind-set of a culture of fast growth based on exploitation saw the rivers as energy not in use. When some rivers in the western United States were dammed, electricity was sold for money worth a tiny fraction of the emergy value in the power received and used wastefully. For instance, the city of Las Vegas, Nevada, developed giant lighted signs, gambling casinos, and entertainment palaces with the cheap electricity. In what way, if any, did the dam reinforce the society and environment compared with the loss of river functions? By 2000 dams with little contribution were being removed to preserve salmon runs. Chapter 13 discusses this and other adaptations for a lower-energy future.

Nuclear Power

In the last half of the 20th century, the nuclear fission process was developed, and more than 400 nuclear plants were built worldwide. These plants control radioactive chain reactions that generate intense heat. Because the temperature at the core is too high, various cooling methods are required to disperse some of the heat until the temperature is low enough to operate machinery without melting the metals. Nuclear fission uses concentrated uranium ores, which are mined in many places, but the supplies are limited. Nuclear electric power is a nonrenewable energy source.

At first nuclear power was taking more emergy from the economy than it yielded because of the large inputs required to enrich the uranium, build the complex installations, pay interest on the capital, and deal with the frequent interruptions in service experienced in the early days of their operation. Later, problems were solved, more efficient enrichment processes were found, and there were years of sustained power output with net benefit (table 7.3). After runaway chain reactions released radioactivity at Three Mile Island, Pennsylvania, and Chernobyl, Ukraine, few nuclear plants were built. But many countries had become dependent on nuclear electric power. Today , many of the more than 400 plants are reaching the end of their 30- to 40-year expected lifetime, and it is not clear whether they will be replaced. Like fossil fuels, nuclear fuels are becoming a little scarcer each year and thus more costly. Some of the radioactive fuels in atomic bombs from the old Soviet Union are being converted for use in nuclear power plants.

Breeder Reactors

Nuclear fission power plants that use enriched uranium generate the chemical element plutonium in their fuel rods as a byproduct. Plutonium can also sustain chain reactions and therefore is usable in atom bombs or, when controlled, to fuel nuclear power plants. The arrangement for processing plutonium from nuclear fission to make more power has been called a breeder. But to get the intensely radioactive plutonium out of the old fuel rods and into the form needed to operate more power plants, very expensive robot operations are needed. Plutonium is highly toxic and if ingested goes directly to the bones, causing bone cancer. Human beings are the most sensitive of all organisms to radioactivity and cannot physically handle the old fuel rods. Because of the costs and toxicities, and because plutonium can be made into weapons, President Carter canceled breeder work in the United States. France operated a breeder reactor for a time, sending some of the rods to Japan.

Electric Power and Information Centers

The global development of electric power led to global communication, computers, space satellites, television, and the information revolution, which are evaluated in chapter 8. The world is now covered with networks of interconnected electrical power lines. These are the basis for the cities in which most people now live. People switching on quiet devices and house lights in the cities are so far from the huge, throbbing, noisy, waste-generating power plants that they take for granted their clean luxury living. Their lifestyle is based mostly on fossil and nuclear fuels, and the reserves of these are disappearing.

EMPOWERED CITIES

The energy systems network of the planet starts with the sun and earth and converges to people and the centers of society in cities. The centers feed back out to the landscape pulses of control action, material recycle, and other high-quality work that

TABLE 7.3   Emergy Yield Ratios of Sources Measuring Net Benefit to Society

ITEM	EMERGY YIELD RATIO[a]	SOURCE
Renewable environmental resources		
Wood biomass, 10–100 yr growth	3–12	[b]
Tidal electric, La Rance, not counting estuary loss	15	[c]
Hydraulic power, elevated basin counted free	10	[c]
Geothermal steam, volcanic area	8	[c]
Ocean thermal electric	1.5	[c]
Photovoltaic grid in commercial use	0.43	[c]
Agricultural products		
Short-cycle wood plantation	2–3	[b]
Shrimp aquaculture	1.8	[c]
Silk from silkworms on mulberry	1.1	[c]
Ethanol from sugar cane or grain	1.1	[c]
Palm oil	1.1	[c]
Nonrenewable reserves		
Natural gas at wellhead	3–40	[c]
Coal and lignite at the mine, thick seams	4–40	[c]
Oil, international prices, 1974–2001	3–12	[c]
Offshore oil	6–10	[c]
Alaska oil pipeline	11	[c]
Peat, sun-dried	3	[c]
Oil shale	0.3	[c]

[a] Empower of yield divided by the empower from the economy for necessary processing (fig. 7.16), a measure of net emergy.
[b] From Doherty (1995).
[c] Various references summarized in *Environmental Accounting* (Odum 1996).

reinforces the empower of the whole system (chapter 4). Human societies could develop higher emergy and transformities and prevail over alternatives by locating their centers in places where environmental energies converge naturally. The convergence of river networks is shown in fig. 5.10. For example, the civilization of ancient Egypt developed at the junction of the Nile River from the south and the distributary channels that connected the river with the Mediterranean Sea to the north.

Agrarian Cities

In agrarian times, without many nonrenewable inputs, important cities developed where environmental energies converged. After a time, accumulations of energy, matter, and information in these city centers were enough to set off progressive pulses of cultural frenzy. Seen from space, the earth was like a Christmas tree with flashing lights, where each light was the surge of civilized progress organizing an area for a time. Some of the information on new innovations from each center, if shared broadly enough, was available to help the ascendancy of later cultures in other areas.

In energy systems diagrams, cities are on the right of the convergence of successive energy transformations. Cities are the source of the feedbacks of control and recycle that disperse to the rural areas from right to left. Figure 7.13a shows some components of an agrarian landscape and the corresponding energy systems diagram, when the items are arranged from left to right according to scale of time, space, and transformity.

Fuel-Based Cities

Since the fossil fuel revolution, human societies have been located mostly in great cities. As explained by energy hierarchy principles of self-organization in chapter 4, the centers contain large emergy storages, high concentrations of activity (high empower density), and high-quality work (high transformities and storms of pulsing activity). Although agrarian landscapes converge emergy from rural sources to support their cities, the cities of the 21st century directly receive the concentrated fossil fuels for processing in the cities. Then the products spread out, with matching interaction over the landscape (fig. 7.14). Figure 7.13 compares the fuel-based city with the agrarian city.[7]

Shu-li Huang (1998) and students evaluated the emergy basis for the city of Taipei, the center of Taiwan (fig. 7.15). It developed in a valley surrounded by mountains and a river delta, a natural convergence of high emergy. Similar study was made of the emergy basis for Jacksonville, Florida (Whitfield 1994), where several rivers converge, connecting with natural harbors suitable for trade.

The Automobile City Lifestyle

The human being is a complex hierarchy of many scales organized together, from biochemical molecular processes to the complex information processing in the brain. *Homo sapiens* evolved as part of social groups interacting with a landscape of ecosystems. Apparently inputs from the green environment are important to maximum function. People seek the input from nature or parks, even when they live concentrated in the cities.[8]

While people were migrating to the cities, fossil fuels were cheap. The psychological need for ecosystems by people working in the centers caused pathological overuse of automobiles in cities. The simple design of people living as part of the central structure was displaced by suburban living and commuting (fig. 7.14b). Individuals sought individual cars for their freedom, their power, access to ecosystems, and

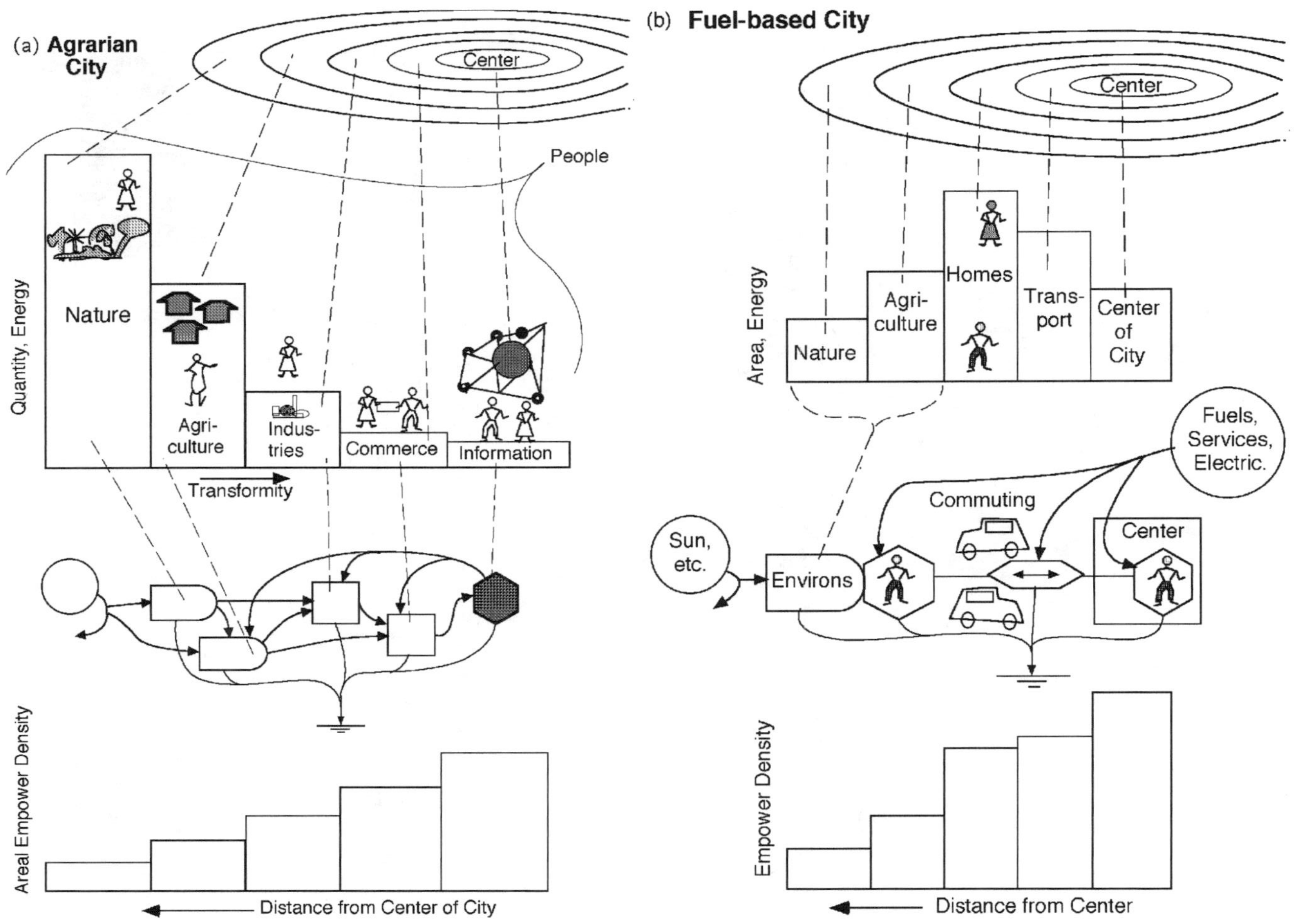

FIGURE 7.13 Comparison of energy distribution, systems diagram, and empower density of cities. (a) Cities at the center of agrarian landscape based on renewable energies; (b) urban landscape based on automobiles, commuting, and fossil fuels.

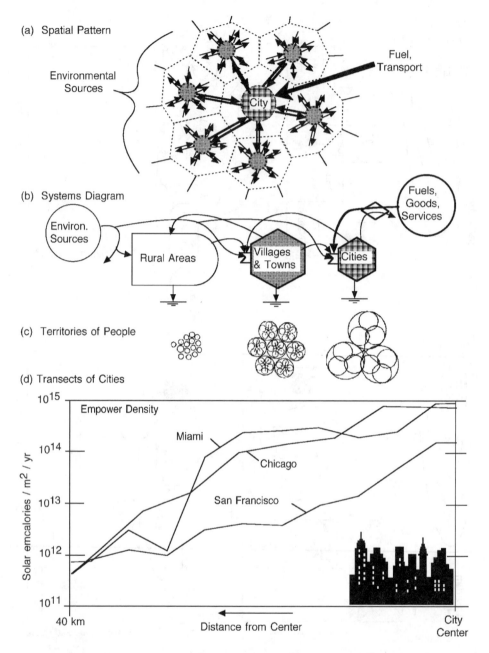

FIGURE 7.14 Characteristics of fossil fuel cities. **(a)** Spatial hierarchy of centers and transportation; **(b)** systems diagram summarizing energy and hierarchy; **(c)** radius of circulation of people indicating their territories (Doxiadis 1977); **(d)** empower densities as a function of distance from city centers. The transects in fig. 7.14d are averages calculated by Robert Woithe as part of the Cities project. For each city, six straight lines radiating out from the center of the city were drawn on land use maps. Empower densities for each type of land use were used from the study of Jacksonville by Douglas Whitfield (1994).

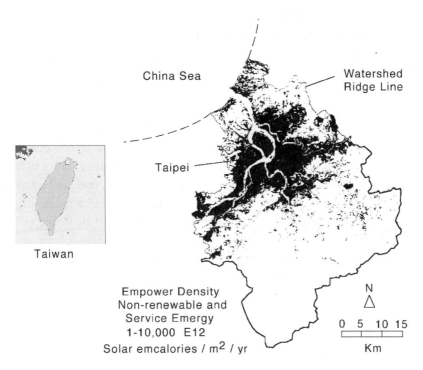

China Sea

Watershed
Ridge Line

Taipei

Taiwan

Empower Density
Non-renewable and
Service Emergy
1-10,000 E12
Solar emcalories / m^2 / yr

N

0   5   10 15

Km

FIGURE 7.15 Spatial organization of the city of Taipei, Taiwan, indicated by the empower density (Huang 1999).

the time they saved. The city was transformed by the great emphasis on transportation devoted to the oscillation of automobile people in and out. Individual cars and the highways to support the daily shuttle took over the city organization, destroying neighborhoods and causing slums to develop with great waste of emergy.

Emergy of Alternative Sources

Since the "energy crisis" of 1970s, millions of dollars have gone for research seeking primary sources to replace the fossil and nuclear fuels. A primary source is one that generates more empower than is used up in its processing. Ultimately, what a society can do depends on the net empower from its primary sources. The renewable empower of sun, tide, and earth was the primary source supporting agrarian society, now mostly integrated into the fossil fuel system. Each alternative proposed as a primary source for the future ought to be evaluated for its potential to yield net benefit.

NET EMPOWER

The idea of net empower evaluation is shown in fig. 7.16. A good source yields more emergy to society than is required from society for the processing. In other words, a good primary source has high net empower yield. In fig. 7.16, yield (Y) is greater than feedbacks (F). The ratio of yield to feedback is the net emergy ratio (Y/F). Net emergy ratios for hundreds of energy sources have been evaluated.[9]

Table 7.3 shows the net emergy ratios of the fuels used by society in the past. Agrarian society used organic products of photosynthesis such as wood and peat, with net emergy ratios of about 2 or 3. After fossil fuels were predominant in much of the 20th century, the net emergy ratio was 40 or more, an enormous increase that set the earth into its growth surge of benefits and calamities. Since the start of the Organization of Petroleum Exporting Countries in 1971, fuel prices have been higher, so that the fuels available to most economies have had a net emergy yield oscillating between 3 and 12, still enough to keep the earth in global economic expansion until now. Table 7.3 summarizes net emergy ratios for many energy sources. More details are given in a companion book *Environmental Accounting* (Odum 1996).

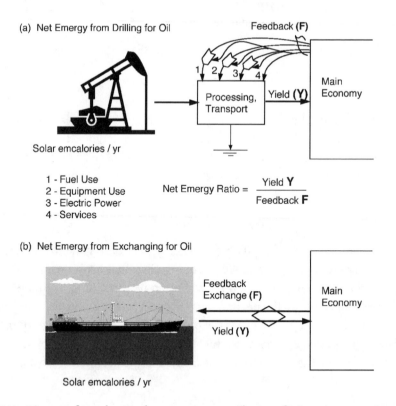

FIGURE 7.16 Diagrams for evaluating the net emergy contribution of primary energy sources (yield and feedback), here using oil as an example. Net contribution is measured with the emergy yield ratio (Y/F). **(a)** Evaluation of oil wells; **(b)** evaluation of the oil obtained by foreign exchange, where F is the emergy that is bought by the oil supplier with funds received in payment.

The net emergy idea is so simple that it received public attention in 1975, and Senator Mark Hatfield helped pass a law intended to require net emergy evaluation of federal projects.[10] In those days emergy was still called embodied energy, and the law confused energy with what is now called emergy. Consequently, net energy and energy yield ratios were calculated instead of emergy ratios. Agencies and industries thus made the error of counting all kinds of energy as if they contributed to work equally (chapter 3). This ignored the small energy but high emergy of human services, strategic metals, equipment, and indirect environmental damage. As a result, energy policy in governments and industries was based on the erroneous calculations that suggested that solar technology, ethanol from intensive agriculture, and short-rotation forest plantations could replace fossil fuels. Several billion dollars were wasted trying to extract oil from shale rock because neither Congress nor the industries paid enough attention to our calculations and public testimony, which showed no net empower (Odum et al. 1976).[11]

Net empower yield of something usually is highest before it is processed and declines as more and more emergy from society is used for mining, transporting, processing, converting, and distributing. For example, fig. 7.17 shows the successive steps in cutting, collecting, chipping, and transporting wood chips to substitute for fossil and nuclear fuels in Sweden[12] (Doherty 1995). Wood in the forest has a net empower yield, but after the many transformations required to make wood powder, almost no net emergy is left. To support society, wood has to be used with less processing, as in the past. Less concentrated energy can support only a simpler society.

Not every source must have net empower, but a society cannot operate at all without at least one primary source that has the net contribution to support everything else. A primary source can help maximize a system's total empower by bringing in other emergy sources, even if they are not a net contribution by themselves. For example, the equipment for solar water heaters is supplied by the fossil fuel economy to convert sunlight into hot water. There is no net empower contribution, but less fossil fuel is used than in direct water heating with gas or electric power.

CONCENTRATING SOLAR ENERGY

As explained in chapter 4, solar energy is inherently dilute. By the time it is concentrated to fuel status, its net emergy yield is small. Because of the success of industrial agriculture, people assume that net empower of solar production can be increased by more intensive farming or forestry practices. This is wrong, as proved by Steven Doherty (1995) in his analysis of forest production in Sweden, Puerto Rico, and the United States (fig. 7.18). The more often a forest is replanted and harvested, the less net yield. Very high yields come from forests allowed to grow a long time without much effort by society. In other words, the net empower of solar energy depends on time of growth.

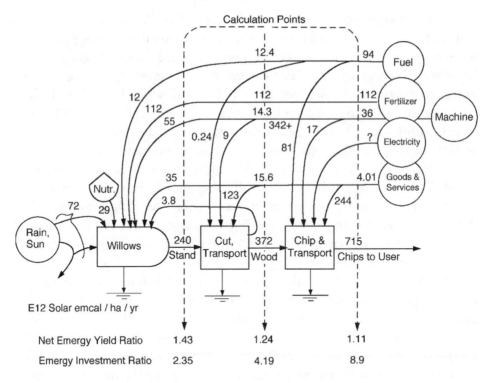

FIGURE 7.17 Energy systems diagram and emergy flows for Swedish willow plantations from Doherty (1995). Emergy yield ratio (yield emergy/feedback emergy) decreases with processing. Emergy investment ratio is the emergy fed back from the economy divided by the local free environmental emergy contribution. It measures intensity of contribution from the economy.

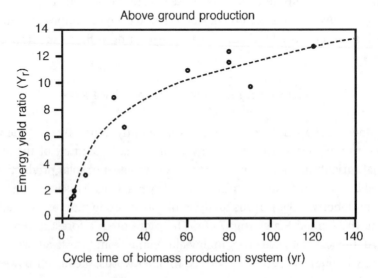

FIGURE 7.18 Net emergy yield ratio (Y/F in fig. 7.16) of forest wood production as a function of the time of growth between replanting cycles (Doherty 1995).

Many transformation steps are required to process and concentrate dilute solar insolation to high-quality electric power using organic photosynthesis of the chloroplast, which is the green plant's photovoltaic cell (fig. 3.3a). With the intent of skipping steps, hardware photovoltaic cells have been researched for decades, trying to generate electric power from solar energy with net emergy, which would make them economical. But these designs ignore the energy hierarchy law (chapter 4) that requires many calories of available energy at one level to make a few calories at higher levels. Figure 7.19 compares electrical current generation from silicon solar voltaic cells with that from a wood power plant operated on old-growth logs in the Amazon. Evaluations that claim net yield from solar cells leave out the huge empower required in the human services for manufacture, distribution, support, connections, operation, management, and maintenance.

The greater the human population, the smaller the area of forests remaining, and the less time is usually allowed for growth. The global net empower of solar energy decreases with population. As populations have increased, times between shifting agriculture farming have decreased, which reduced yields.

Where a dilute renewable energy has to be concentrated to support society, either emergy is used to concentrate the energy spatially or time is allowed for the energy to accumulate in a broadly distributed storage. There is an emergy equivalence between accumulation of available energy over time and the work of concentrating energy in space. Self-organizing systems do both (chapter 4).[13]

Life Support Carrying Capacity for Humanity

In wildlife management the phrase *carrying capacity* sometimes is used to describe the ability of an ecosystem to sustain a species population without damaging its functions. Carrying capacity for wildlife is the population level for long-range survival. It depends on the power supplied of the types needed by that species. For example, a landscape supplies seeds used by doves. The resource needs of a species of dove are mostly a characteristic of the species and are genetically determined.

The carrying capacity of an area for people includes not only the emergy supplied from that area but the additional emergy that can be brought into that area by exchange or purchase. The resource required per person depends on the standard of living, which is best expressed in emergy units because it includes direct inputs from the environment and purchased inputs. Estimates of human needs should include all the works that provide stability, reserves, protection, and so forth. Note the wide range in emergy per person of people in different countries and cultures (table 7.4).

Any state or nation has its budget of environmental emergy plus the emergy that its economy is bringing in from nonrenewable sources within the nation or from without. The local carrying capacity (*C*) for humans can be estimated as

$$C = Pe + (R)(Pe)/E,$$

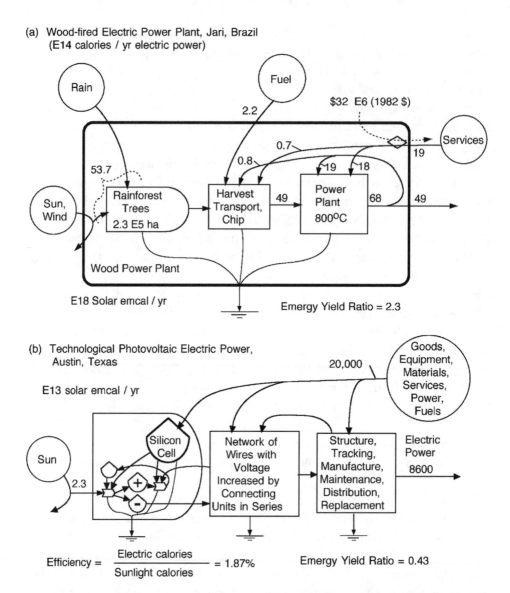

(a) Wood-fired Electric Power Plant, Jari, Brazil
   (E14 calories / yr electric power)

(b) Technological Photovoltaic Electric Power,
    Austin, Texas

$$\text{Efficiency} = \frac{\text{Electric calories}}{\text{Sunlight calories}} = 1.87\%$$

FIGURE 7.19 Comparison of methods of conversion of solar energy to electrical power. **(a)** Electric power plant operating with net emergy yield using rainforest logs at Jari, Brazil (Odum and Odum 1983). **(b)** Grid of photovoltaic cells generating electric power in Austin, Texas, with no net emergy (Odum 1996). Emergy yield ratio is the emergy yield divided by the emergy inputs from the economy.

TABLE 7.4 Human Emergy Support

Item	Empower Density[a] (E10 secal/m²/yr)	Empower Share[b] (E14 secal/person/yr)
Pygmies in deep forest[c]	5	1,240
Farmers in monsoon agriculture[d]	1.4	0.58
Modern society, dense population[e]	2	489
People averaged over the earth's land[f]	4.8	12
People in the United States[g]	22	81
People in Jacksonville, Florida, based on fossil fuel[h]	36–13,000	239
People in a closed system, Biosphere 2[i]	48,000	10,200
Person in space capsule supported from Earth[j]	1,900,000,000	6,500,000

Note: secal = solar emcalories; 1 acre = 4,047 m².

[a] Anual empower in solar emcalories per year divided by the land area.

[b] Annual empower in solar emcalories per year divided by the human population.

[c] Supporting forest empower, 1 person/mile², 640 acres/mile²:

   (131,000 calories/m²/day)(365 d/yr)(transformity 1,000 secal/cal) = 4.78×10^{10} secal/m²/yr.

   (4.78×10^{10} secal/m²/yr)(2.6×10^6 m²/mile²) = 1.2 E17 secal/person/yr.

[d] Monsoon agricultural support, 1 person per acre:

   (80,000 cal/m²/d)(180 d/yr)(1,000 secal/cal) = 1.44 E10 secal/m²/yr;

   (1.44 E10 secal/m²/yr)(4,047 m²/acre)(1 person/acre) = 5.83×10^{13} secal/person/yr.

[e] 1.61×10^{23} secal/yr for India (Odum 1996) divided by 3.29×10^6 m² = 4.89×10^{16} secal/m²/yr; empower share, table 10.1.

[f] World empower from fig. 9.4: 7.25×10^{24} secal/yr divided by 1.51×10^{14} m² land area = 4.8×10^{10} secal/m²/yr; 7.25×10^{24} secal/yr divided by 6×10^9 people = 1.21×10^{15} secal/person/yr.

[g] 2.1×10^{24} secal/yr (Odum 1996) divided by 260×10^6 people = 8.1×10^{15} secal/person/yr; 2.1×10^{24} secal/yr divided by 9.4×10^{12} m² area = 2.23×10^{11} secal/m²/yr.

[h] Evaluation of Jacksonville by Douglas Whitfield (1994).

[i] Evaluation of Biosphere 2 (Beyers and Odum 1993): 1.02×10^{18} secal/person/yr. Annual support: (6.18 E18 secal/yr)/1.28 ha/10,000 m² = 4.8 E14 secal/m²/yr.

[j] 6.5 E20 secal/person/yr calculated for NASA Skylab by Noyes (1977); multiplied by 3 astronauts and divided by 100 m² = 1.9×10^{19} secal/m²/yr.

where C is the number of people supported per area, Pe is the local environmental empower per area, R is the regional investment ratio (ratio of outside empower to local environmental empower), and E is the annual empower use per person.

Vegetarian Diet and Human Nutrition

Vegetarian diets were important in many cultures in the past, and they are suggested for the future as a way of conserving available energy. The biological systems of human beings apparently evolved at the apex of food chains and require a very high diversity input of chemicals such as vitamins and minerals and good sources of carbohydrates and fats for basic calories. Figure 7.20 summarizes the way the food energy and diversity have been supplied to humanity. The assembly of the necessary substances usually was provided by the food chain to wild or domestic animals that combined the requirements together in so-called protein sources. In urban society the supply of food diversity to vegetarians is also provided by a complex network of businesses collecting and distributing specialty foods. This means gathering a great diversity of the necessary chemical molecules in ratios suitable for the support of humans. Many support systems have two types of food input, one flow a staple that supplies calories in a carbohydrate form such as potatoes, the other a protein supplement that supplies the daily requirement for diversity of molecules. The high-quality foods require more processing work combining varieties of molecules.

For tribal cultures in forest ecosystems, support came from a complex network of hundreds of species of plants and animals processing diverse products for the complex needs of humans. Now urban people receive support from a network of hundreds of people using fossil fuels directly and indirectly to process, transport, and sell a similarly high diversity of needed products. Either way, emergy is re-

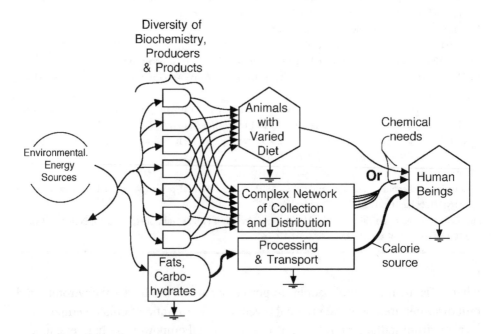

FIGURE 7.20 Diagram of the necessary use of power to converge products through food chains or collection networks to develop diverse, high-quality nutrition for humans. Solar protein generator system without fuel energy subsidy, compartmentalized according to the principal necessary functions: producer, diversifier, controller, combining selector, and people. Each block function may involve many species and their energies for coordination.

quired for collecting and processing. For the industrialized system, combining and transporting are aided by fossil fuels.

A person can eat adequately without meat if someone does the extra work of gathering enough kinds and diversity of plant foods. As an alternative, he can eat the more expensive meat products, which already provide a nutrition close to his need. In meat, the combining has already been achieved by the integrating and mixing actions of the animals, microbes, and their systems of work and sustenance. Thus, the transformity of protein is higher than that of carbohydrates.

Human Support

The carrying capacity for a person of earlier cultures based entirely on renewable environmental energy was about 1 to 250 people per square kilometer (0.36 square mile), with 99.5% or more of the organic matter production metabolized in the work of life support. Higher densities occurred in special places where the energy hierarchy of nature transported the production of larger areas by means of a stream, such as salmon passing in waterways.

Civilizations support dense populations if they are able to match the fossil fuels with other environmental resources to be amplified. Primitive areas with rich oil deposits could only sell their fuel at first because they lacked technology and interacting inputs such as water.

The standard of living is best estimated from the annual emergy use per person (table 7.4). Notice the wide range of values. Rural people with little income but in low density can have as high empower use as some people in cities. Global standards may be decreasing.

Empower Signature

Any area can be characterized by its pattern of resource use, which is an energy signature of its activity (fig. 7.21a). Energy sources are arranged from left to right in order of transformity. Figure 7.21a, showing the energy flows in Florida, might be very misleading if used to infer the basis for society. But inflows are correctly compared on an emergy basis in fig. 7.21b. This is an empower signature, which shows the amounts and kinds of inputs on a common basis. It predicts the kinds of energy users that will predominate in a system that adapts to these resources. The example shows that the state of Florida is supported mainly by the fossil and nuclear empower plus that of purchased goods and services, which are also fossil fuel based. This signature is typical of developed areas, which are becoming more alike.

Many thoughtful leaders have long warned of the future shortages of fossil fuels, although the timetables of disaster are continually being revised as new sources of these fuels are found. But the earlier warnings are pertinent even if the timing of crisis is still unknown. Oil and coal will not run out, but the ratio of energy found

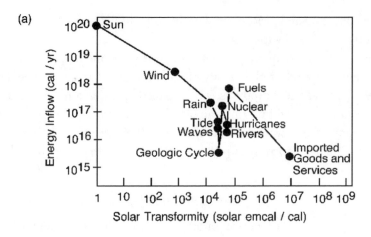

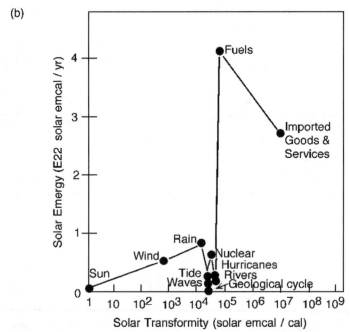

FIGURE 7.21 Energy signature: principal sources of energy and emergy used by the state of Florida, plotted according to their transformities (Odum 1983). **(a)** Annual energy flow; **(b)** annual emergy flow (empower).

to energy spent in obtaining them will continue to decrease until net emergy ratios of remaining deposits are less than alternative renewable sources. If the net yield of potential energy begins to approach that of wood, we will have returned to the solar energy–based economy, and by that time the standards of living of the world will have retrogressed to those of two centuries ago. Whether such changes will come suddenly in a catastrophe or slowly as a gradual trend is one of the great issues of our time (chapter 13).

Life Support in Space Capsules

The carrying capacity question also arises in the discussions of people in space outside the earth. The biosphere really is an overgrown space capsule, and the questions about carrying capacity are similar. What is the area necessary to support a person on the surface of the moon? How much area is needed when moon energy is supplemented by some fuel from Earth? The carrying capacity of an environment for one person depends on the transformity of the work the person has to perform. It is not just the inputs to keep a body alive. The emergy of support required per person in a space capsule is many times higher than that on Earth and much more than can be obtained from solar energy captured by a space ship (table 7.4).[14] Emergy required for support in the NASA Skylab was 94,000 times larger than required for a U.S. citizen on Earth (Noyes 1977; see also review of efforts to supply life support in closed systems Beyers and Odum 1993).

Insight on what is needed for people doing complex work was provided by Biosphere 2, the 3-acre glass enclosure in the mountains of Arizona containing ecosystems and self-contained agriculture (Marino and Odum 1999). Eight people were supported in a minimal way for 2 years, despite large inputs of electric power and engineering. See table 7.4. The emergy required to operate the rainforest life support area in Biosphere 2 was 2,300 times larger than that used by the natural rainforest from which many trees were seeded (Leigh 1999). See chapter 12.

The carrying capacity for people in the future depends on the rate of depletion of the nonrenewable resources and the potentials of alternative energy sources. We postpone that discussion to chapter 13.

SUMMARY

The energy supporting humanity was examined with systems overviews beginning with simpler societies using renewable resources. Small tribes were supported by hunting, gathering, and shifting agriculture within complex ecosystems. Larger populations were supported in seasonally pulsing monsoon climates with rice paddies, aquaculture, and domestic animals. Without energy reserves societies are limited by the dilute nature of solar energy, which has to be concentrated many times to support high-transformity humans. Because of the fourth energy law (maximum empower principle), the efficiency of solar energy transformation is inverse to light intensity. Net emergy yield of biomass depends on time of growth.

Two centuries of fossil fuel use increased the production of food, clothing fiber, housing, heat, electric power, and information, expanding populations and organizing society in hierarchical urban centers with a wasteful automobile culture. Developed societies at the beginning of the 21st century are based on fossil fuels interacting with renewable energy, but much of the energy processing is decimating natural capital and is not sustainable. An index of real wealth, the emergy use per person, has a wide range between people crowded in areas without resources and those enriched in economic centers. The emergy required for astronauts in space is too large for self-sufficient space colonization. Net emergy evaluations did not find any alternative energy sources that can replace the declining fossil and nuclear fuels. Patterns of resource use by past societies help us see what is possible for global humanity as the use of nonrenewable resources fades.

Bibliography

Note: An asterisk denotes additional reading.

Alkire, W. H. 1965. Lamotrek atoll and inter-island socioeconomic ties. *Illinois Studies in Anthropology,* 5.

*Armstrong, N. E. and H. T. Odum. 1963. Photoelectric ecosystem. *Science,* 143: 256–258.

Beyers, R. J. and H. T. Odum. 1993. *Ecological Microcosms.* New York: Springer-Verlag.

Birdsell, J. B. 1966. Some environmental and cultural factors influencing the structuring of Australian Aboriginal populations. In J. B. Bresler, ed., *Human Ecology,* 51–90. Reading, MA: Addison-Wesley.

Brody, S. 1945. *Bioenergetics and Growth.* New York: Reinhold.

Brown, L. R. 1965. *Increasing World Food Output.* Foreign Agricultural Economics Report No. 25. Washington, DC: U.S. Department of Agriculture, Economic Research Service, Foreign Regional Analysis Division.

Brown, L., C. Flavin, and H. French. 2001. *State of the World 2001.* New York: Worldwatch Institute.

*Bjorndal, T. 1990. *The Economics of Salmon Aquaculture.* Oxford, UK: Blackwell.

Carter, W. E. 1969. *New Lands and Old Traditions.* Latin American Monograph, No. 6. Gainesville: University of Florida Press.

Clendenning, K. A. and H. C. Ehrmantraut. 1951. In E. T. Rabinowitch, ed., *Photosynthesis,* Vol. 2, Part 2. New York: Interscience.

Conover, R. J. 1964. Food relations and nutrition of zooplankton. *Proceedings of the Symposium on Experimental Marine Ecology, Occasional Publication* No. 2. Chapel Hill: University of North Carolina Graduate School of Oceanography.

Doherty, S. J. 1995. *Emergy evaluations of and limits to forest production.* Ph.D. dissertation, University of Florida, Gainesville.

Doxiadis, C. A. 1977. *Ecology and Ekistics.* Boulder, CA: Westview.

Fischer, A. W. 1961. Economic aspects of algae as a potential fuel. In F. Daniels and J. S. Duffie, eds., *Solar Energy Research,* 185–189. Madison: University of Wisconsin Press.

*Giles, G. W. 1967. Agricultural power and equipment. In *The World Food Problem,* Vol. III, 175–208. A Report of the President's Science Advisory Committee, Washington, DC.

Hagen, E. E. 1966. *Man and the Tropical Environment* [mimeo]. Symposium on Biota of the Amazon, Belém, Peru.

Harris, M. 1965. The myth of the sacred cow. In A. Leeds and A. P. Vayda, eds., *Man, Culture and Animals*, 217–228. Publication No. 78. Washington, DC: American Association for the Advancement of Science.

*Hickling, C. F. 1961. *Tropical Inland Fisheries.* New York: Wiley.

Hoon, V. 1996. *Living on the Move: Bhotiyas of the Kumaon Himalaya.* Walnut Creek, UK: Altamir Press, Sage Publications.

Huang, S.-L. 1998. Urban ecosystems, energetic hierarchies, and ecological economics of Taipei metropolis. *Journal of Environmental Management,* 52: 11, 39–51.

——. 1999. Spatial hierarchy of urban energetic systems. In S. Ulgiati, ed., *Advances in Energy Studies, Energy Flows in Ecology and Economy,* 499–514. Proceedings of the International Workshop, Porto Venere, Italy, May 26–30, 1998. Rome: Museum of Science and Scientific Information (Musis).

*Jenkins, D. W. 1968. *Biogenerative Life Support Systems.* Washington, DC: National Aeronautics and Space Administration, SP-165, 1–6.

Johnson, M. J. 1967. Growth of microbial cells on hydrocarbons. *Science,* 155: 1515–1519.

Kleiber, M. 1961. *The Fire of Life.* New York: Wiley.

Leigh, L. S. 1999. *Diversity limits and Biosphere 2.* Ph.D. dissertation, University of Florida, Gainesville.

Mann, K. H. 1965. Energy transformations by a population of fish in the River Thames. *Journal of Animal Ecology,* 34: 253–375.

Marino, B. and H. T. Odum, eds. 1999. Biosphere 2: Research past and present. Special issue of *Ecological Engineering,* 13(1–4): 1–358.

McPherson, A. T. 1964. *Food for Tomorrow's Billions. Proceedings of Food in the Future, Concepts for Planning.* Washington, DC: Dairy and Food Industries Supply Association.

Noyes, G. 1977. Energy analysis of space operations. In H. T. Odum and J. Alexander, eds., *Energy Analysis of Models of United States,* 401–422. Gainesville: Report to Department of Energy, Department of Environmental Engineering Science, University of Florida.

Odum, H. T. 1962. Man and the ecosystem. Proceedings of the Lockwood Conference on the Suburban Forest and Ecology. *Connecticut Agricultural Experiment Station Bulletin* 652: 57–75.

——. 1967. Energetics of world food production. In *The World Food Problem,* Vol. III, pp. 55–94. A Report of the President's Science Advisory Committee, Washington, DC.

——. 1983. *Systems Ecology.* New York: Wiley. Reprinted in 1994 as *Ecological and General Systems.* Boulder: University Press of Colorado.

——. 1984. Energy analysis of the environmental role in agriculture. In G. Stanhill, ed., *Energy and Agriculture,* 24–51. Berlin: Springer-Verlag.

*——. 1986. Enmergy in ecosystems. In N. Polunin, ed., *Ecosystem Theory and Application,* 337–369. New York: Wiley.

——. 1996. *Environmental Accounting, Emergy and Decision Making.* New York: Wiley.

——. 2001. Emergy evaluation of salmon pen culture. *Proceedings of the International Institute of Fishery Economics.*

Odum, H. T., C. Kylstra, J. Alexander, N. Sipe, and P. Lem. 1976. Net energy analysis of alternatives for the United States. In *U.S. Energy Policy: Trends and Goals. Part V: Middle and Long-Term Energy Policies and Alternatives,* 254–304. 94th Congress 2nd Session Committee

Print. Prepared for the Subcommittee on Energy and Power of the Committee on Inter-
state and Foreign Commerce of the U.S. House of Representatives, 66-723. Washington,
DC: U.S. Government Printing Office.

Odum, H. T., M. T. Brown, D. F. Whitfield, S. Lopez, R. Woithe, and S. Doherty. 1995. *Zonal
Organization of Cities and Environment*. Unpublished report to the Chiang Ching-Kuo International Scholar Exchange Foundation, Taipei, Taiwan.

Odum, H. T. and E. C. Odum, eds. 1983. *Energy Analysis Overview of Nations*. Working Paper
WP-83-82. Laxenburg, Austria: International Institute for Applied Systems Analysis.

*Odum, H. T. and E. C. Odum. 2000. *Modeling for All Scales. An Introduction to Simulation*. San
Diego, CA: Academic Press.

Paddoch, C. 1972. An analog simulation of a simplified model of New Guinea people, their
pigs, ritual feasts and warfare. In *Report to Atomic Energy Commission*, 649–658. Contract
At-940-10-4156. Gainesville: University of Florida.

Parker, F. W. 1963. *Food for Peace*, 6–20. Madison, WI: American Society of Agronomy, Special
Publication No. 1.

Petrides, G. A. and W. G. Swank. 1964. Population densities and the range-carrying capacity for
large mammals in Queen Elizabeth National Park, Uganda. *Zoologica Africana*, 1: 209–225.

Pimentel, S. and C. W. Hall, eds. 1984. *Food and Energy Resources*. San Diego, CA: Academic
Press.

President's Science Advisory Committee on World Food Production. 1967. *The World Food
Problem*. Washington, DC: U.S. Government Printing Office.

Rappaport, R. A. 1971. The flow of energy in an agricultural society. *Scientific American*, 224:
116–133.

Romitelli, M. S. 1997. *Energy analysis of watersheds*. Ph.D. dissertation, University of Florida,
Gainesville.

Ryan, S. 1990. Diurnal CO_2 exchange and photosynthesis of the Samoa tropical forest. *Global
Biogeochemical Cycles*, 5(1): 69–84.

Smith, R. F. 1970. The vegetation structure of a Puerto Rican rain forest before and after short
term gamma irradiation. In *A Tropical Rain Forest*, D-103–D-135. Oak Ridge, TN: U.S.
Atomic Energy Commission, Division of Technical Information.

Turnbull, C. 1916. *The Forest People*. New York: Simon & Schuster.

Turnbull, C. 1963. The lesson of the Pygmies. *Scientific American*, 208(1): 28–37.

Whitfield, D. 1994. *Emergy basis for urban land use patterns in Jacksonville, FL*. M.S. thesis, University of Florida, Gainesville.

Wilson, E. O. 1984. *Biophilia*. Cambridge, MA: Harvard University Press.

*Wolman, A. 1965. The metabolism of cities. *Scientific American*, 213: 179–190.

Notes

1. Systems that are loaded to run slowly have a high conversion efficiency, but when loaded
to accomplish more energy laws require a decline in efficiency. In fig. 3.3c the output load of
the photosynthetic machinery is the same (in the short run), but the input force increases with

light intensity. Whereas the optimum efficiency for maximum yield in one energy transforma-tion step is 50% (appendix fig. A3), photosynthesis has a series of successive energy transfor-mations, so that the combined conversion is less. When loaded for maximum power, the light intensity is intermediate, and the efficiency of the first step of light conversion is moderate (10%). In the bright light of summer, efficiencies are necessarily lower. Light in excess of the optimum loading is allowed to pass to other receptor pigments below.

2. The systems diagrams of resources in support of societies in this chapter are highly aggre-gated to help readers visualize causes. *Environment, Power, and Society* included energy flows for the Dodo tribe of Uganda and for peoples of the savanna based on big game herds. Diagrams for two dozen other societies were published in detail elsewhere (Odum 1962, 1967, 1984).

3. Hagen (1966) gave a 1940 population of 1.4 people per square mile for the Amazon, in-cluding the towns. Turnbull (1916, 1963) gave a population of 40,000 pygmies for a rainforest area of 50,000 square miles, or 0.8 person per square mile. In Australia, Birdsell (1966) found aboriginal populations in densities up to 550-person villages per 600 square miles in high-rainfall areas and down to 500 people per 40,000 square miles in dry central areas.

4. Graphs of agricultural yield as a function of fossil fuel use from *Environment, Power, and Society* were updated and redrawn as a composite using empower. Global limitations to agriculture have been summarized by Lester Brown and associates in annual publications of the World Watch Institute (Brown et al. 2001).

5. McPherson (1964) reported cost as $0.55 (1964 $) per kilogram. Percentage of fuel con-verted was from a report of Shell Development Company pilot plant in which 10^7 grams of food (50% protein) was produced from 2 million cubic feet of methane (4.6×10^8 kilocalories) (Rappaport 1971). Emergy money ratio in 1964 was 11×10^{12} solar emcal/dollar (Odum 1996). Higher efficiencies were claimed in slow batch experiments by Johnson (1967).

6. Sylvia Romitelli (1997) evaluated the Coweeta Watershed of North Carolina and Brazilian watersheds, showing how transformity and empower density increasing toward the mouth measured the ability of river floodwaters to generate productivity of floodplains, valleys, and deltas. She evaluated the ways in which hydroelectric dams divert this empower.

7. Comparisons of cities were conducted as part of a joint project titled "Ecological Ener-getic Evolution of Urban Systems: Cross Comparison of Chinese and American Societies" by the Graduate Institute of Urban Planning, National Chung-Hsing University and Center for Environmental Policy, University of Florida, Gainesville. The work was sponsored by the Chiang Ching-Kuo International Scholar Exchange Foundation, Taipei, Taiwan (Odum et al. 1995).

8. The author's suggestion regarding inherited ecological affinity was made in the Lock-wood Conference on Urban Forests (Odum 1962). The word *biophilia* has been suggested by Wilson (1984).

9. Values are scattered in the literature, some assembled in *Environmental Accounting* (Odum 1996), which has a bibliography of emergy evaluation.

10. The law is Public Law 93-577, The Non-Nuclear Energy Research and Development Act of 1975.

11. The results were given and published in congressional testimony (Odum et al. 1976). A conference was held sponsored by the TRW corporation, Stanford University, and the Na-

tional Science Foundation. Embodied energy calculations were rejected by a summarizing committee that recommended implementing oil shale development.

12. Work supported by a project of the Swedish Power Board exploring potential to substitute spruce and other forest production of Sweden for nuclear power. Calculations included conversion to wood powder, a very intensive fuel, but there was no appreciable net emergy yield, which means it could not support itself.

13. There may be an opportunity to relate this principle to the space–time concepts of cosmology.

14. The emergy per person required `for highly educated human work (high transformity) is much greater than that sometimes estimated to support the metabolism of a passive body in space. Space capsules in continuous sunlight beyond atmospheric absorption and the shadow of the rotating Earth receive about four times more light than any place on Earth, but the area needed to support a person on sunlight follows from the transformity.

An astronaut needs: (1 E8 solar emcal/cal)(22.5 E6 gcal/day metabolism) = 2.2 E15 solar emcalories per day;

Each square meter receives sunlight: (solar constant 2 E4 gcal/m^2/minute)(1,440 minutes/day) = 2.88 E7 solar calories/m^2/day

Absorption area needed per person: 2.2 E16/2.88 E7 = 7.6 E8 square meters (760 km^2), which is an area about 28 km by 28 km.

THE GEOBIOSPHERE builds and maintains structural storages with productive work. Some structural storages are nonliving, such as those of a mountain or a thunderstorm. When inputs decrease, nonliving structures dissipate, and later if energy inputs are again available, self-organization has to start over. With information the products of self-organization carry over from one episode of growth to another, making life, progress, and evolution possible (see chapter 4). The configurations of structural storage are extracted, copied, duplicated, and tested in new applications. This chapter considers structure, information, evolution, and their energy basis.

ENERGY AND STRUCTURE

The storage tank symbol (fig. 8.1) represents accumulations of energy and emergy above the surroundings.[1] To be sustained, structure has to be continually supplied by processes of productive buildup. As already described with the second energy law, there is spontaneous loss of structure that is called depreciation.

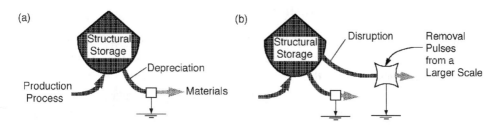

FIGURE 8.1 Energy support and loss from a structural storage **(a)** with depreciation from small-scale fluctuations; **(b)** including the disruption losses from large-scale pulses.

The Perpetual Shakedown

The jumping motions of warm molecules in all matter, plus the many kinds of frictions in the world of larger systems, invariably shake structure into disorder. The higher the temperature, the faster the little structures shake apart; and the more weathering processes, mechanical wind processes, destructive radiations, and earth movements there are, the more the large structural patterns become disrupted. The loss of structure is another manifestation of the principle of energy degradation.[2] When some of the small molecular structures shake apart as the molecules wander and vibrate, the effects may be very large energetically because there are so many molecules in a small volume.

The classical concept of the second law is too simple because it concerns only one small scale. The destructive pulses from higher levels in the energy hierarchy involve less energy, but they have higher transformity and large effects (fig. 8.1b). There is a second law action at every scale in which the oscillations of the next lower scale wear away structures while the infrequent pulses of the larger scale dismantle structure in bursts. Thermodynamic teaching will remain incomplete until the energetics of different scales is recognized with the fifth law of energy hierarchy (chapter 4).

Self-organization of Nonliving Structure

As explained in chapter 3 (fig. 3.6), productive processes recycle materials and accelerate output with autocatalytic feedback reinforcement (fig. 8.2). Nonliving examples of structural storage include the cold ice in a glacier, a heated house in winter, the elevated rocks of a mountain, the accumulations of water in a lake, or the clouds in the sky (fig. 5.3). Figure 8.2 shows depreciation losses driven by the available energy of the structural storage and energy-aided decrease caused by infrequent actions of larger units. Destruction or consumption by larger-scale units on longer time scales may seem catastrophic to those on a smaller scale. There is net gain and growth when there is a positive balance of inputs and outflows between episodic removal pulses.

Structure and Operation

Structures exist because of the necessary part they play in operations useful to other units and processes. Clouds may be part of operations that generate rain. Mountain storages facilitate the earth cycles; leaf structures are necessary to photosynthetic operations. As drawn in fig. 8.2, the building of structure and the contribution to other operations are mutually reinforcing outputs from the same production process. Whether we are discussing nonliving or living structures, where autocatalytic processes are involved, units may be represented in overview with the hexagonal symbol of fig. 2.10i.

(a) Nonliving Autocatalytic Structures
    Examples: Storms, Mountains, Flames

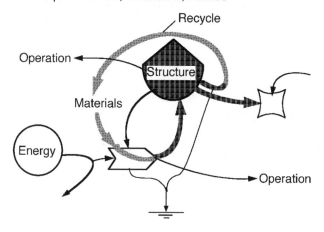

(b) Living Structure

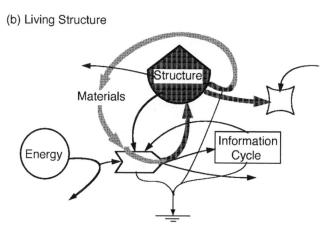

FIGURE 8.2 Comparison of nonliving and living structures, including the main features of self-organizing processes: a renewable energy source, energy transforming interaction, autocatalytic reinforcement, recycling of materials, depreciation, input of disruptive pulses from a larger scale, and contributions to operations upscale and downscale. (a) Nonliving structure; (b) living structure containing an information maintenance loop (see fig. 4.15).

Selection by Reinforcement in Nonliving Systems

Whenever substantial power supplies are available, a selection from among variable energy flows[3] rewards and reinforces the pathways that draw the greater power. A familiar example is cumulus cloud formation on sunny summer days (fig. 5.3). The heating of the ground develops variation in many small rising currents that form small competing cumuli that draw on the energy of the stored moisture. The ones that receive the most loop reinforcement draw more air flow and thus more energies, causing the demise of the others. The process of selection may develop a few large-scale thunderstorms by the end of the day. The rising columns are parts of loops of air that pass in downcurrents back to the ground, constituting a mechanism that reinforces the cloud over the rising column and helps it move higher and increase in volume. This is a natural selection process that builds structure by selecting from a variety of contributors the pattern that reinforces energy input.

Another example is the development of branching stream networks (Leopold and Langbein 1962) driven by the potential energy of rain. At first, choice is generated by random development of many small tributary branches. Later, as these varied developing drainage patterns join each other, some are reinforced and others pirated. The cutting of the ground allows the big ones to be reinforced and takes energy away from the smaller ones. The drainage pattern in the land represents structure that is carried to the future as landscape memory.

Many features of fluid dynamics may be explainable in these general systems terms. The ratio of forces required to start eddies is described by Reynolds's number for two adjacent flows at different speeds and by Richardson's number for one flow rising from a heat surface relative to another fluid. These numbers indicate energy thresholds for autocatalytic reinforcement for maximum empower.[4]

INFORMATION AND IMMORTALITY

Apparently missing on many other planets, information developed on earth in the evolution of life and is now operating with the social learning systems of humans and their computers. The development of information vastly improved the ways of building and sustaining structure. As plans are copied, duplicate structures and operations are made and widely shared. By means of information, systems structures could be made, improved, and replaced indefinitely.

In chapter 4, information was introduced as the description of components, configurations, and programmed sequences of functional systems that is sustained by extracting, copying, and reapplication to operations (fig. 4.15). Life is made possible by its information maintenance cycle (fig. 4.15), which is shown as a box in the structural unit in fig. 8.2b.

An item is useful information if using it to reconstruct a system is easier than generating the system from scratch. For example, the electrical system of a radio is a network of parts, connections, and adjusted temporal responses. The essence of the system can be written on paper, including specifications and wiring diagrams. These plans are information, which can be copied, distributed, and shared. With the plan, the radio can be reconstructed without being reinvented.

Thus to repair and replace we can use templates and plans, just as is done in the building industry and in biological reproduction. But even the templates and plans shake apart and develop errors according to the second law.

Carrier Depreciation, Information Loss, and the Correction of Error

Even the information written as ink on paper carries a small content of potential energy. The ink pattern is a chemical concentration capable of depreciation and dispersal (Brillouin 1962), and the paper carrying the information has enough potential energy to drive its own depreciation. If the paper and its ink depreciate, potential energy disperses. Although the potential energy content is tiny, the quality of this information may be great (high transformity) and the loss important.

Figure 8.3 illustrates how error accumulates in structural damage over time. First there are damages that can be repaired, but later there are irreparable errors, and the unit fails (fig. 8.3a). To continue the function, errors have to be repaired by replacement of the structure using the information template (fig. 8.3b). But later, after errors develop in the information template, replacements don't work, and the function is lost. To sustain the structures and functions indefinitely requires making many copies, using them to make many structures, selecting and keeping those that function best.

Readers will recognize the concepts of natural selection introduced by Darwin to explain the maintenance and evolution of species. The making of more offspring than can survive provides the opportunity for those most suited to surrounding conditions to last and in turn produce more offspring. If all the offspring survive, there is no selection; a choice must be provided by excess creation and mechanisms of selection. The simplest is self-selection by the best adapted.

The concept of natural selection involves the choice by nature from among alternatives provided. In energetic terms, reproducing excess offspring is a power-demanding process that can be measured in terms of the energy flow required for the reproductive process and the offspring.

Thus on one hand we have the shaking and other disordering forces, and on the other hand we have the process of proffering more duplicates than are needed and selecting the best to serve as the templates for the next round. As fast as errors develop in the templates, the selective process takes them out. If too little power is put into super-duplicating and selection, the errors will exceed the repair, and the stock will lose structure and fail.

Information Cycle

To sustain information without error requires continual processing in a circle (figs. 4.15 and 8.4). The plans (information) for successful systems are extracted, copied, distributed, and used to make more structure to operate systems. To be sustained, a system must maintain enough copies without errors, shared widely enough and tested continuously enough to keep the structures functional and competitive. Large-scale patterns of society reproduce and maintain their information with education one generation after another, a process of copying and selecting.

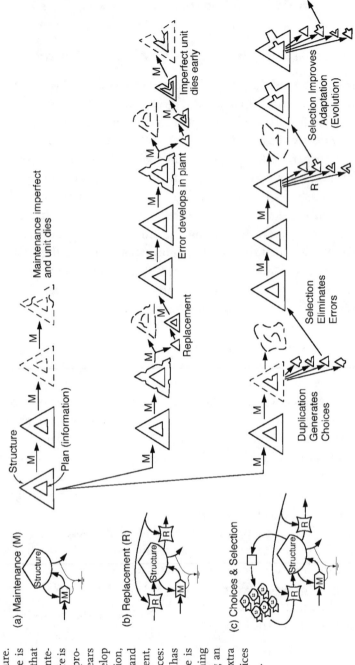

FIGURE 8.3 Sustaining structure.
(a) Maintenance work: Structure is maintained, but imperfectly, so that eventually the unit fails. **(b)** Maintenance and replacement: Structure is maintained, and one unit is reproduced to replace the one that wears out, but eventually errors develop in the plan used for reproduction, and the unit becomes imperfect and dies. **(c)** Maintenance, replacement, and loop selection of extra choices: Structure is maintained and also has reproductive replacement. There is the additional work of maintaining an error-free plan or generating an improved plan by combining extra choices for reproduction, generating choices and selection that benefits them.

(a) Maintenance (M)

(b) Replacement (R)

(c) Choices & Selection

Structure

Plan (information)

M

Maintenance imperfect and unit dies

Replacement

Error develops in plant

Imperfect unit dies early

Duplication Generates Choices

Selection Eliminates Errors

Selection Improves Adaptation (Evolution)

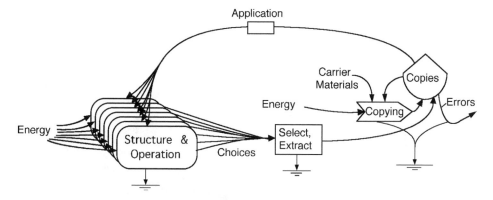

FIGURE 8.4 Cycle of information and operational testing. The cycle includes selection of operations, extraction of information, duplication of plan, distribution for sharing, and reapplication to operations.

Biological examples of the cycle of information and operational testing are the life cycles of species reproduction and dispersal. Thousands and thousands of reproductive life cycles are known to biological science; all are examples of the cycle that maintains genetic integrity. There is a timely need to put all these diagrams on a common basis, arranging the stages from left to right in order of transformity. The empower of a closed loop is the sum of the contributing inputs, which is a constant for the entire loop in any given averaging interval (i.e., one life cycle). Transformities are obtained by dividing the loop empower by the energy flow in each stage.

An example is the salmon life cycle in fig. 8.5, calculated for the Umpqua River in Oregon.[5] In the course of one cycle, part of the empower is supplied by the mountain river and part by the oceanic food chain. Transformity is highest in the eggs, which is the stage of extracted information before its dispersal and reapplication to form juvenile fish.

To sustain the system of structure and information, the cycle has to generate and test template copies faster than errors develop. In the rapid evolution of computer software, as in biological evolution, there are many examples where designs were eliminated because the population of users sharing the information was too small or the energy of support too little.

Whatever the scale, fluctuations on a smaller scale generate variations and errors during copying and operation. These variations provide the choices for selection according to the performance of operations. When this process develops an improvement in functions, the progressive change is called evolution.

Structural Life Cycles

To sustain structures indefinitely requires a system of life cycle replacement, whether it is cars, microbes, or landscapes. Main pathways are shown in fig. 8.6.

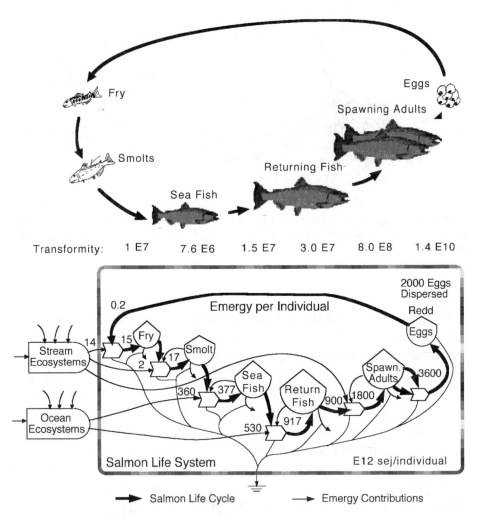

FIGURE 8.5 Life cycle and transformities of salmon, illustrating the cycle of copying, application, selection, extraction, and recycle that sustains a species. NOTE: Emergy units are solar emjoules.

The life cycle of structure includes construction, depreciation, repair, and removal for replacement with new construction. A self-maintaining population pumps in flows of materials and potential energy; combines materials to form parts; throws old parts out, sometimes reusing them as materials; rearranges new parts and disarranged situations; and transforms fuel energies into structural storages of potential energy with the form needed to operate the system.

Figure 8.6 shows the system for sustaining and testing information and structure. In populations of living organisms the system is tightly packaged, whereas the same functions are separate and more flexibly combined in society's structures. Living structures with capacity for self-repair hold their form and function using a continuous cold fire (biochemical reactions with the energies coupled to do

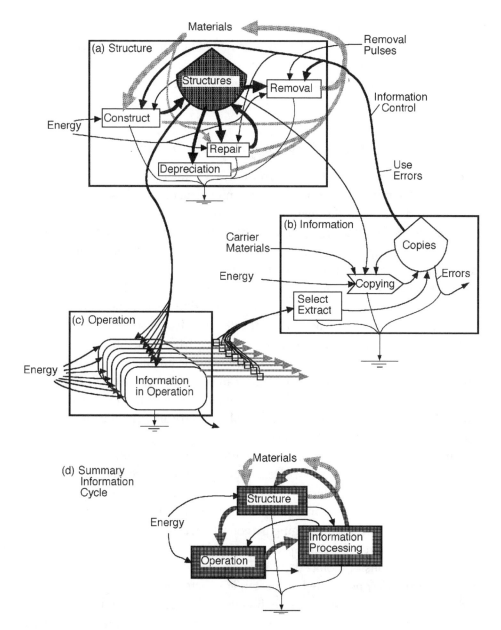

FIGURE 8.6 System of structure, operation, and information necessary to sustain and evolve structure. **(a)** Structural life cycle with pathways of repair and replacement; **(b)** information process with selection, extraction, copying, and reapplication for error correction and evolution; **(c)** set of parallel operations, each organized with a copy of the structural and operational information; **(d)** summary of the system of structure, information, and operation, with the information circle darkly shaded.

work). Living maintenance is continuous. When an animal or plant tissue is doing nothing except resisting the shakedown, its rate of burning fuel and using oxygen is called basal metabolism. The more structure there is, the more metabolism is required for its support (Mayumi 1991). However, the metabolism per unit structure decreases with size and scale, as explained with fig. 6.6.

Repair, Senescence, and Replacement

Among higher organisms, the pattern of survival includes some temporary repairs (fig. 8.3a), but generally, a deterioration of structure is allowed, followed by a discard at death, with the new individuals being made with mass production processes (fig. 8.3b). What is repair and what is replacement is a matter of scale. A structure often is repaired by replacing a smaller part. What is replacement on one scale is a part of repair of the next larger scale.

As a structure of connected parts gets older, accumulating damages, making a new one becomes energetically cheaper than repairing the old one. For organisms and cars, the parts are attached and intricately connected so that repair by rearrangement, or by removal and replacement of a part, is difficult. Repair is not a mass production procedure. However, a set of separate unconnected units does not show such collective senescence, and units can be replaced as needed. The pattern is similar in this respect for people, oaks, and automobiles. In the open ecological system the parts are not intimately connected geometrically; therefore, they may be separated and repaired one at a time easily.

Thus senescence and replacement of the whole may be economical when the parts are intimately connected, whereas repair by part replacement and potential immortality may be the pattern when the parts are not intimately connected. Senescence apparently occurs only in physically attached units of such complexity that the requirements of disengaging parts for replacement becomes too high. Senescence is a property of the compact and complex.

Loose associations of mobile organisms are found in lower levels of the energy hierarchy, whereas concentrated, intricately connected, monolithic rigid structures tend to develop where power flows are highly concentrated in hierarchical centers. Examples of unconnected associations are the organisms in plankton of the sea, the scattered plants in a new field, or the houses in suburbs. Examples of connected associations are the reefs of oysters (Copeland and Hoese 1967), heavy root networks of some tropical forests, or the continuous apartment complexes of cities. It is the latter group that develops senescence at the group level as well as in its parts. We are used to the idea in urban renewal that some high-density, continuous structures are more cheaply replaced than repaired. An example in a simpler ecosystem is the senescence of barnacle associations illustrated by Barnes and Powell (1950) and shown in fig. 8.7. When old and top-heavy, they break off or are broken off by animals that serve an urban renewal role in the animal city. New growth and succession refill the gaps.

(a)

(b)

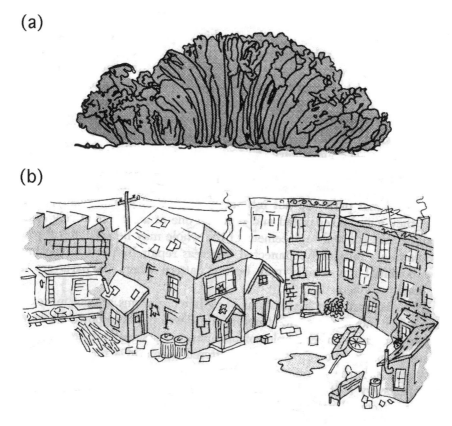

FIGURE 8.7 Associations that have monolithic structure and develop senescence. **(a)** Cross-section through crowded barnacle growths from Barnes and Powell (1950); **(b)** urban ghetto where houses have common walls.

Little Shaking in a Big Corral

To maintain order of a tiny structure, we must repair it more often, for the molecular shaking can break it apart readily if the structure is only slightly larger than the molecules doing the shaking. Many readers may remember seeing demonstrations of Brownian movement through the microscope. Visible particles smaller than bacteria can be seen to shake because of the bombarding and wiggling of the invisible molecules around them.

A structural form can be built on a larger scale and maintained with less if it is made up of large blocks (fig. 8.8). The molecules may wander and disorganize within the small blocks without ruining the desired structure, which is on a larger scale or is in the relationship of the big blocks. Thus, we see such different structures as buildings of bricks or bodies of elephants, with less maintenance required per pound than those of microscopic structures such as electrode metal surfaces and microorganismal bodies. The energy hierarchy (chapter 4) explains why energy transformations and dispersal at microscopic levels can maintain structures of greater dimension but less energy.

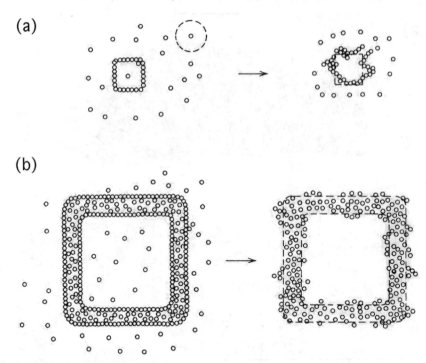

FIGURE 8.8 Comparison of the relative loss of form of large and small squares through random molecular motion. **(a)** Small square in which the action of one molecule constitutes a large proportion of the form; **(b)** large square in which the random motions of single molecules are less important, and the form lasts longer without repair.

Levels of Selective Reinforcement in the Energy Hierarchy

As explained by the energy hierarchy in chapter 4, units of each size support and are controlled by units of a larger size. The larger units can select the smaller structures and operations that maximize power contribution, as in fig. 8.9. Rapid processes and pulses on the smaller scale generate variations and choices that are available for selection by larger-scale units to the right. The energy used performs the work of generating the choices and supporting the chosen. Duplicates not chosen may be consumed for energy and materials.

The classification of organisms is a hierarchy with many individuals in a species, many species in a genus, many genera in family, many families in an order, many orders in a class, and many classes in a phylum. Territory and turnover time increase along this scale. The information in the higher categories spreads and distributes the smaller categories while accumulating the products that result from the more successful.[6]

In plant succession, multiple seeding and initial colonization constitute the choice generator, whereas mineral cycles and developing food chains provide choice machines for rewarding the plants by developing feedback loops.

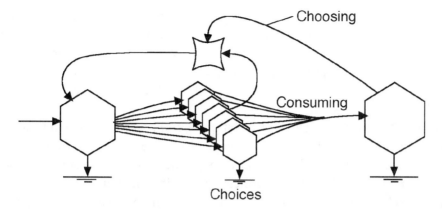

FIGURE 8.9 Information maintenance using 2 levels of the energy hierarchy. For an ecological example, high-level consumers provide choosing services that reinforce maximum power of the prey and predator.

In the embryological development of blood vessels or enzyme systems, the initial proliferation of many possible connections, largely at random, constitutes choice generation; the reward of those that manage to find a reinforcement loop by establishing successful circular flow provides the selection that draws resources from the others and reinforces the chosen. Much evidence of adaptive growth within organisms is given by Goss (1964).

In animal populations, the choice generator is excess reproduction, and the choice machine may be the carnivores selecting the best adapted by eating those less adapted (fig. 8.9). In organized reproductive systems, the mutations and reassortments of genetic mechanisms are the choice generator. In development and progress in industry, the choice generator is the research division, and management serves as the selector.

In higher organisms, selection may be programmed into the behavior of managerial species such as carnivores. In humans, the introduction of will, consciousness, and group-favoring motivations (e.g., religious, economic, egoistic) further specializes the selection of subsystems. These mechanisms help maximize empower.

Reinforcement in Information Networks

Even within the networks of information processing there is organization, with choice generating and selective reinforcement. For example, creative thinking often consists of hours of mental testing and trials that constitute the choice generator. Most of the ideas fail to receive the encouragement of striking a rewarding sensation of truth, but a few are reinforced. Creative thinking therefore relies on an effectively motivated thought generator and a well-developed truth-reward response. People who are not creative may be impatient with the necessary unrewarded probing, may lack a random thought generator, or may not have enough pleasure sensation when the ideas ring true. Creative children, with their half-organized thought generators, often devastate the teacher's desired classroom order and are disapproved, and their potential for innovation may be discouraged.

Creative special-purpose learning computers have been made, each with a varying device followed by a selection process that fixes the pathways that match some outside criteria (Nilsson 1965).

Figure 8.9 also illustrates the normal process of science. Many kinds of trials called hypotheses are attempted, but the choosing system is empirical measurement that makes certain the concept is loop rewarded by some agreement with nature. In this way, the network of theory grows, joining empirical facts. It is the empirical choosing system that sets science apart from philosophy. Ideas by themselves that do not receive loop reinforcement from real measurement wander off into interesting but unreal patterns. Both working units are required: the idea generator and the chooser.

Coon (1954) describes humans as connectors who transform potential energy into social structure. The shared information of social structure is an emergent property at a higher level based on the contributions of people.

INFORMATION BITS AND ENTROPY

There is a large body of quantitative knowledge concerned with the complexity of information, which is called information theory.

Information Bits

To most people *information* is a broad term covering words, data, messages, codes, and other inputs to the human mind that are stored in the memory or in libraries and are the basis for effective actions. Shannon and Weaver (1949) provided a quantitative measure of complexity intended to define information. This information index is the *logarithm of the number of possibilities*. The number of combinations increases as the product of added possibilities. Because logarithms are exponents, adding exponents multiplies the possibilities. In other words, this logarithmic information index is an additive measure of the complexity of possibilities.

This index of information measures the number of possible combinations of components or interconnections, expressing the result in bits. A bit is defined as the amount of uncertainty in the situation of one decision between two possibilities. For example, there is one information bit in flipping a coin, which has two possibilities, heads or tails, requiring one decision to make a choice.[7]

In any system, some of the possible organization may exist, some may not exist, and some may be unknown. In the following equation, total information bits are the sum of that defined and recorded, that uncertain, and that not organized.

$$I = I_{\text{organized and known}} + I_{\text{uncertain}} + I_{\text{known to be unorganized}}.$$

For example, consider that there are 16 possible pathways in the system in fig. 2.8, including a path from each unit to itself. A count of these is a measure of the uncertainty inherent in a situation with 16 pathway possibilities. But fig. 2.8 only has eight known pathways. The other possible connections are either missing or unknown.

In networks with one interconnection between each unit, the number of connections increases as the square of the number of units. The energy required for organizing and maintaining the network may be proportional to the number of pathways and increase similarly. The information bit measure indicates the decisions and work necessary to arrange a system, to keep it organized, or to transmit the message to another place or to the future using memory (fig. 8.10).

The information bits in a system are not only in the connections but also in the number of functional units. These information bits are the number of decisions necessary to specify the units.[8] Most systems have different kinds of units with different numbers of each kind. There is uncertainty, and thus information bits, in the opportunity for various ratios in different species quite apart from pathway organization, which we have already discussed. The information bit content associated with species composition was introduced by Margalef (1957) and widely used in ecology. Species information is calculated as if the species were letters in a message, which they actually are in the message of inheritance by which ecosystems are maintained and developed. Along with graphic expressions of species diversity, the information bits in species counts are used as a diversity index.[9] See chapter 10.

Information Content of Molecular Patterns

At the molecular level with billions and billions of molecules, the information content becomes very large. A gram of water, for example, has about 3 E22 molecules, with an immense number of possible configurations. The information bit content is huge.

Traditionally it is assumed that the molecules are disordered, distributed at random as they move and shake. The heat energy (which is molecular motion energy) keeps the molecules vibrating and shuffling.[10] The information bits are regarded as measuring the disorder. However, fig. 4.13 suggests there is a hierarchical organization in the distribution of energized molecules. Let us not assume that molecular complexity is all disorder.

Earlier and independently from the Shannon–Weaver use of information, the logarithm of the number of molecular configurations was found to be a measure of the complexity of chemical substances and, if multiplied by Boltzmann's constant, identical with entropy.[11] Entropy is the information content (uncertainty) of the complexity of the molecular states.

When water evaporates, the molecules move about more freely as vapor, producing even more complexity than before, with more combinations of states possible. The entropy has increased and, in this sense, so has the complexity. After transport

(a)   More emergy required per added species

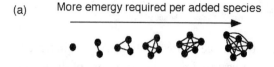

FIGURE 8.10 Emergy required to organize diversity. (a) Emergy and the number of possible connections increasing
with number of kinds; (b) power required to connect different populations.

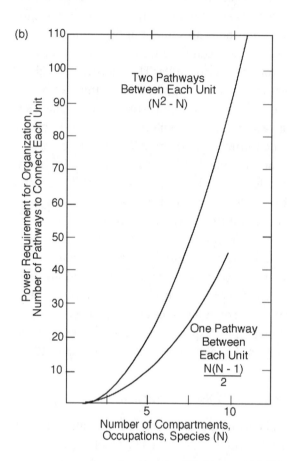

(b)

Two Pathways
Between Each Unit
$(N^2 - N)$

One Pathway
Between
Each Unit
$\dfrac{N(N - 1)}{2}$

Power Requirement for Organization,
Number of Pathways to Connect Each Unit

Number of Compartments,
Occupations, Species (N)

to other areas there may be potential energy in the relationships of the water vapor to
the surroundings. The energy in the vapor state is called latent heat and is the main
fuel of atmospheric storms, where the vapor condenses into cloud droplets (liquid
phase), releasing the latent energy, heating the air, and causing winds (fig. 5.6).

Entropy and Life

Maintaining living structure can be compared to maintaining ice because both
have structure that tends to be dispersed by heat. Refrigerators work by removing
the disordering heat from the box while storing potential energy in the process.
While the refrigerant fluid circulates, the refrigerators store ice at low temperature
with available energy relative to surroundings (fig. 8.1a). The entropy is less in the
cold ice. As required by the second law, the cooling is possible because much more
available energy from the power source is degraded into heat of the environment,

an entropy increase. That there has to be an overall increase in molecular entropy is one way of stating the second law.

Using entropy, Schrödinger (1944) put basal metabolism of living organisms into similar language. He said that the high degree of ordered structure (low entropy) in biological tissue is maintained by removing disorder with processes that degrade into environmental heat (high entropy). For example, there is much more available energy in food than is stored as biological, ordered structure. Refrigerators, crystals, tissues, and societies have an ordered structure (improbably low entropy content), which is maintained by continually pumping out into the environment large quantities of dispersing heat so that the overall entropy (storage and environment) always increases. In the nonliving refrigerator system, we maintain a low-temperature, low-entropy storage with less shaking (crystalline ice). In the living system, nonrandom structure is maintained at normal temperature by pumping out the shaking actions that would disrupt it.

However, the Schrödinger description is only half the story because not everything useful has low entropy. Useful storages are maintained at higher temperature than their surroundings, as in the warmth of mammals or heated houses in winter. At high temperatures, the molecular complexity (entropy) is higher than that of the surroundings. Whereas some biological structures such as bone crystals have low entropy, some aspects of living structure have higher complexity than their surroundings. Energy transformations support structural storages that are a mix of low-entropy materials and high-information networks, both maintained away from equilibrium.[12] R. E. Ulanowicz and S. Jorgensen initiated other approaches for relating the energy used to the entropy and information supported.[13]

EVOLUTION

In evolution, not only are the structures and patterns of the system maintained against the disordering influences, but, by definition of the term, novel changes are made, which may be progressive. The origin of life was the start of an information cycle.

The Ecological System Precedes the Origin of Life

Many have wondered how the great complexity of present life could have started from a nonliving world of unorganized chemical substances. The energetic history of the earth suggests that a long time is required to achieve the present complexity of life through step-by-step additions of information. Natural selection organizes by its choices a few bits at a time. As already explained, natural selection operates on nonliving, self-organization as well as on living systems.

Earth energy flows circulate matter and hold ordered patterns of structure by pumping their potential energy budgets down the energetic drain. These flows include those in the world's wind systems, hurricanes, the water systems of rivers and oceans, and the cycle of erosion, deposition, and mountain building (chapter 5). As long as there is a steady power inflow of potential energy such as sunlight, and as long as there are closed circuits of minerals, a system has the ingredients necessary for duplication, selection, and reinforcement. As shown in fig. 8.11, the circulation of seas provided a route for materials to pass from an upper photo-chemical zone where sunlight energy is received to a dark zone where chemical reactions can reorganize using stored energy and activated molecules energized by the light. Figure 8.11 suggests some steps in the chemical evolution toward life.

Differences in solar heating circulate waters (fig. 5.1b), which contain mineral elements and organic molecules generated by ultraviolet radiation. At the surface photons cause oxidation–reduction separations, which drive dark reactions. Thus the first part of the modern system of P and R existed before life. While circulating between light and dark zones, the molecules persist that grow longer polymer strands. Natural selection chooses the strands that break, in a kind of primitive reproduction. Thereafter selection favors those that last longer and develop more complex structures. There probably never was an exact moment life started to exist, for the energetics and cycles were already there. The specialization, capsulization, and subdivisions into living units occurred step by step. Information was increased in small increments one choice at a time, bit by bit.

The scheme to explain the origin of life shown in fig. 8.11 is like the production–consumption pattern of the modern biosphere (figs. 1.3f and 2.2) and like the model of choice–loop selection in fig. 8.9. The photochemical action was a choice generator; the circulating fluid provided a reinforcing loop, and competition in the dark selected the units capable of evolutionary change.

Power and Evolution

The rate of evolutionary change depends on the available power. More energy flow generates more choices and selections. Innovation can occur if outside conditions have changed and more power is available than is required to maintain and replace structures and information. For example, more food resource generates extra off-spring and more of the larger organisms that select. Without power excess, not much change of structure and information is possible, and losses of structure may occur.

When the environment is changing, the allocation of power may go toward mechanisms of adaptation. Conversely, in times of stability more power goes to make smaller-scale processes efficient. Examples are plants and animals that re-produce asexually when conditions are uniform but apply energy to sexual processes when the environment is changing.

The energy control of evolution applies to any reproducing system on any scale, including primitive organisms and our complex society. The evolutionary changes on a large scale may require emergy inputs of high transformity such as those from the deep earth, astronomical impact, or innovative information.

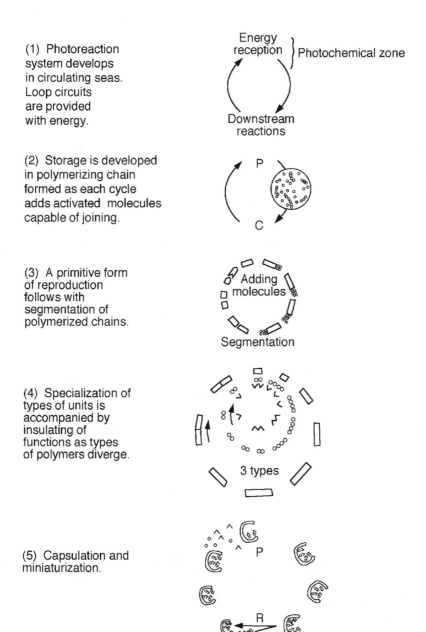

(1) Photoreaction system develops in circulating seas. Loop circuits are provided with energy.

Energy reception } Photochemical zone

Downstream reactions

(2) Storage is developed in polymerizing chain formed as each cycle adds activated molecules capable of joining.

P

C

(3) A primitive form of reproduction follows with segmentation of polymerized chains.

Adding molecules

Segmentation

(4) Specialization of types of units is accompanied by insulating of functions as types of polymers diverge.

3 types

(5) Capsulation and miniaturization.

P

R

FIGURE 8.11 Diagram of the step-by-step origin of life with the choice–loop–selector mechanisms acting on circulating molecules energized daily with photochemical reactions.

Miniaturization

Many mechanical and electronic systems that are initially large and clumsy later evolve into miniaturized, economical systems. Engines, radios, and computers have gone through such stages of development. Miniaturization provides more function for less power and less space, ultimately allowing more function and complexity for the same resources.

We may suspect a similar history in biological evolution. Many groups of animals and plants first appear in the fossil record as large simple structures and later are replaced by smaller, possibly more efficient species. For example, at first there were large armored fishes, ferns, trees, birds, dinosaurs, giant foraminifers, insects, and so on.

If the origin of life was a step-by-step capsulation of the loose circulations of molecules in the sea, as suggested in fig. 8.11, the earliest forms must have been so large and had so little structure that they left little in the fossil record. Gradually, with miniaturization, some component structures must have become recognizable after millions of years of chemical production and consumption. Such a history would help explain the sudden appearance of advanced fossils in Paleozoic rocks. Life from earlier periods may have been too large and loose to be recognized.

If early life formed in this way, with large loops preceding the small capsules, some of our best-miniaturized modern world dominants such as the bacteria may not have been the earliest forms.

Speciation and Insulation

In chapter 3 we introduced the concept of species as functional and specialized energy flows out of which complex networks were self-designed. The formation of species has been called microevolution, and an extensive literature shows examples of species generation. As summarized by Mayr (1963), a population that becomes divided by space or other factors generates genetic differences by mutations and genetic recombination phenomena. Differences in stocks are then exposed to differences in natural selection so that populations develop dissimilarities. In the language of the previous paragraphs, speciation, like other creative work, involves a variety-generating process followed by a choosing mechanism. However, each species depends on the rest of the ecosystem and its support. Speciation is best considered as the process of adapting parts (populations) to evolving ecosystems.

The development of specific behavioral and chemical means for keeping the species functions separate constitutes insulation of the circuits and is expensive in work drains on the available energy budget. If insulating mechanisms are absent, energies are lost through leaks between circuits. The energy reinforcement for developing species-insulating mechanisms is greater when networks are complex. See, for example, the energy network diagram for a section of a rainforest (fig. 6.18).

In human affairs, the evolution of complex industries with specialized occupations is an equivalent process. Each industry receives selective reinforcement and economic rewards for contribution to the larger economy. Whereas maximizing expansion takes precedence during times of growth on excess resources, contributing to the rest of the system takes priority at other times. See chapter 9 on economic systems for more on macroeconomic designs for maximum empower.

Emergy of Information and Its Hierarchy

Information has the emergy content of developing the first copy, the emergy of making and distributing copies, and the emergy of the users involved. Widely shared information has high transformities and broad influence. Like other series of energy transformations (chapter 4), information networks form hierarchical series in which information entities have increasing transformities, territory, and turnover times. Figure 8.12 shows the transformity of information increasing with increasing territory and replacement time to the right. These bars were calculated by dividing the emergy of formation by the energy carrying the information (DNA for the genetic examples and paper for the book examples).

The emergy per book copy is smaller when there are 1,000 copies but much larger when there is only one copy left. The emergy to make duplicate leaves is small compared with the emergy required to maintain a sustainable population of that tree species. The whole emergy required to evolve a species from its precursor species is assigned to a few copies when a species is endangered. The "Species DNA" on the right assigns the whole emergy of biological evolution to the estimated number of species that have ever existed. The highest transformity in fig. 8.12 is the emergy of life assigned to a culture of algae as if it was the last life in existence.

Apparently greater differences in genetic information require more evolutionary time for development. Thus, the emergy content increases along the scale of biological classification from individual to species to genus to family to order to class and phylum.

INFORMATION SUPPORT

Information Storing and Memory

The development of computers has shown how large quantities of information can be stored as memory by putting it in small, microscopic form on tiny computer chips. The entire set of coded genes for a human being is stored in the microscopic genetic strands in each biological cell. Brains also have a huge capacity for learned information, which is stored in nerve connections, also in microscopic form. But the ability to retrieve and use information is rapidly diluted as the number of stored information items increases. To accumulate information without selection is to lose its use.

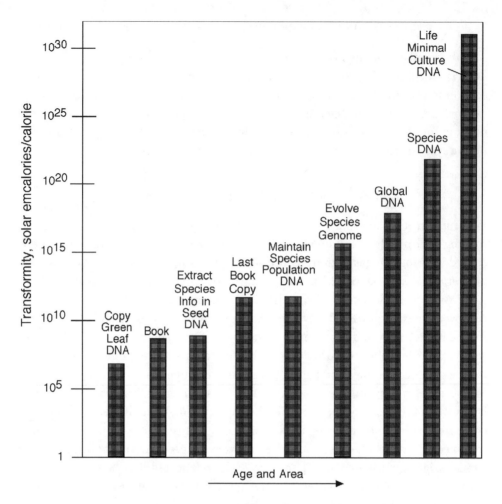

FIGURE 8.12 Transformity of categories of information arranged according to increasing scale of territory and replacement time. Duplicating a leaf, maintaining a forest tree species population, extracting information into a seed, and evolving a tree species were from table 12.1 in Odum (1996). Data are given in joules (J) and solar emjoules (sej).

Emergy of global DNA:

(9.44 E24 sej/yr)(2 E9 years evolution) = 1.9 E34 sej.

Energy of the world's genes:

(1 g dry live biomass/m²)(5.1 E14 m²/earth)(2 E-3 g DNA/g bio) = 1.0 E12 g DNA/earth;
(1.0 E12 g DNA/earth)(5 kcal/g DNA)(4,186 J/kcal) = 2.1 E16 J DNA.

Transformity = (1.9 E34 sej)/(2.1 E16 DNA/earth) = 9 E17 sej/J.

Emergy of the minimal DNA to sustain the essence of life in a minimal culture was estimated from the time to evolve divided by the minimum sustainable DNA, such as that of a microcosm culture of (1,000 g/m³)(1 E-3 g DNA/g) = 1 g DNA;

(1 g DNA)(5 kcal/g)(4,186 J/kcal) = 2.1 E4 J DNA in minimum life.

Transformity: 1.9 E34 sej to form life/2.1 E4 J DNA min life = 9 E30 sej/J.

(*continued*)

Brains deal with excess information by processing it through short-term and long-term memory. Items worth saving from the short-term memory are selected and stored in the long-term memory, while the rest is dumped. A good student learns how to learn by training his or her mind to save the right information. The information society is developing an analogous process in which the huge amount of information on the Internet seems to be society's short-term memory. Heretofore, libraries and scholars have been society's long-term memory. But books decay and have to be replaced. Storage with air conditioning to reduce the silverfish and dry rot is expensive. A challenge for new technology is to find a better long-term memory for the world's information explosion.

Although putting information on a microscopic scale increases memory capacity, it also makes the bits more vulnerable to the errors of molecular shaking, as explained with fig. 8.8. Making many copies and distributing them widely protects information (fig. 8.4). Apparently the long-term information in the brain is not local but distributed over the whole. Information shared over the global society also becomes long-term memory with a long replacement time. But the items globally shared are few, and maintaining such broad status requires continuous expensive recopying and replacement. The Bible, the Koran, and the textbooks of elementary education are examples.

Information Transmission and Power

In the transfer of information, there must be some kind of power flow to carry the information, such as the electricity in the telephone line, the sound waves from the radio, or the energy of a flying bird carrying seeds. Although small, information pathways are energy flows and should be shown on the energy diagrams, usually from right to left. For example, see figs. 8.4 and 8.6. In our society, human thoughts are transmitted, stored, restated, and used to control.

Because potential energy self-organizes autocatalytic circuits and energy-dispersing eddies, noise, and short circuits, the amount of byproduct complexity in-

---

Assume size of one individual of a species is about 100g and DNA 0.001 of that, or (0.1 g)(5 kcal/ g)(4,186 J/kcal) = 2.1 E3 J;

(1.26 E25 sej/spec)/(2.1 E3 J DNA) = 6 E21 sej/J.

Emergy of a book that took 2 years to write: (2,500 kcal/person/day)(365 d/yr)(2 yr)(4,186 J/kcal)(500 E6 sej/J) = 3.8 E18 sej/book 1st copy manuscript.

1,000 copies and $20 each to produce each copy: ($20,000 service)(1 E12 sej/$) = 2 E16 sej/1,000 books.

If book is (500 g)(4 kcal/g)(4,186 J/kcal) = 8.4 E6 J/book copy.

Transformity per book: (3.82 E18 sej)/(1,000 books) = 3.82 E15 sej/copy; (3.82 E15 sej/copy)/(8.4 E6 J/book copy) = 4.54 E8 sej/J.

If a book is the last copy: (3.82 E18 sej)/1 book = 3.82 E18 sej/last copy and (3.82 E18 sej/copy)/(8.4 E6 J/book copy) = 4.54 E11 sej/J.

creases with the power. Power lines, highways, factories, jet planes, and fast rivers are noisy. The word *noise* often is used not only for sound but for all kinds of variations and losses in energy flows.

Because high-power flows are noisy and full of complex variations, they are not the best media for sending information messages. A power administrator does not put a telegraph key on a 100,000-volt power line transmission to the next city. Instead, she might send a message over a low-voltage telephone line. Similarly, we observe that power circuits and information circuits are separated in ecosystems, although both are energy transmissions.

The transmission of information is an important part of any complex system. A plant manager makes his company respond on the basis of a stock market report. A cell makes its biochemical machinery respond on the basis of codes received from its genes. An ecosystem makes its power flows respond on the basis of its memory storages, some of which are biological and some of which may be physical or in libraries, records, rocks, or wood structure. Although there is vastly more energy flow in a power line, a phone message may have very high transformity and be capable of controlling the power line. Items of higher transformity are more easily transmitted and often provide a means to send flows of large empower. Birds flock together before migration, increasing their empower concentration, transformity, and information transmission.

Historical Information and Archaeology

In the earlier history of human evolution little emergy was available for transmitting information to the future. In our current emergy-rich society, we have the luxury of seeking clues in the material remnants of the past for information about early humans. These studies are the field of archaeology. Figure 8.13a shows how the number of these "artifacts" of the past decreases over time with the disordering influences described in this chapter. The emergy of a fossil or pottery remnant is that of its original formation, plus the annual empower of the deposit that protected it, multiplied by the time of preservation, plus the emergy used by archaeologists in collecting it (fig. 8.13b). The information it carries is the emergy of the original number of items (i.e., pottery items) divided by the copies remaining. As time passes, the emergy increases, and the number of copies decrease, so that the transformity and scarcity increase.

Irvin (2000) placed an emergy value of about 100 million emdollars on the Old Spanish Fort Matanzas in Florida by summing the emergy of the coquina rock used, the empower of the builders multiplied by the years required for construction, plus the annual empower of the land times the years since its construction.

Information Limits

Communication engineers describe limits in transmitting messages by the capacity of the channel and the ratio of signal to noise. We have a noisy channel of in-

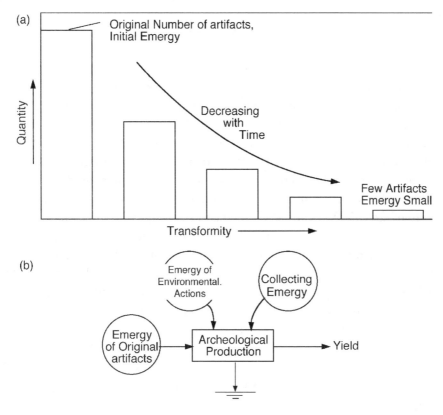

FIGURE 8.13 Emergy of fossils and archeological artifacts. **(a)** Survival curve of past objects with increasing transformity of its information; **(b)** emergy inputs to surviving evidence.

formation transfer from the past to the future. Can we put too many messages in it? What happens when the fuel base of our society decreases and less information can be maintained?

The information explosion is a perplexing property of our public decision processes, for energy requirements are large, and great costs are involved. Information development on the west coast of the United States helped create an electrical power shortage in the year 2001. Society is so concerned with the electronic web and information processing that it may be neglecting long-term information storage. Even libraries are using their resources to foster retrieval of Internet information on servers that may not be there in 10 years. Faculty in universities are pressured to do contract research to support students, whereas their traditional role and the long-term need is to consolidate knowledge and develop simplifying principles.

Particularly as resources decline, tough decisions are required as to what books and disks should be saved. What information is most important? There is a limit to what can be supported because access to information decreases as it becomes more complex. Perhaps decisions can be made on what information to save according to its transformity.

SUMMARY

Energy transformations sustain the structures of environment and society through repair and replacement and by sustaining a linked system of controlling information. Similar patterns and processes of organization are found from the molecules to the global scale of society and the economy. On each scale, the energy hierarchy adaptations for maximum power are sustained by generating choices that are selected by the next level. Integrity of information is maintained by a circle of selecting, extracting, copying, and reapplication to operational use, where performance is the basis for further selection and maximum empower. Some structural storages are in the form of high and low complexity, as measured in entropy units, but the emergy per bit increases with scale. A plausible gradual scenario for the origin of life is stepwise selection of emergent chemical structures in circulating fluids energized by pulses of solar energy over geological time. Emergy values are largest in globally shared genetic and learned information. Rates of biological and cultural evolution and the development of structural storages are greatest in times of high power and are expected to decline with the rising scarcity of concentrated fossil fuels.

BIBLIOGRAPHY

Barnes, H. and H. T. Powell. 1950. The development, general morphology and subsequent elimination of barnacle populations *Balanus crenatus* and *B. balanoides,* after a heavy initial settlement. *Journal of Animal Ecology,* 19: 178–179. Sketch of original photograph from Kendigh, S. C. 1961. *Animal Ecology.* Englewood Cliffs, NJ: Prentice Hall.

Brillouin, L. 1962. *Science and Information Theory,* 2nd ed. New York: Academic Press.

Coon, C. S. 1954. *The Story of Man.* New York: Knopf.

Copeland, B. J. and H. D. Hoese. 1967. Growth and mortality of the American oyster *Crassostrea virginica* in high salinity shallow bays. *Publications of the Institute of Marine Sciences,* 11: 149–158.

Goss, R. J. 1964. *Adaptive Growth.* New York: Logos Press, Academic Press.

Irvin, C. 2000. *Emergy Evaluation of an Estuarine Reserve in Florida.* M.S. research paper, University of Florida, Gainesville.

Jorgensen, S. E., H. Mejer, and S. N. Nielsen. 1998. Ecosystem as self-organizing critical systems. *Ecological Modelling,* 111: 261–268.

Leopold, L. B. and W. B. Langbein. 1962. *The Concept of Entropy in Landscape Evolution.* Washington, DC: U.S. Geological Survey Paper 500A.

Margalef, R. 1957. La teoria de la informacion en ecologia. *Memorias de la Real Academia de Ciencias y Artes de Barcelona,* 32(13): 373–449. Translation in *Society of General Systems Yearbook,* 3: 360–371.

Mayr, E. 1963. *Animal Species and Evolution.* Cambridge, MA: Harvard University Press.

Mayumi, K. 1991. A critical appraisal of two entropy theoretical approaches to resources and environmental problems, and a search for an alternative. In C. Rossi and E. Tiezzi, eds., *Ecological Physical Chemistry*, 109–130. Amsterdam: Elsevier.

Nilsson, N. J. 1965. *Learning Machines*. New York: McGraw-Hill.

Odum, H. T. 1970. In H. T. Odum and R. F. Pigeon, eds., *A Tropical Rainforest*. Division of Technical Information TID 24270, U.S. Atomic Energy Commission. Springfield, VA: Clearinghouse for Federal and Technical Information.

——. 1989. Emergy and evolution. In S. C. Holberg and K. Samuelson, eds., *Preprints of 33rd Annual Meeting of International Society for the Systems Sciences*, Vol. 1, 211–218. Ostersund, Sweden: International Society for the Systems Sciences.

——. 1996. *Environmental Accounting, Emergy and Decision Making*. New York: Wiley.

——. 2000. *Emergy of Global Processes. Folio #2, Handbook of Emergy Evaluation*. Gainesville: Center for Environmental Policy, Environmental Engineering Sciences, University of Florida.

Rossi, C and E. Tiezzi, eds. 1991. *Ecological Physical Chemistry*. Proceedings of an International Workshop Held in Sienna, Italy, November 8–12, 1990. New York: Elsevier.

Schrödinger, E. 1944. *What Is Life?* Cambridge, MA: Cambridge University Press.

Shannon, C. E. and W. Weaver. 1949. *Mathematical Theory of Communication*. Urbana: University of Illinois Press.

Ulanowicz, R. E. 1986. *Growth and Development, Ecosystems Phenomenology*. New York: Springer-Verlag.

Notes

1. In *Environment, Power, and Society*, the word *order* was used for accumulations away from equilibrium. However, that word is ambiguous because it may be used to refer to such regular structures as atoms in a crystal and to very irregular patterns, such as the gases in a hot flame. Nor is *entropy* a good word for storage with potential. A cold crystal is a low-entropy state, and a hot flame is a high-entropy state, but both may have potential energy relative to the surroundings.

2. This principle is also stated as follows: The universe tends to run down; molecular patterns trend from the less probable to the more probable. In dispersion of energy as discussed in chapter 1, the wandering of molecules from some less probable situation to one more probable is a heat dispersion. The ratio of heat dispersed (Q) to the temperature (T) is called the change of entropy. The molecules wander so as to even up the intensity of their motion (temperature), and the ratio in the section from which the molecules wander goes down more than the ratio goes up in the section into which the molecules wander. Consequently, the overall ratio always increases, counting the whole process of wandering. Thus, this entropy ratio is often said to measure the amount of degradation going on.

3. In *Environment, Power, and Society*, the source of variation was attributed to randomness. However, small-scale variation was explained in chapter 3 as the normal pulses of smaller scale phenomena rather than inherent randomness. On each size there is a smaller scale that generates choices for the next larger scale to selectively reinforce.

4. The Reynolds number is a dimensionless number in fluid dynamics providing a criterion for dynamic similarity. It is named after Osbourne Reynolds (1842–1912). Typically it is given as follows:

$$\text{Re} = \rho v_s L / \eta,$$

where

v_s = mean fluid velocity,
L = characteristic length,
η = (absolute) dynamic fluid viscosity,
ρ = fluid density.

The Reynolds number is used for determining whether a flow is laminar or turbulent.

Richardson's number (R_i) is used to predict the instability of air. It is the ratio of the static stability (N^2) to the square of the wind shear (dU/dz):

$$R_i = N^2 / (dU\,dz)^2,$$

where $N^2 = (g/q)(dq/dz)$.

Here U is the wind speed, g the gravitational acceleration (about 9.8 m/s^2), q the potential temperature, and z the height.

5. The following are notes on the emergy evaluation of salmon in the Umpqua River, Oregon. Data are based on joules, and emergy is expressed in solar emjoules (sej). There are 4.187 joules per calorie and 4,187 joules per kilocalorie.

Emergy of headwater streams from rain, 32.7 E20 sej/yr; and from geologic emergy of the mountains, 32.0 E20 sej/yr; therefore,

$$(64.7 \text{ E20 sej/yr}) / (1.10 \text{ E10 m}^3/\text{yr runoff}) = 5.9 \text{ E11 sej/m}^3 \text{ H}_2\text{O}.$$

Emergy of rain at the surface = 2.9 E4 sej/J from table 4 in Folio #2; for

1,000 m, 25.1 E4 sej/g; for 500 m, 19.8 E4 sej/g.

For each cubic meter of rain per year on 1 square meter at average elevation 500 m, emergy from rain is (1.98 E5 sej/g)(1 E6 g/yr) = 2 E11 sej/m^3 water/yr.

Emergy from the mountain into that water from a square meter is

$$(1 \text{ m}^2 \text{ area/m}^3 \text{ rain})(32 \text{ E20 sej/yr}/3.1 \text{ E10 m}^2) = 1 \text{ E11 sej/m}^2/\text{yr}.$$

Sum of rain and geologic input = 2.0 + 1.1 = 3.1 E11 sej/m^3 rain on land;

(3.1 E11 sej/m^3 rain on land)/(1 E6 g/m^3) = 3.1 E5 sej/g rain on land;
(350 m^3/sec)(3.15 E7 sec/yr)(1 E3 kg/m^3) = 1.10 E13 kg/yr discharge;
(64.7 E20 sej/yr)/(1.10 E16 g/yr) = 5.9 E5 sej/g;
(1.10 E10 m^3/yr)/(1.3 E10 m^3 rain) = 84.6 % runoff.

Emergy in headwater streams is concentrated from 3.1 E10 g/yr rain to 1.1 E10 g/yr of runoff:

(3.1/1.1)(3.1 E5 sej/g water) = 8.7 E5 sej/g.

Emergy added per redd from share of stream water times its transformity:

(1 m width)(0.1 m depth)(0.1 m/sec)(60 days)(8.64 E4 sec/day)(5.9 E11 sej/m³) = 3.1 E16 m³;

(3.1 E16 m³)/(2,000 eggs/redd) = 1.5 E13 sej/egg emerging;

(1.5 E13 sej/fry)/(1,000 J/ind) = 1.5 E10 sej/J.

Emerging fry:

2 months egg developing in gravels = 20% of eggs (0.2)(2 E8) = 4 E7 eggs remaining. Smolts = 10% of eggs: (0.1)(4 E7) = 4 E6 individuals; 40 g each.

Energy in individual:

(40 g)(0.2 dry)(5 kcal/g)(4,186 J/kcal) = 1.67 E5 J each; emergy of the added growth: 10% conversion of food energy with transformity: 1 E6 sej/J; energy input used;

(1.67 E5 J/ind)(10)(1 E6 sej/J) = 1.7 E12 sej/ind;

(1.7 E12 sej/ind)/(1.67 E5 J/ind) = 1.0 E7 sej/J.

Emergy of global ocean area:

15.83 E24 sej/yr from Folio #2 (Odum 2000), table 1 divided by area of ocean: (15.83 E24 sej/yr)/(3.61 E14 m²) = 4.38 E10 sej/m²/yr.

Fish at sea = 1% of eggs: (0.01)(2 E8) = 2 E6 fish.

Energy per fish:

(12,000 g/fish)(0.20 dry)(5 kcal/g)(4,186 J/kcal) = 5.0 E7 J/fish.

Fish food per square meter of ocean as 1% of the net primary production:

(4 kcal/m²/day)(365 days)(4,186 J/kcal)(0.01 efficiency) = 61,116 J/m²/yr.

Transformity: Emergy flow/area divided by the food energy flow:

(4.38 E10 sej/m2/yr)/(61,116 j/m2/yr) = 7.16 E5 sej/J.

Emergy per fish based on 10% efficiency and transformity of food:

(10)(5.0 E7 J/fish)(7.16 E5 sej/J) = 3.6 E14 sej/fish;

(3.6 E14 sej/fish)/(5.0 E7 J/fish) = 7.2 E6 sej/J for sea fish;

Emergy of returning fish, 7.2 E20 sej/yr.

For Umpqua watershed: (408,000 returning fish/yr)/(1.3 E4 km²) = 31.4 fish/km²;

(4.1 E5 ind/yr)(15,000 g/ind)(0.20 dry)(5 kcal/g)(4,186 J/kcal) = 2.57 E 13 J/yr or 6.3 E7 J/fish.

Emergy of sea fish to generate returning fish:

(3.6 E14 sej/ind)(2 E6 fish)/4.1 E5 fish = 1.76 E15 sej/fish, an increment of (17.6 – 3.8 = 13.8 E14 sej/fish); (1.76 E15 sej/fish)/(6.3 E7 J/fish) = 2.8 E7 sej/J.

Spawning adults=0.1% of eggs; (15,000 g)(0.20 dry)(5 kcal/g)(4,186 J/kcal)=6.3 E7 J stored.

(4.1 E5 returning)(1.76 E15 sej/ind)/(2 E5 spawning)=3.6 E15 sej/ind, increment (3.6 E15 sej/ind −0.9 E15 sej/ind =2.7 E15 sej/ind); (3.6 E15 sej/ind)/(6.3 E7 J/ind)=5.7 E7 sej/J.

Eggs from spawning adults: Half females; eggs half of body weight:

(2,000 eggs/adult/yr)(1 E5 female spawning fish)=2 E8 eggs.

Egg volume:

(0.1)(15,000 g/fish)(1 ml/g)/(2,000 eggs)=0.75 ml/egg.

Egg energy:

(0.75 ml/egg)(0.10 g dry/ml)(7 kcal/g)(4,186 J/kcal)=2,197 J/egg.

Emergy:

(3.6 E15 sej/spawning adult)(2 adults)/(2,000 eggs)=3.6 E12 sej/egg before release. (3.6 E12 sej/egg)/(2.2 E3 J/egg)=1.6 E9 sej/J (a value of the isolated information of an egg).

6. A computer simulation model of taxonomic evolution, based on 10,000 years to evolve a species, was used to estimate the emergy of higher categories of systematics (Odum 1989, 1996).

7. Information complexity (I) is defined as the logarithm to the base 2 of the possible combinations (C). Thus, $I = \log 2\ C$. A logarithm to the base 2 is the exponent one applies to 2 to get C. When one combines two parts of a system, one may compute the total possible combinations in the new complex by multiplying together the possible combinations in each part. Because logarithms are exponents, one accomplishes this by adding them. Information is thus an additive measure of complexity.

8. The number of decisions needed to specify one unit depends on the number of possible choices. Having defined one unit among a number of units leaves one less possibility for the next decision, two less for the next decision, etc. Therefore, the number of decisions to define the set of units is the factorial of the number. The information bit content is the logarithm of the factorial. To define the four units in fig. 2.8, the number of yes–no choices is $4 \times 3 \times 2 \times 1 = 24$. Its logarithm to the base 2 is $3.32 \log 10\ 24 = 4.6$ bits.

9. Where a system has many kinds of items, each present in different ratios, such as letters of the alphabet in a message or species in a forest, there is a form of the information formula that gives the bits per individual (H) due to the composition in the system or in information messages used to describe and transmit it. Thus,

$$H = -\sum_{i=1}^{n} pi \log_2 pi,$$

where pi is the probability of each type of item in the system and n is the number of kinds.

10. Heat is the molecular shaking motion; the level of molecular shaking is called temperature. When a person feels hot, his molecules have energy, are shaking more, and indicate this to the person's conscious self through a sensory system. As the shaking decreases, we say the temperature goes down. At a certain temperature, which is almost reached in some laboratory work, the shaking is zero and the thermometer reads −273°C; temperature cannot go below this point, called *absolute zero*. The molecules are still and are said to be ordered. The *absolute temperature scale* starts at absolute zero and thus is at +273°K when water freezes.

If we start adding some of the shaking energy (heat) to material initially at absolute zero, the temperature rises. If we sum up little by little the heat calories added and divide each amount by the temperature at which it was added, according to the entropy quotient formula described earlier, we obtain a gradually increasing quantity that measures how much shaking would have to be removed again to restore the complete order of absolute zero. Thus, the final quantity determined by such calculations is the *entropy content* of the structure.

11. The Boltzmann–Planck formula is $S = k \log m$, where m is the number of unorganized molecular states, k is Boltzmann's constant, 33 E-27 kcal/(°C)(molecule). A mole is 6.02 E23 molecules. Here log to the natural base e is used (2.718). To convert units, use the following:

1 cal/degree = 8 E23 bits. There is about 1 bit per molecule for each cal/(°C)(mole).

12. The Schrödinger booklet implied that there is some law about the amount of power flow that has to go into the drain (increasing external disorder and entropy) to sustain each unit of structure and storage against its thermal shaking and other disorganizing tendencies. An appropriate Schrödinger ratio is the ratio of entropy increase required per unit of entropy maintained. This ratio was calculated for the tropical rainforest at El Verde, Puerto Rico, by expressing the energy of metabolism and the potential energy of forest storage in entropy units by dividing each by the environmental temperature (Odum 1970). Where temperatures are similar and constant, the ratio is the replacement rate. Recognizing the concepts of energy hierarchy, Tiezzi suggested that entropy of different scales should be put on a common basis with the concept *emptropy*, in which each of the heat changes is expressed on a common basis as emergy divided by the Kelvin temperature (Rossi and Tiezzi 1991). In some articles the definition of information and the definition that was independently made for molecules so as to yield entropy were done with different algebraic signs. Either is correct if used consistently. We can talk about positive and negative information bits and positive and negative entropy without really changing anything except the agreed-upon convention. As disordering proceeds, negative entropy increases in magnitude, as does positive entropy. They both would become zero when all disorder is converted into crystalline order at absolute zero.

13. Ulanowicz (1986) developed the concept of *ascendency*, in which the logarithm of the complexity of pathways (macroscopic entropy) was calculated after the pathways were weighted according to their energy flow. This is an index of the complexity of power circuits. Jorgensen et al. (1998) assigned exergy values to genes by estimating Gibbs free energy weighted according to the probabilities of formation of complex molecules.

THE SYSTEM of humanity and environment is aided by the circulation of money, a special form of information. The acts of buying and selling appeared as human beings developed the mental capacity and social tendencies that allowed the exchange of goods. The circulation of money helped networks self-organize for maximum power by reimbursing people for their work immediately. Money is now the principal control mechanism for several levels of the energy hierarchy. However, the influence of economic values is expected to decrease as the global shortage of nonrenewable resources reduces annual empower. This chapter considers ways in which energy and money indicate good public policies, including policies for microeconomics, macroeconomics, and international exchange.[1]

ENERGY SYSTEMS AND MONEY

Money is something that can be exchanged for goods, services, materials, information, or other items of real wealth. Money may include material objects (e.g., gems or coins) or information objects (e.g., paper money or promises to pay), which a society recognizes as generally valuable for purposes of exchange. Money circulates as a countercurrent between people and in systems diagrams is represented as dashed or purple pathways. Money flows, can be stored, and is imported or exported to other systems. In the short term, money is conserved, which is a way of saying that the money flowing into a system equals that in outflows, plus the change in storages and money lost.

Exchange

The process of exchange is represented in fig. 9.1 using the diamond-shaped exchange symbol from fig. 2.10k. The symbol represents the swapping of two commodities in fig. 9.1a and the exchange of money for a commodity in fig. 9.1b. In fig. 9.1c, human service interacts with a natural resource inflow to make a product for

FIGURE 9.1 Representation of exchange. **(a)** Barter; **(b)** purchase of a service or commodity using money; (c) money flow to the human part of a production process.

(a) Barter

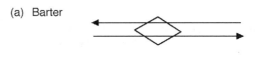

(b) Purchase

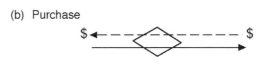

(c)

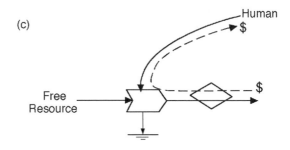

sale. The money is shown going to the human only because no money ever goes to nature for its contribution.

Price and Market Value

The ratio of the money paid to the item bought is the price. If the price is determined by the surrounding markets, it can be indicated on the systems diagram with an additional input arrow (fig. 9.2a).[2] Market value often is described as what people are willing to pay.

When something people want is scarce, they are willing to pay more, so the price rises. Having a higher price results in more money going to the limiting process, which helps increase the supply. In this way, human spending behavior helps remove limits to production. Figure 9.2b is a systems view of the price–scarcity relationship. Where the amount of money flow (J_m) is constant and the real wealth (commodity J_c) varies, the price is inverse to the commodity flow. In other words, the inverse relation of price and scarcity is a system property to which human behavior has become trained to respond.

Countercurrent Circulation of Money

The circulation of money in an economic system is shown for the coupled system of production and consumption in fig. 9.3. Money evolved to allow the production of one person to be rewarded by a compensation from some other part of society.

FIGURE 9.2 Systems view of market price controlling a purchase. (a) Systems diagram and equation for price; (b) resulting graph of price vs. commodity flow when the flow of money is held constant.

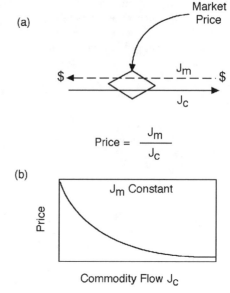

We receive food from the grocery store by passing money in the opposite direction, to the grocer (fig. 9.1b). We receive money when we deliver work (fig. 9.1c) that makes an energetic contribution to the function of at least one other unit. Money circulates with little loss, whereas energy flows diminish as they flow from sources and disperse into the heat sink as used energy (fig. 9.3d).

An index of the annual circulation of money is called the *gross economic product*. It is the index most used to measure the size and condition of the economy. In the simplified model of fig. 9.3d gross economic product is the same money flow that pays the producers and also is the income of the consumers.[3]

Consistent with its position in the energy hierarchy, the energy flow in the work of a downstream consumer and its recycle back upstream are small compared with the energy of production flowing downstream (flowing to the right in fig. 9.3d). But the money flow to the production sector and the consumer sector are similar. Therefore, the ratio of money to energy increases downstream and is a measure of the increasing amplifier value of the energy there.

Although Boulding[4] compared money circulation to the circulation of nutrients, the direction of material flows is different, going in the same direction as the available energy (fig. 9.3b). The circulation of money in the opposite direction (fig. 9.3d) helps keep the flow of inputs to production and the flow of products balanced so that neither production nor consumption is limiting. Computer simulations of these relationships were given elsewhere (Odum 1987; Odum and Odum 2000).

Money as a Loop Reward Selector

In ecosystems, production and consumption are mutually reinforcing by the circular loop of products to consumers and materials, goods, and services to production

FIGURE 9.3 Exchange in a system of agricultural production and consumption. Work by consumers aids farm production by supplying labor, equipment, and fertilizer. **(a)** Simple loop system reinforcing without formal payment; **(b)** cycle of materials moves in the same direction as the available energies; **(c)** exchange controlled by barter arrangement; **(d)** control of exchanges by the countercurrent circulation of money.

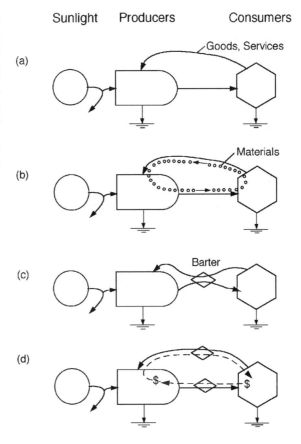

(fig. 9.3a). Through such reinforcing loops, the system selects species and pathways that contribute most to the function of the whole system.

In economic systems, money keeps people engaged in reinforcing the network. When money is paid for useful work, the reinforcement to a worker is immediate. A higher price stimulates people to produce more; a lower price stimulates consumers to use more. Thus, price and its associated human behavior patterns are self-regulating to keep available commodities optimal. Money helps maximize power and empower.

Emergy, Emdollars, and the Emergy/Money Ratio

Whereas emergy measures the real wealth produced and consumed, money buys the emergy as it circulates. The buying power of the money depends on the empower. The more emergy flow, the more the money buys. The ratio of the emergy used to the money circulating in each state or nation indicates, on average, the money's ability to buy real wealth there.

The *emergy/money ratio* can be used to estimate *emdollars* (em$), *the economic equivalent of emergy*. Emdollars are calculated by dividing an emergy storage or flow by the emergy/money ratio. The emdollars are the part of the buying power of the gross economic product due to a contribution of real wealth (emergy). Emdollars indicate what part of the total buying power a resource is contributing to the economy. Conversely, the emergy/money ratio indicates the real wealth equivalent of a unit of circulating money.

Figure 9.4 shows the annual global emergy budget of the earth (introduced earlier with fig. 5.5). The global money circulation is shown with dashed lines and labeled with the 1997 value of annual gross economic product. The ratio, 1.9 E11 secal/$, was the global emergy/money ratio in that year. The emergy in use by the world's economy is the basis for the buying power of world money flows.

In the simple model of production and consumption (fig. 9.3), the emergy of the product flow and the flow of consumer services is the same. At steady state, when neither changes, the emergy/money ratio is constant. Therefore, the real wealth buying power of the money as it circulates is constant. We might say, then, a unit of money is an *emergy certificate* (not an energy certificate).

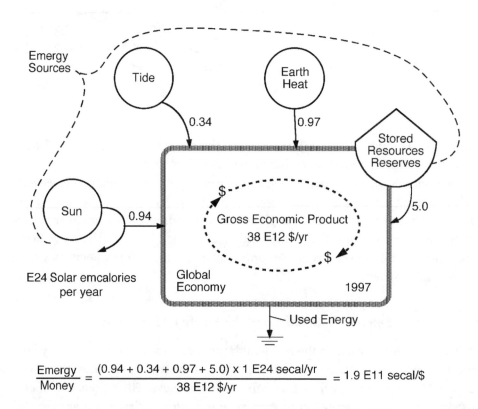

$$\frac{\text{Emergy}}{\text{Money}} = \frac{(0.94 + 0.34 + 0.97 + 5.0) \times 1 \text{ E24 secal/yr}}{38 \text{ E12 \$/yr}} = 1.9 \text{ E11 secal/\$}$$

FIGURE 9.4   Emergy in use by the system of humanity and nature of planet Earth, which is the basis for the buying power of the world's gross economic product. The ratio of annual emergy to money circulation is useful for evaluating emdollars. This evaluation is done using the global emergy base of 9.44 E12 sej/yr (Odum 1996).

Economic Condition

Changing either the emergy use or the money circulation changes the emergy/ money ratio. When more money is circulated but the supply of real wealth is not increased, then the buying power is less, a condition called *inflation*. Inflation also results if the money circulation is unchanged but the real wealth is reduced. In 1973, when world fuel prices increased sharply, less fuel emergy was used, and the annual inflation rate increased between 5% and 25% in most of the world.

Economic growth exists if the gross economic product, corrected for inflation, increases. After the inflation is estimated from changes in the *consumer price index* and subtracted, the economic product is said to be in *constant dollars*. There is economic *recession* if the money circulation decreases (expressed as gross economic product in constant dollars).

In 1929, when the money circulating was sharply reduced and the stock market crashed, the emergy/money ratio increased. The buying power of a dollar increased. This was *deflation*. However, incomes were so reduced and the economy so disrupted with unemployment that few people had much money to spend, a condition called *depression*. Plenty of resources were available, but they were not being processed.

Wheel Analogy and Monetary Policy

Adding or subtracting money can accelerate or decelerate the economic system. A system of production and consumption is represented as a wheel in fig. 9.5. Note the circulating flow of money, which is passing in a reverse direction from the flow of available energy and emergy that it controls. The money circulation is a kind of lubricant. Like ball bearings, the money turns in the opposite direction at the points of exchange between the energy flows and work processes. Monetary policy tries to control the economy by affecting the money supply, which affects the money circulating.

Adding some money in order to cause money to circulate faster increases the energy and emergy flows if unused resources are available (fig. 9.5a). Of course, adding money also causes some inflation. During the last century of growth in the United States, it was the general economic policy to accelerate the economy, encouraging a little inflation. Because there were rich nonrenewable resources available to develop, the growth in emergy use more than compensated for the inflation, so the standard of living generally increased.

In depression, with the slowdown in money flows, there is less money in relation to energy flow at each point of exchange. Prices are low, which reduces the stimulus to production. With less emergy, the standard of living declines. Reducing money circulation drags on the whole system (fig. 9.5b). Many kinds of factors can contribute to depression, such as the disarrangement of society

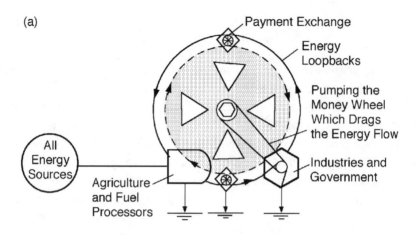

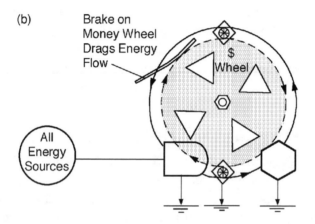

FIGURE 9.5 Circulation of money, drawn as a wheel coupled to emergy processing. **(a)** Time of inflation with the money circulation being pumped to stimulate energy flow; **(b)** depression, in which a weak money circulation limits emergy flow and use.

that follows some wars and the disruption of society when large populations are displaced from agriculture to urban jobs. During the depression in the 1930s, tax money was used to support public works, make jobs, and stimulate money circulation.

In early pioneer times colonists learned to be frugal and save food, clothes, houses, wood, fuels, and imported goods for winter and other hard times ahead. As the economy became more complex and money became more important, ordinary people transferred their realistic attitudes about real wealth over to money, even though money is something different: mere symbolism traveling in the opposite direction. When economic conditions were threatening, people saved money in a tin can for later use as if money were potatoes or seeds. Behavior that predisposed people to use money less when bad times threatened was the opposite of what was needed. The money wheel was retarded (fig. 9.5b). Restricting money flow helped to intensify depressions. Later policy followed the economic model of Keynes, in which depression was battled by spending more. The money wheel was pumped (fig. 9.5a).

Production Functions and Diminishing Returns

With examples from ecosystems (fig. 3.9) in chapter 3, production was explained as the output of processes in which more than one necessary input was combined to form a new product or service. Increasing the availability of one input factor usually causes one of the other necessary inputs to be locally limiting so that there are diminishing returns (fig. 3.10 and appendix fig. A5). Production in the economic system follows these principles, with output proportional to the product of the local concentrations available. At one time and place or another, each of the factors in fig. 9.6 has been important and limiting to an economy. In this diagram we don't show the money flows, although the data usually are in the form of money paid for each of

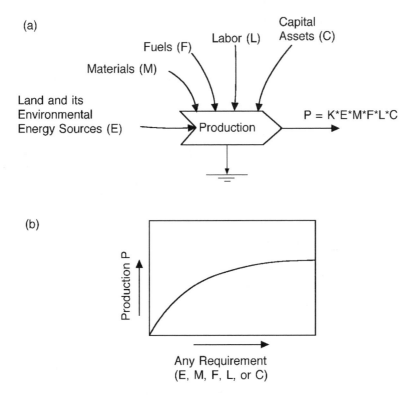

FIGURE 9.6 Diagram of the input factors usually contributing to economic production. **(a)** Production function *P*, a product of the forces and concentrations locally available; **(b)** diminishing returns on production when increase of one factor causes others to be limiting. For equation see appendix fig. A5.

the inputs. Two centuries ago, when there was still an agrarian economy, the land–environment factor was considered most important for production by the economists of the time. In current textbooks those economists are called physiocrats.

When a factor is abundant and not limiting, it becomes a part of the constant in the production equation and thus omitted from thought. As the fossil fuel–based Industrial Revolution developed, even agricultural production was stimulated by machines using fuels. For a time, resource inputs were not as limiting as labor (proportional to population) and capital (capital assets and the money to build them).

If the emergy of one factor is abundant enough, it can offset and be substituted for other factors of lesser quantity in the production function. When rich fossil fuels and mined materials became available, they dwarfed the renewable environmental contributions. While they were nearly unlimited, they substituted for other inputs, and their effect on production was constant and ignored.

The economy accelerated growth, and economic production was regarded as dependent only on labor and capital. Abundant energy kept materials from becoming limiting by supporting recycle. Nothing was believed capable of limiting growth. With the shocks of energy shortage in the 1970s, fuels were added to production functions again. But by the year 2000 people were convinced that something would always substitute for potentially limiting factors.

ENERGY AND MONEY IN MACROECONOMIC OVERVIEWS

Macroeconomic models overview an economy with a larger-scale view. Here systems diagrams show how the action of money depends on energy resources. Figure 9.4 evaluates the emergy sources of the global economy for 1997, showing the most important source to be fossil fuels.

Money Circulation and the Energy Hierarchy

Figure 9.7 is an aggregate overview, with sectors of nature and society arranged left to right according to energy hierarchy.For simplicity, the energy sources (sun, tide, and earth energy) are aggregated as one input on the left. Labeled "nature," the environmental sector without people or money is on the left, including the oceans, freshwater, and land ecosystems. On the right, with high transformity, is the emerging sector of globally shared human information including public opinion, mostly without money. Money is shown in the center circulating as a counter-current between a sector with agriculture, aquaculture, and forestry producers, the mostly urban sector of human settlements, and the sector of economic industries that use the water, air, land, and minerals of nature. What money can buy depends on the productivity and information of the connecting systems that operate without money.

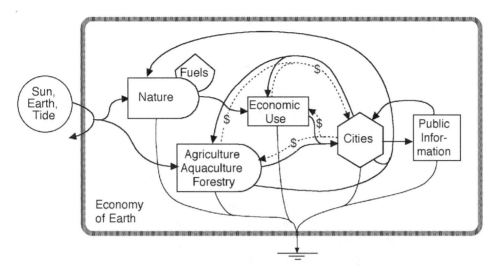

FIGURE 9.7 Main sectors of the global economy of society and nature showing the circulation of money. Renewable energy sources to the environment are aggregated as one source.

Macroeconomic Equations and Dynamic Simulation

Textbooks introduce concepts of macroeconomics with a set of equations relating production, savings, income, investment, consumption, and labor. These relationships can be combined as a single system (fig. 9.8). The flows of money are dashed lines circulating as a countercurrent to the real wealth flows of energy and resources to which they are coupled by prices. There are two storage tanks. The momentary store of money for investment is labeled M (capital). The capital assets tank (Q) represents the structural stores and reserves of real wealth.

In the figure, savings go into the storage tank, becoming capital (M), and money flows out as a countercurrent driving the resource flow into structural storage (Q). Part of the production goes into structure and part to the consumers in countercurrent payment for their labor. Income (Y) receives and supplies the savings–investment (S–I) loop and the consumer (C) loop. It is a countercurrent to the real production (P_R).

The traditional economic equations have no provision for energy limits. As shown in fig. 9.8a, a constant energy availability usually is assumed. In other words, there is no energy limit. By referring back to fig. 3.6, the reader will recognize an autocatalytic design that grows exponentially (fig. 9.8c). However, if the energy source is a limited flow type (fig. 9.8b), the system builds storages of capital and capital assets and levels off as shown in fig. 9.8d.[5] Macroeconomic concepts are incomplete if the effects of energy limitations are not included.

The turnover time of the capital (money supply) is 16 years or less, whereas the turnover time of the real wealth capital assets of the structural storages of society is much longer (50–200 years).[6] Using money to represent real wealth storages in

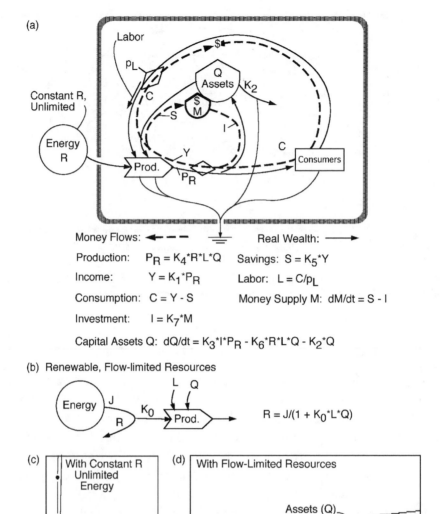

FIGURE 9.8 Systems diagram of the equations of macroeconomics representing the circulation of money shown coupled to pathways of real wealth flowing in the reverse direction. Reprinted from Odum (1996) with permission from Academic Press. **(a)** Systems diagram and macroeconomic equations shown with unlimited energy (constant force); **(b)** more realistic energy arrangement with limited, flow-controlled energy source; **(c)** result of simulating the model with unlimited energy; **(d)** growth curves of capital and capital assets with flow-limited source (Odum and Odum 2000).

simulation models is incorrect, causing simulations of growth and oscillations to be too rapid. If money is separated from real wealth, dynamic simulation models can be realistic.

Input–Output

Flows of money through an economy sometimes are represented with input–output tables. In chapter 2, fig. 2.8, two ways of representing the circulation of materials were represented, one as an input–output table and one as a network in a systems diagram. The model includes photosynthetic production, consumers using the food, and materials recycling. In fig. 9.9a a similar model with the same input–output table is shown again, but this figure shows agricultural production with the circulation of money in the opposite direction from the flow of real wealth that the money is purchasing (fig. 9.9b).

Future money flows sometimes are predicted when the flows in the input–output table are increased by the percentage growth expected for the whole economy. Coefficients relate one flow to another. Because such simple extrapolation does not change the coefficients, all the flows keep the same proportions.

To make the model real, energy sources must be added and the producer and consumer units given the ability to grow individually and autocatalytically according to the energy available to each. Then the relative flows change. The revised model with energy flows is shown in fig. 9.9c with the countercurrent of money in dashed lines and diamond-shaped exchange symbols indicating the links. The input–output model is thus transformed into a dynamic model. The model indicates the principle that economies are mutually controlled by the available energy and by the money paid according to price. If the energy source is unlimited (constant force), then growth of the model is explosive, as in fig. 9.8c. But if the source is limited by an externally controlled flow source such as sunlight, the system levels off.

Money and Growth

When unused resources are available, people self-organize in an overgrowth frenzy that maximizes empower, like that of colonizing ecosystems (chapter 3). In human society with a free economy, the mechanism of overgrowth is capitalism. When energy sources are not yet limiting and new businesses compete, the ones that start first outgrow the others (chapter 3, fig. 3.8). The successful businesses start first by borrowing. They can pay back interest because of their growth.

Figure 9.10 shows the autocatalytic growth for an isolated agrarian economy, like those two centuries ago. If new money is added to the money supply as the system grows, and the ratio of money to emergy remains constant, then there is no inflation.

Borrowing adds to the money supply, as diagrammed in fig. 9.10a. A person who loans money or buys stocks or bonds gets a promise-to-pay note, stock, or

(a) Input-output Table

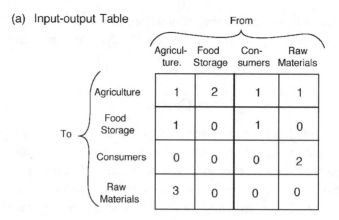

	From			
	Agricul- ture.	Food Storage	Con- sumers	Raw Materials
Agriculture	1	2	1	1
Food Storage	1	0	1	0
Consumers	0	0	0	2
Raw Materials	3	0	0	0

To { (applies to the rows: Agriculture, Food Storage, Consumers, Raw Materials)

(b) Money Circulation

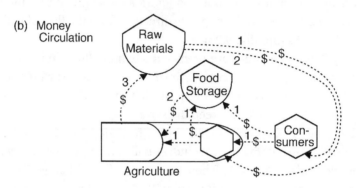

(c) Energized Model

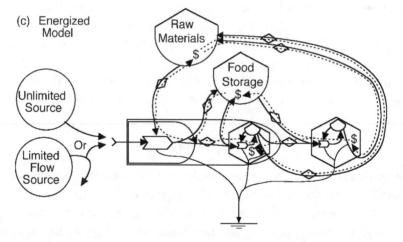

FIGURE 9.9 A 4-sector input–output model. **(a)** Flows between sectors in tabular form; **(b)** countercurrent of money for the same table; **(c)** systems diagram with 2 alternative energy sources and dynamic relationships added that go normally with the flows of money and real wealth (Odum 1987).

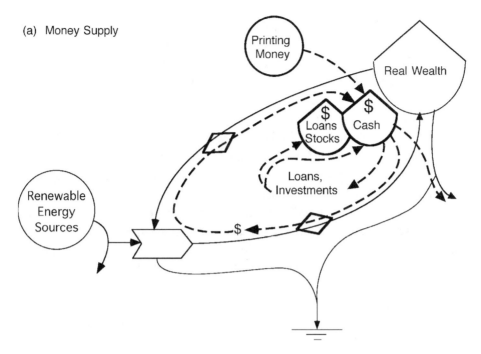

(a) Money Supply

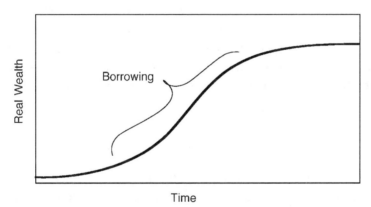

(b) Growth on Renewable Resources

FIGURE 9.10 Systems view of money supply and growth. **(a)** Circulation of money and the way loans and investments generate new money, especially during growth; **(b)** growth of assets and money supply in an agrarian economy based on renewable resources.

bonds. The money he loans keeps circulating. The loan papers (promissory notes, mortgages, stocks, bonds) also are a form of money because they can be exchanged for real wealth. Thus during growth, the money supply is increased by the borrowing process and by printing of money. The dashed lines in fig. 9.10a show the ways investments and loans add to the money supply. The increased circulation of money as a countercurrent helps accelerate the production of real wealth.

At times in the past, money was printed faster than the growth in real wealth, causing great inflation and leading the public to lose faith in the currency. At such times the limited supplies of gold and silver were more acceptable and became the basis for economic exchange, either as metallic coins or as paper that was redeemable at banks for gold and silver. There was a gold standard. During the period of Spanish colonization of the new world in the 16th century, large volumes of gold were sent back to the economy of Europe, causing a kind of inflation in which the buying power of gold was temporarily less.

In 17th-century Sweden, methods of mining and refining silver were developed with hydropower augmented by wood smelters. Silver coin became the exchange medium among the upper class in towns and part of the means for supporting Swedish armies controlling northern Europe[7] (Sundberg et al. 1994).

The paradox about gold and silver is that the real wealth value of the refined metals is higher than its buying power as a currency (chapter 4). In other words, its emdollar values are greater than its dollar value,[8] properties that go with its uses in the chemical industry and jewelry. By the end of the 19th century the growth in real wealth within the human economy was faster than the mining of new gold and silver for currency. Real growth without enough circulating money contributed to depressions. The gold standard was abandoned, but the money supply was controlled to limit inflation or deflation. One way to do this was to control borrowing by having national central banks set the interest rates.

At times in history when growth of an agrarian economy was not possible, borrowing and high interests fell out of favor. Lending money was called usury and regarded as evil as people adapted their cultural mores to the energy system. But in times of growth such as the year 2000, interest and dividends were taken for granted. Few people currently think of interest as unearned income or question its justice.

Growth on Nonrenewable Resources

A simple overview of the economy in fig. 9.11a has production based on renewable environmental energy sources, concentrated nonrenewable resources, and their interaction. The countercurrent circulation of money flows in proportion to the money supply. The ratio of money in circulation to production is an index of the average price. A typical computer simulation result is the S-shaped growth curve in fig. 9.11b.

As growth progresses, the innovative technology, information, and other structural assets increase. These assets contribute to autocatalytic processes, stimulating further growth. Therefore, production increases. If the money circulation is constant or increasing only slightly, the prices fall. Later, the decrease in concentration of nonrenewable resources causes the production to decrease and the prices to rise again.

Some authors (Simon and Kahn 1984) used the decrease in the prices of resources early in economic growth as evidence that resources do not causally control the economy. They thought that people would pay higher prices as soon as

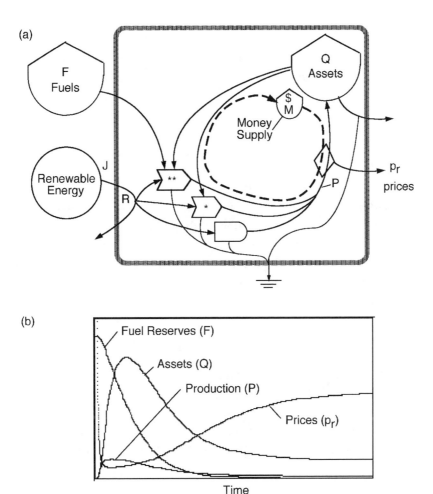

FIGURE 9.11 Overview model of economic growth on nonrenewable resources interacting with renewable resources. **(a)** Energy systems diagram of source, assets, and production; **(b)** result of simulation where money supply (small tank) is constant or increasing only slightly. Equations are as follows: Renewable resources, $R = J/(1 + k1FQ + k0Q + Pw)$; production, $P = PwR + k4RQ + k3RFQ$; for assets Q, $dQ/dt = P - k5Q$; price, $pr = kM/P$; fuels F, $dF/dt = -k2RFQ$.

resources began to decrease. This is an example showing how prices are system controlled, not a simple function of individual behavior.

ECONOMICS AND ENVIRONMENT

As explained earlier, the monied economy is embedded in and dependent on environmental systems for its resources (fig. 9.7). Conversely, environmental systems that prevail are those that are symbiotic with the economy and can benefit from human society. Although some politicians still speak of jobs versus environment, the public is beginning to understand that the economy is maximized only when the works of the environment are sustained. Principles for maximizing the economies of society and nature concern the interface, traditionally the field of *microeconomics*.

Interface of Environment and Society

The environment affects human society and its economy by many pathways. Figure 9.12 shows that much of the contribution of environment to society has no corresponding circulation of money. Some of these indirect interactions, not identified with payments of money, provide for people's aesthetic needs, which appear to be necessary for a high quality of life. However, environmental processes can be given their emergy values through evaluation of their inputs. Then the emergy value is divided by emergy/money ratio to obtain emdollars. The emdollars usually are much larger than the dollars paid for environmental contributions.

Economic Use of Environmental Products

Environmental systems generate wood in forests, fish in lakes, minerals in geologic processes, food and fiber in crops, and other resources that are collected by humans engaged in businesses of processing products for economic use. The interface of economic use is shown in fig. 9.13a. At the interface, real wealth is joined by the circulation of money. Money obtained from sales goes to pay for the

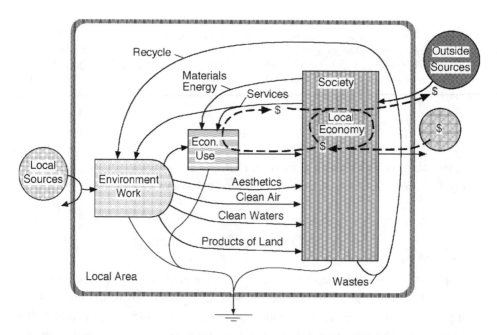

FIGURE 9.12 Interface between environment and economy showing some of the indirect contributions of environment without monetary exchange.

(a) Exploiting Interface

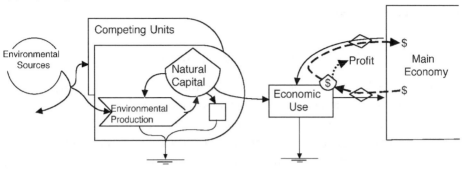

(b) Sustainable Interface

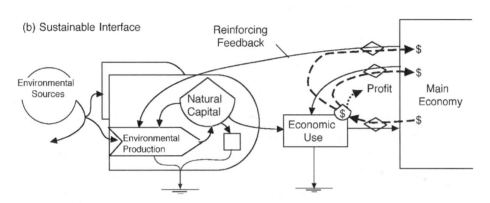

FIGURE 9.13 Systems diagrams of the economic use of environmental products. **(a)** Exploiting interface that drains and displaces useful production; **(b)** sustainable interface with feedback reinforcement.

necessary inputs of goods, labor, services, materials, and fuels and, in a growing society, for profit.

A monetary evaluation of an environmental product includes the *costs* paid for the inputs to the processing work and the market values (prices). Because money is paid only to people for their costs and profit and not to the environment, money paid for the first step of economic use is not a measure of the environmental work contributed. When environmental products are abundant and contribute most to the economy, costs are low and markets give them low prices. When environmental products are scarce, costs of processing are high, prices on the market are high, and more money is circulated. But the net contribution is small and the standard of living is lower. In other words, *money cannot be used to evaluate the real contribution of nature.*

However, the emergy of the flow of an environmental product can be evaluated (dividing by the emergy/money ratio) to obtain the flow of emdollars. Processed products used by other parts of the economy have the emergy contribution of the environment plus that of each of the inputs fed back from the economy. These products often have higher transformities than those created without economic feedbacks.

Natural Capital Evaluation and Discounting

Storages of environmental products are natural capital and have large emergy and emdollar values. Most are storages and stocks maintained by environmental systems to operate their own autocatalytic production processes (fig. 9.13). One of the concepts driving environmental destruction is the idea of selling environmental storages now and putting the money in a bank or other investment to earn interest. Especially during times of economic growth, a substantial interest rate can be assumed to be available. That interest rate is called a discount rate. Having the money now rather than later makes a big difference in the compound interest earnings.

Whereas "exploit now" may seem to make sense for short-term business profit, it is incorrect for maximizing total emdollars for public benefit. By leaving the stock of natural capital, the environmental system may be growing emdollars faster than the bank is paying interest dollars. Then the exploitation is a loss to the economy. For example, a fishery might be producing 20% per year, whereas the bank was paying only 5% interest. Furthermore, if the exploitation causes the desired populations to be displaced by unusable weed species that dominate for years, then very great losses result. Also, natural capital may be contributing unrecognized emdollars indirectly, which are lost with the stock.

Sustainable Reinforcement by the Economy

As drawn in fig. 9.13a, the yields are unsustainable because the economic user is draining environmental products such as fish populations without the reinforcing feedback that is required in all systems (chapter 4). Furthermore, using the stock (natural capital) reduces the ecosystem's own autocatalytic reinforcement, and its production is reduced. As the product gets scarce, prices rise and the drain continues, driving the stock to even lower concentrations. The environmental system may respond with overgrowth of *competing populations*, displacing the desired stock. Unfortunately, a half-century of destructive environmental exploitation of fish populations was encouraged by efforts to manage environmental systems with single-species models for maximum sustainable yield.

However, fig. 9.13b includes the necessary feedback from the economy to reinforce the symbiotically desirable environmental production. Note the loop of monetary expenditure required to pay for the reinforcement. There have been many efforts to use tax moneys and tax incentives to encourage environmental systems, but there was no means to determine how much reinforcement was necessary. According to the energy hierarchy concepts, feedback reinforcement should be given the same emergy as the pathway of environmental use.

To compete in an unregulated economy, business must engage in short-range exploitation or have the resources taken by others. Sustainable economic use requires that either the users organize necessary reinforcements or use public money to pay for them. Free market forces alone cannot reinforce the environment. For example, during its most prosperous years New Zealand provided free phosphate fertilizer to farmers to reinforce soil production.

It would be desirable to have an automatic response mechanism to exchange reinforcement for products of the environment. One idea is a special emdollar currency that could be circulated between environmental managers and environmental users. A landowner or government land manager would exchange some of her product for the emdollar coupons. This special currency would then be sent to the local environmental agency to purchase inputs that reinforce the environment. The economic users of the environment would buy emdollar coupons from the agency as needed with regular money that would operate the agency. The emdollar coupons could also be bought and sold like other commodities.

Appropriate Pulsing Frequency

As explained in chapter 4, each scale in the energy hierarchy has an appropriate frequency of pulsed consumption that maximizes empower in the long run. The recommendations for economic use in the discussion of fig. 9.13 were made for the long-run steady state. On a smaller scale of time and space there are pulses of consumption. A tree is cut, a fish is harvested, or a peat deposit is mined.

On the smaller scale, short-term removals of natural capital in pulses can generate maximum empower if they are followed by enough reinforcement and if the frequency of pulsing use is optimal. Fisheries and forests were destroyed when excess capital investment created harvesting industries on the wrong scale. Fish and logs were harvested on too large a scale for optimal sustainable replacement.

Hierarchy of Circulating Money

Although money does not circulate to nature, its circulation among people is part of the hierarchical organization that maximizes empower. When money pays for a product, that money circulates back to the economy to pay for an input of higher quality (fig. 9.1c). As products are processed in successive steps, from environment, to processor, to manufacturer, to wholesaler, to retailer, to consumer, each step involves additional inputs from the environment and additional cost. As shown in fig. 9.14, each step has additional money added to the circulation, which is often called *value added*. Notice how each step to the right has additional energy transformations, decreased energy, increased transformity, and increased money circulation.

There is no money circulating in the sectors on the far left because the work there has little or no human participation to pay for. As already explained, money

(a) Energy Transformation Series

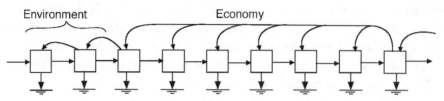

Environment   Economy

(b) Money Circulation

(c) Prices and Density of Money Circulation

(d) Spatial Organization   Center of City

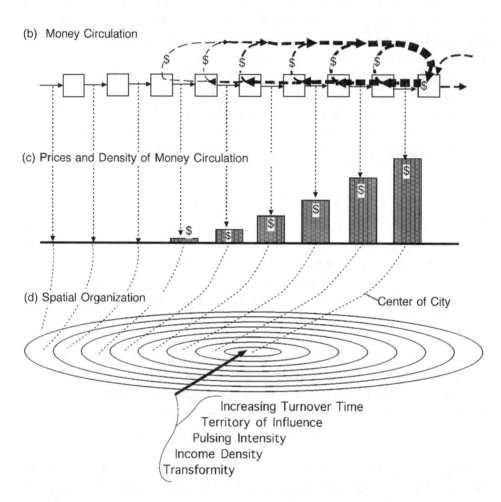

Increasing Turnover Time
Territory of Influence
Pulsing Intensity
Income Density
Transformity

FIGURE 9.14 Circulation of money in the energy transformation hierarchy. **(a)** Aggregated view of the series of energy transformations from rural environment to the center of cities; **(b)** increasing circulation of money with each transformation in which emergy and monetary value are added; **(c)** increase of concentration of money circulation and prices toward the center; **(d)** sketch of the zonal spatial distribution of processes.

is not a measure of environmental work. In the energy transformation chain to the right, after people and money are involved, increased dollar value correlates with increased emergy.[9]

As suggested with fig. 9.14d, the circulation of money converges toward the center and disperses on its return outward. This flow is similar to those of the materials and information described in previous chapters, but the circulation of money is a countercurrent to those flows. The concentration of money flow per area increases toward the center of cities, where everything has to be bought. In rural surroundings, more inputs to people are free. For example, water and waste disposal cost less, so less money circulates there.

GLOBAL ECONOMY AND INTERNATIONAL EXCHANGE

Trade between nations and areas is one of the mechanisms of self-organization that maximizes empower. Resources and products that are abundant in some areas but limiting in others are mutually stimulating when exchanged. Equitable trade also joins areas and nations into a more unified system that helps maintain peace. Whenever people have tried to isolate their economies with taxes on trade (tariffs) to prevent outside competition, their economies have become depressed because they did not manage for maximum empower.

Thus free trade became public policy, an ideal based on the assumption of equitable exchange. It was assumed that buying and selling with a common monetary unit, such as the international dollar and global market prices, represented equal value in exchange. But free trade made developed countries rich, with high standards of living, leaving less developed countries devastated because money does not evaluate the real wealth in raw products correctly.

Whereas real wealth benefits of trade are not proportional to the money paid, trade balances can be evaluated with emergy and emdollars. As emphasized in fig. 9.14, more money circulates but buys less in cities. In developed nations that are more urban, a dollar buys less than in the undeveloped areas, where people get more of the real wealth life support directly from the environment without circulating money. Figure 9.15 compares the exchange of real wealth when $1 circulates between the United States and countries such as Ecuador and Brazil. A trade or a loan that is repaid returns four times more real wealth to the United States than is received in exchange by the more rural nation. For example, shrimp from estuarine ponds of Ecuador enrich the United States while draining the emergy that used to benefit the local economy (Odum and Arding 1991). No country can pay 400% interest on a loan, and consequently debts often are defaulted, which upsets banking and trade. If past loans are reevaluated using the emergy/money ratios of each country, it may be found that the returns on loans sometimes were overpaid. Efforts to exchange debt for setting aside conservation lands also paid too much

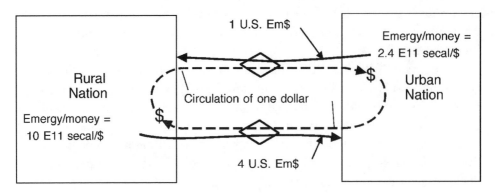

FIGURE 9.15 Unequal exchange of real wealth when a dollar circulates between a rural nation and an urban developed nation with lower emergy/money ratio.

to lending banks because the calculations of equity were based on money, which vastly underestimated the real worth of the lands being conserved.

Another example of uneven trade is the purchase of fuels on international markets in fig. 7.16b (table 7.3). Countries selling oil supply 3 to 12 times more real wealth emdollars than they receive in buying power. Oil-supplying countries would develop their economies, capital, and circulating money faster by using more of the fuels at home rather than selling them.

The present system of trade is not maximizing global empower because so many countries are losing their natural and economic capital and producing much less than they might. Instead of free market policy, which allows extra financial investment power of developed nations to strip less developed nations, international equity treaties could evaluate all aspects of exchange, such as trade, loans, foreign aid, and exchanges of people, adjusting emergy equity between countries each year. All countries would become more prosperous and symbiotic and help maximize global empower. Chapter 10 has more on the global hierarchy and the distribution of powers.

International Interface of Nations

Each nation exchanges real wealth, money, people, and information with other nations. Figure 9.16 shows the international interface of the United States. The money going overseas tends to balance money returning. As soon as dollars accumulate overseas, their value to international currency exchangers decreases. Then more dollars are spent to buy U.S. products, which tends to use up the excess. In fig. 9.16 the balance of trade between the United States and the outside countries is negative (0.69−0.86=−0.17 trillion $), but in emergy terms, there is a large positive balance (9.3−1.5=+7.8 E23 secal/yr). Dividing by the emergy/money ratio, the benefit is (+7.8 E23 secal/yr)/(2.7 E11 secal/$)=2.88 trillion emdollars. Developed nations receive much more real wealth than they export or pay for.[10]

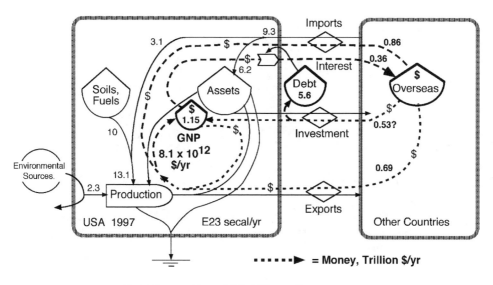

Total Emergy Use = 21.6 E23 secal/yr

$$\text{Emergy/Money} = \frac{(10 + 3.1 + 2.3 + 6.2) \times 1 \text{ E23 secal/yr}}{(8.1 \text{ E12 \$/yr})} = 2.7 \text{ E11 secal/\$}$$

FIGURE 9.16 Systems diagram of the economy of the United States in 1997 showing flows and storage of real wealth. Money circulation (dashed lines) and money supply (storage tank symbols) are boldfaced.

Economic and Emergetic Roles of Nations

There is now a single global economy, with money speeding electronically around the world through a complex network of business and finance. Figure 9.17a shows a simplified summary of the three main sectors of the global economy in 2000. Each nation is different in its emphasis on one or more of the global functions shown. On the left is the production from oceans, natural areas, agriculture, aquaculture, and forestry based on use of renewable resources. Countries that have large contributions of this type are New Zealand, Argentina, and Australia. In the center of fig. 9.17a are contributors of nonrenewable reserves of fuels and mineral materials, such as the oil-supplying nations.

To the right are urban centers of information and finance that use the resource emergy to generate high technology and services for the other areas. Examples are the developed nations of Switzerland, Netherlands, and Japan. Very large countries such as the United States, China, and Russia have some areas with each kind of operation. However, no nation is self-sufficient. Forty-three percent of the emergy basis for the U.S. economy comes from outside (fig. 9.16).

Each country develops an economy tuned to its own resources plus what can be added through trade. Because much of the world's emergy is based on fossil fuels, the

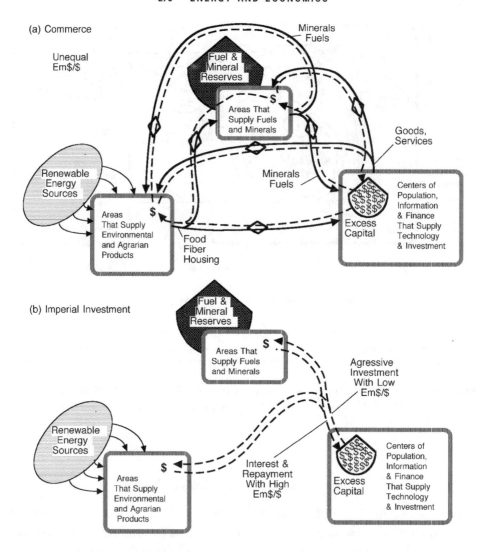

FIGURE 9.17 Main kinds of functional areas of earth's global economy, including the environmental–agrarian sector using renewable energy, the fuel and mineral mining sector, and the centers of accumulation of real wealth, population, information, and finance. **(a)** Mutual exchange of products and the countercurrent of money; **(b)** pathways of aggressive investment causing surge of inequitable economic development, a form of weedy overgrowth accompanied by inequitable emdollar exchange and defaulted loans.

first trade strategy for all nations should be to trade for and use the high net emergy yield of fossil fuels.

Imperial Capitalism

Because more real wealth flows into urban areas, capital is also generated there without loss of money's buying power. By the year 2000, with global corporations and international financial organizations, global capitalism was free to maximize profit. National money of developed nations, when converted to international dollars, has low emergy/money ratios (high em$/$), which gives the money greater buying power when converted to money of the less developed areas. Aggressive investments (fig. 9.17b) develop waves of enterprises using local resources, exporting the raw products, or using labor without paying full costs. Because of the larger emergy/money ratios of the less developed countries, the interest and repayments in emergy terms were several hundred percent higher than the original investments.

The emergy flows to developed urban centers, through favorable terms of trade and by means of interest payments, accelerate capital development, from which aggressive imperial investments are directed into the other areas by agencies of global capitalism. Global investing bleeds net emergy benefits from less developed areas to developed areas because of the imbalance in emergy/money ratios explained in fig. 9.15.

By analogous comparison with ecosystems, the global economy in 2000 was like a massive weed overgrowth using up undeveloped reserves. Because of the emergy matching principle (chapter 4), maximum global empower based on fossil fuels requires the fuels to be matched with emergy of lower transformity, which helps strip the rural areas. However, the excess of unused fuel reserves in the year 2000 was beginning to disappear, causing fuel and electric power prices to rise. Thus the world economy began to shift to a more equitable, cooperative mode that maximizes economies as the overgrowth mode loses its advantage.

Some very small states and nations have policies encouraging international tourism. These areas use natural and historical assets as attractions to draw people who have money to spend on vacations. The money they spend can buy goods, services, fuels, materials, or other real wealth. However, in many cases the visitors pay only for some local services at minimum wages, whereas they carry out, dilute, or destroy more real wealth of the products they consume and the environmental aesthetics they encounter. With decreasing emergy in the area, prices rise, and local people lose control of their lands and livelihoods. See chapter 12 for guidelines on making tourism mutually beneficial.

The emergy–monetary evaluation of each nation, like that summarized for the United States in fig. 9.16, suggests what international policies are desirable to maximize national emergy use and that of the world as a whole. When emergy evaluation is better understood, such policies could help make the world energy–monetary hierarchy be more appropriate, without the extremes of rich and poor nations seen at present.

Taiwan made a remarkable transition to a fully developed economy in 25 years with policies maximizing its empower use. Figure 9.18 shows the energy hierarchy of the island centered in the urban district of Taipei. Agricultural products were 34% of the economy on an emergy basis. These environmental products were consumed at home, thus benefiting Taiwan with high emergy/money ratios. Only high-value manufactured products were sold. The foreign exchange was used to purchase fuels and other raw materials with high emergy/money ratios. People were sent abroad for education, bringing back information and technology. Mon-

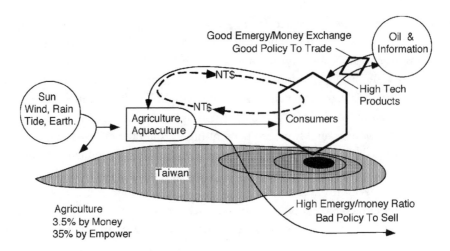

FIGURE 9.18 Sketch of Taiwan summarizing the economy and its energy basis in rural resources, imported oil, and information. Phrases indicate the international polices that caused rapid economic development.

etary capital and capital assets were generated faster by increasing home emergy use than by selling basic products.

Global overview models such as fig. 9.11 show that the economic system's growth is a frenzied and unsustainable pulse of achievement based on nonrenewable resources. Simulations of these models show that descent to lower levels of emergy use and money circulation will follow. See chapter 13 for suggestions for the prosperous way down.

SUMMARY

This chapter introduces economics by showing how markets and the countercurrent of money are mechanisms for self-organizing society for maximum empower. Rather than deriving economic cause from human choices, a systems view suggests that human culture adapts its attitudes and choices to fit the reality of the energy transformation system and its hierarchy. Circulation of money is efficient and useful in the intermediate scale between the less concentrated environmental sectors and the widely shared information sectors of public opinion. Classic models of microeconomics and macroeconomics are given additional dimensions by showing how they process real wealth measured in emergy–emdollar units. Growth capitalism is identified as a large-scale analog of weed overgrowth. The destructive aspects of free trade are explained by the different emergy/money ratios in nations with different degrees of economic development. Emergy evaluation provides the information needed to set policies that may result in more international equity and at the same time more total production and use of real wealth over the earth as a whole. The challenge ahead is to prepare policies for economic descent without crashing (see chapter 13).

BIBLIOGRAPHY

Note: An asterisk denotes additional reading.

Boulding, K. 1962. *A Reconstruction of Economics*. New York: Science Editions.

*Costanza, R. 1981. Energy analysis and economics. In H. E. Daly and A. F. Umanak, eds., *Energy, Economics, and the Environment*, 119–146. Selected Symposia of American Association for the Advancement of Science, 64. Boulder, CO: Westview.

Costanza, R., R. D'arge, R. de Groot, S. Farber, M. Grasso, B. Hannon, K. Limburg, S. Naeem, R. O'Neill, R. Paruelo, J. Raskin, R. G. P. Sutton, and M. van den Belt. 1997. The value of the world's ecosystem services and natural capital. *Nature* 387: 253–260.

Galbraith, J. K. Lecture notes, LBJ School of the University of Texas.

Jansson, A. M. and J. Zucchetto, eds. 1978. *Energy, Economic and Ecological Relationships for Gotland, Sweden: A Regional Systems Study*. Ecological Bulletin No. 28. Stockholm: Swedish National Science Research Council.

Jevon, W. S. 1865. *The Coal Question: An Inquiry Concerning the Progress of the Nation, and the Probable Exhaustion of Our Coal Mines*. London: Macmillan.

Odum, H. T. 1983. *Systems Ecology*, chapter 23. New York: Wiley. Reprinted as *Ecological and General Systems*. Boulder: University Press of Colorado, 1994.

——. 1987. Models for national, international, and global systems policy. In L. C. Braat and W. F. J. van Lierop, eds., *Economic–Ecological Modeling*, 203–251. Amsterdam: North Holland.

——. 1991. Emergy and biogeochemical cycles. In C. Rossi and E. Tiezzi, eds., *Ecological Physical Chemistry*, 25–65. Proceedings of an International Workshop, Nov. 1990, Siena, Italy. Amsterdam: Elsevier.

——. 1996. *Environmental Accounting, Emergy and Decision Making*. New York: Wiley.

Odum, H. T. and J. E. Arding. 1991. *Emergy Analysis of Shrimp Mariculture in Ecuador*. Working Paper. Narragansett: Coastal Resources Center, University of Rhode Island.

Odum, H. T. and E. C. Odum. 2000. *Modeling for All Scales*. San Diego, CA: Academic Press.

Odum, H. T., E. C. Odum, and M. Blissett. 1987. *Ecology and Economy: "Emergy" Analysis and Pubic Policy in Texas*. Policy Research Project Report #78. Austin: Lyndon B. Johnson School of Public Affairs, The University of Texas.

Simon, J. I. and H. Kahn. 1984. *The Resourceful Earth: A Response to Global 2000*. Oxford, England: Basil Blackwell.

Sundberg, U., J. Lindegren, H. T. Odum, and S. Doherty. 1994. Forest emergy basis for Swedish power in the 17th century. *Scandinavian Journal of Forest Research*, Supplement No. 1.

Zucchetto, J. and A.-M. Jansson. 1985. *Resources and Society*. Ecological Studies #56, New York: Springer-Verlag.

NOTES

1. Traditionally the contribution of energy was studied in the field of resource economics, often emphasizing mining, agriculture, forestry, and fisheries. More recently, the field of eco-

logical economics, with additional journals, has included the contribution of ecosystems (*Journal of the International Society for Ecological Economics*). The author's development of energy economics since *Environment, Power, and Society* was published involved energy evaluation, analysis, and computer simulation of models with the circulation of money coupled to energy transformations (Odum et al. 1987; Odum 1983, 1987, 1991, 1996; Odum and Odum 2000).

2. When a diagram of the exchange process represents most of the flows of money and commodities of that type that exist, the ratio of flows generates the price, and the arrow can be shown going from the exchange symbol out, delivering the price to control transactions elsewhere. Jevon (1865) discovered that the price of coal was inverse to the quantity available.

3. The gross economic products calculated annually by statistical bureaus of government agencies are made with many definitions and conventions to represent depreciation, trade, and other factors. Some authors have modified the gross economic product by adding in economic evaluations of the natural capital being consumed (Costanza et al. 1997). However, this is double counting. The global economy is already intricately coupled to the renewable and nonrenewable source use that is its basis. The global buying power of money is based on all of the global emergy of renewable resources and natural capital being used.

4. Boulding (1962), an economist, found many similarities between ecological systems without humans and human economies, including the comparison of currency to mineral cycles. He suggested *ecosystem* as a suitable name to include the economy and other aspects of humanity and nature.

5. Simulation of economic minimodels was given in our simulation book, which includes the programs in BASIC and EXTEND on an accompanying disk (Odum and Odum 2000).

6. Galbraith (lecture notes) estimated the turnover of money in the 1980s to be 16 years, which was before the development of faster Internet transactions. The replacement time of Texas highways was estimated as 50 years. Aspects of society of larger scale, such as major structures of cities, skyscrapers, bridges, and the information storages in libraries, have longer replacement times.

7. Sundberg arranged for emergy evaluation of the original silver mining and processing of the 17th century (Simon and Kahn 1984).

8. Gold was concentrated by the rivers in South Africa in two successive geological uplift cycles over a long period so that the earth emergy inputs were large (Odum 1991). Calculations were based on an unpublished 1987 report of a policy research project by Ragu Bhatt with H. T. Odum, *Policy Implications of Gold Emergy* (see Odum et al. 1987).

9. Ann-Marie Jansson and J. Zucchetto (1978; Zucchetto and Jansson 1985) proposed that embodied energy and market values of goods and services might correlate. As explained with fig. 9.13, the relationship is inverse at the interface where free environmental inputs first enter the economy.

10. Analyses such as that summarized in fig. 9.16 have been made on many states and nations, with emergy ratios ranging from 12 E12 secal/$ for New Guinea down to 1 E12 secal/$ for the United States. Published values (Odum 1996) are soon out of date because the emergy/money ratios decrease each year.

CHAPTER 10

ENERGETIC ORGANIZATION OF SOCIETY

WITH THE ability to learn and cooperate, people occupy the top of the hierarchy of earth energy. Networks of humanity include human biology, individuals, families, populations, occupations, businesses, and governments. People process resources (chapter 7), information (chapter 8), and money (chapter 9). In its structure and functions human society has many similarities with ecosystems. Both require inputs from many scales of the energy hierarchy and feed their work back to reinforce their sources. Figure 10.1 shows some of the necessary pathways for people in order of transformity from left to right. This chapter considers human populations, the organization of human society, and its energy basis.

POPULATIONS

Even though control comes from higher levels of social organization, human society first depends on its population and their life cycle to provide order. The number of people is a balance between reproduction and mortality. In any local area there is also immigration and emigration.

Models of Population Number

Figure 10.2 diagrams some of processes affecting a population in different combinations. Each model includes a typical computer simulation for that design.

In fig. 10.2a there are no energy limits. Population growth is autocatalytic and exponential and can be represented as a constant gain amplifier, with an intrinsic reproductive tendency. In 1798 Thomas Robert Malthus predicted that if reproduction continued at a constant rate per individual, the acceleration would cause the food resources to fail abruptly, and the population would crash as a result of poverty, malnutrition, and disease.[1] The exponential growth graph is sometimes called Malthusian growth.

Figure 10.2b is a "wildlife" model of population supported by an inflowing renewable energy source that limits growth. The population draws from a small food

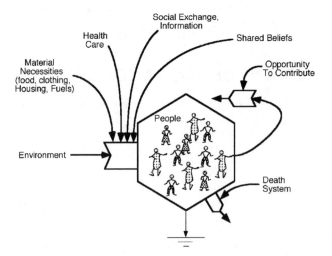

FIGURE 10.1   Necessary inputs to people.

reserve that is being renewed by regular energy inflow. Simulation of growth start-ing with a few individuals may overshoot, causing an oscillatory transient.

In fig. 10.2c energy for growth is unlimited (availability constant), but the popu-lation becomes limited by crowding. For example, dense populations of microbes or animals developing in a limited space can accumulate toxic wastes and inhibit reproductive behavior. Mathematically, these interactions increase quadratically (in proportion to the square of the number, kN^2). The model is called *logistic* and levels off.

Crowding accelerates the spread of contagious disease (e.g., the hoof and mouth disease epidemic in Britain in 2001). The very large energy used in modern health care, preventive medicine, and veterinary medicine has raised the threshold of population concentration at which diseases become epidemic, but catastrophic population crashes can result when these measures are interrupted.

The model in fig. 10.2d adjusts reproduction to fit the available energy. Some mechanism senses the available energy. For example, birds reduce the number of eggs laid when food levels are low (Lasswell and Kaplan 1950); rotifers put out fewer eggs when algal concentrations are less (Edmondson 1965). In short, many natural populations in undisturbed systems practice a kind of birth control (Wynne-Ed-wards 1962) in which reproduction is a function of the energy budget.

Figure 10.2e is a pulsing model that uses a slowly renewed energy source and reproduction stimulated by interactions within the population. When resources are initially abundant, the autocatalytic feedback to increase reproduction is co-operative and quadratic (proportional to the square of the number, kN^2). Growth in the United States was quadratic (Von Foerster et al. 1960) after the onset of the fossil fuel–based Industrial Revolution. The simulation in fig. 10.2e pulses, pass-ing rapidly from rapid growth to descent as the immediately available resource is consumed.

FIGURE 10.2 Models relating population growth and maintenance to available energies. **(a)** Unlimited exponential growth; **(b)** growth on a storage supplied by steady energy flow; **(c)** autocatalytic growth on unlimited source with quadratic losses; **(d)** model with reproductive rate programmed to be proportional to available energy; **(e)** model of pulsed population growth based on previous accumulation of slowly renewed energy reserves; **(f)** population growth on a nonrenewable energy reserve with energy of assets used to reduce mortality and inhibit epidemics. Equations for the models follow. Many others were published elsewhere (Odum 1983).

(a) $dN/dt = k_1N - k_2N$.

(b) $dF/dt = J - k_1FN$; $dN/dt = k_2FN - k_2N$.

(c) $dN/dt = k_1SN - k_2N^2$.

(d) $dN/dt = f(A)k_1N - k_2N$.

(e) See appendix fig. A10.

(f) $dN/dt = k_1(A/N)N - k_2N(1 - k_3A) - k_4N^2(1 - k_5A)$; $dF/dt = -k_0FNAR$; $dA/dt = k_6NAR + k_7AR - k_8(A/N)N - k_9N(1 - k_3A) - k_{10}N^2(1 - k_4A)$; $R = J/(1 + k_{11}FNAR + k_{12}AR)$.

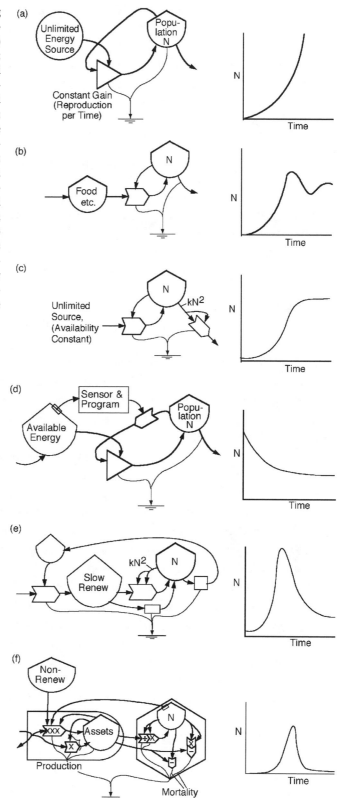

The population model in fig. 10.2f has energy pathways to reproduction and two pathways that represent the action of medicine and public health to inhibit death rates. As resources become less available, reproductive rates decrease, and mortality rates increase, including the quadratic pathway representing epidemics. The simulation shows how the shift from population growth to population descent can occur in a short time. A current example is the AIDS epidemic affecting many countries in Africa.

Human Populations

When most of the energy available to society came from agrarian food supplies, reproduction often was curtailed by famine and the diseases that accompany physiological weakness. Now that the energies available to developed societies are in great excess, food has not been limiting in much of the world, and in poorer countries welfare programs help ensure adequate nutrition. As a result, population growth curves have been steep and Malthusian.

Maximum reproduction rate sometimes has been considered the criterion for natural selection's survival of the fittest (e.g., in Darwin's early writings). But maximizing reproduction is useful only in the overgrowth stage of early succession. For later stages of organization, resources are diverted from reproduction to develop organization, diversity, information, efficiency, and large-scale needs. This is exemplified in Darwin's later writing. Lotka's criterion for fitness (chapter 3) is contributing most to the maximum empower of the *system,* and this depends on the stage of the growth cycle.

As explained with figure 3.13 and chapter 7, the economy as a whole may be in an energy-controlled pulse that may turn down soon (chapter 13). The big question is whether population reproduction rates will decrease in proportion.

Human demographers found that human reproduction in the last two centuries rarely seemed to be related to locally available resources. While society as a whole was accelerating energy use and development, reproductive rates, even among the poorest, were responding to growth opportunity images. However, in later stages of urban development, people become so involved in occupations and complex interactions that their reproductive energy is diverted away from reproduction. In other words, people may reproduce according to energy availability and the image of growth of their economy, not according to their individual resources. Because our energy supply was still expanding in the final years of the second millennium, it may be that the system of society had an energy sensor operating, which was still indicating unlimited growth because of high levels of cheap fuels on the world markets.

World birth control initiatives now being undertaken lack the automatic energy sensor for quantitative control found in many natural populations. For human society, some regularly calculated statistics on empower input could be used to influence the number of offspring in any year, as suggested in the model in fig.

10.2d. One suggestion is to use educational financing, with reproduction allowed only after resources have been assembled for the education of the new offspring. To make population regulation work in a way that is consistent with democratic principles may require innovations in social institutions.

Any policies or events that bring population down as fast as the resources decline will sustain the present standard of living (emergy per person). See chapter 13.

Human Life Cycle

Millions learn from the study of Shakespeare in school that "all the world's a stage, and all the men and women merely players" (*As You Like It*, II, vii, 139). The roles that people play from birth to death are progressive steps in an energy hierarchy (fig. 10.3a). While territory of support and influence gradually increase, the number of

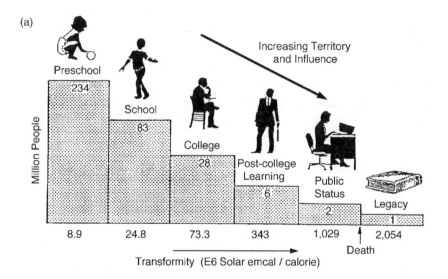

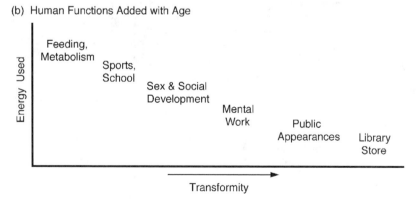

FIGURE 10.3 Life cycle of human beings passing through the hierarchy of age and experience. **(a)** Distribution of population of the United States by stage, age, and transformity; **(b)** stages in the human life cycle in which functions of higher transformity are added.

people reaching each stage of experience and education decreases. Notice the behavioral functions that appear with age that are superseded by larger-scale roles at later stages (fig. 10.3b). After learning about food and waste processing, there is elementary education and making friends, followed by passion about sex, and learning an occupation before beginning a later life stage of net societal contribution.

Transformity of people and their main interactions with society also increase with age. The emergy contribution of a person at each stage may be estimated by the energy metabolism multiplied by the hours of effort times the transformity of the person's level of education and experience from fig. 10.3a.

Health Care

The public health and health care sector of society consists of doctors, nurses, and other health care workers, hospitals, and health maintenance organizations (fig. 10.4). The costs are borne by a mosaic of sources: charges to patients, health insurance, payments from employers to insurers, and appropriations based on taxes from local, state, and federal government. The energy and monetary costs of keeping a population healthy have risen because the populations are larger, their interpersonal exposures have increased, the age of the population has increased (more maintenance required with age), new progress with complex and costly medical technology, new drugs, increasing number of people with birth defects kept alive,

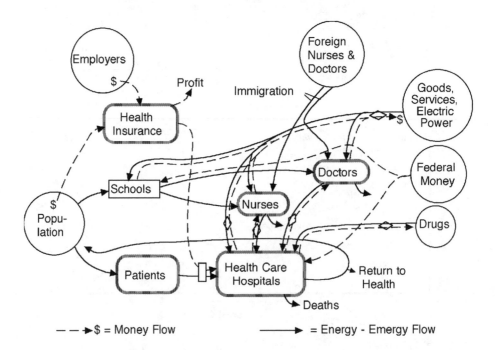

FIGURE 10.4 Model of the U.S. health maintenance system, modified from Logan (1998) (heat sinks omitted).

rising expectation for care and length of life, and capture of health institutions by for-profit capitalism.

Institutions that provide regular services can be government owned, private profit, or private nonprofit. From chapter 9, recall that profit making is appropriate where there are unused resources that can support expanding development. Efficiency of return on cost is sacrificed to accelerate growth. But once established, health maintenance systems could sacrifice profit for efficient service. Recently, there was an international struggle over the balance between profit and efficiency in health care. Although sometimes justified as the means for research, profit tends to become the unearned income of stockholders. Charging an extra percentage of cost to pay for medical and drug research without profit might make more sense.

The Future of Health Care in the United States[2]

The current system of health care in the United States is one of entrepreneurship. Although federal funding and support of the health care system have steadily increased over the past 50 years, the drivers of the system still consist primarily of profit-making entities such as insurance companies, hospitals, and physicians. The profit-making motivation of these key entities drives the system to encourage growth in treatment and cost, whether it is warranted or not.

A simulation model of these aspects of the health care system in fig. 10.4 (Logan 1998) reinforces the idea that continuation of the current system over the long term will result in a rapid growth pattern in treatment rates and costs. The model suggests that the greatest inflation will be in the growth of insurance capital reserves and insurance company profits, perhaps because of the position of the insurance industry as an intermediary, where it is first to receive inflows of financial resources that allows it to siphon off profits easily. The model also suggests a large drain on the financial resources attached to hospitals, perhaps because of their inferior position in the system in relation to inflows.

Simulation results of this model suggested that the introduction of managed care, in which physicians and hospitals may receive incentives not to treat patients, resulted in no substantive decrease in the rate of health care inflation. These results have been supported in the last five years. Health care inflation slowed down briefly during the 2 years after President Clinton's attempt at health care reform in 1994 but has since escalated again; it is currently running at approximately 6% (Altman and Levitt 2002). This pace cannot continue, given the limits to growth suggested previously (chapter 7).

Unfortunately, the crowded policy arena for health care currently consists of a very diverse number of organizations and entities who compete heavily. This large number of competitors tends to result in the inability of legislators to effect anything but very small, incremental change in the system. Medicare allowed older adults to receive federal benefits that approximate universal health care; recently children have also been included in state initiatives that provide federal support.

But the majority of people either have insurance-funded health care or are uninsured. Repeated efforts at major health care reform, including the last attempt in 1994, have resulted in failure.

To be successful now and in the future, health policy must address three issues—cost, quality, and access—before some significant change in the system brings catastrophe. Perhaps the future will include a dramatic increase in demand, as might be caused by a plague or bioterrorism that will make radical reform of the health system necessary. Without reform, the present rapidly growing system may overshoot, causing its eventual collapse. The incrementally advancing federal coverage of patients at both ends of the age spectrum may soften the degree of the overshoot. One can only hope that when the need for reform comes, the United States can adapt gracefully and improve on models from other countries using socialized health care or universal coverage.

A better health care system would include more funding for wellness and preventive care and less emphasis on tertiary care. Health disparities based on gender and race would be absent. The financial incentives that drive the current system would be removed. Treatment of patients would be based on relative outcomes rather than profit incentives. Computerized guidelines for each disease process and treatment would streamline care and provide efficiency. There would be a greater focus on public health and more attention given to aggregate community health outcomes. Financial incentives to teaching hospitals to expand graduate medical education would be reduced to slow growth of the physician supply. Financial incentives to other health care professions such as nursing would be expanded to accommodate growth in population with less costly alternatives to the highly educated physician. This would also allow for a multidisciplinary collaborative approach to patient care and would give the patient more autonomy to become the decision maker for the health care team.

Age and Replacement in Ecosystems

Healthy ecosystems have mechanisms for removing the living parts when they begin to falter because of the depreciation of structure and information. Thus the plankton of the sea and the antelope of the plains both are consumed by carnivores, resulting in the recycling of minerals back into the system, which sustains production. For many populations of animals, the next level of carnivores provides a death system that is programmed so as to help maintain the genetic quality and appropriate distributions and density of the prey population (fig. 8.9).

Slobodkin and Richman (1965) brought plankton into the laboratory without carnivores and found symptoms of individual senescence and lower production, privations that would be selected against in the natural environment. The energy required to maintain an effective population is the energy required to support the carnivores. The energy of the units selected against and eaten supports the quality-choosing function (fig. 8.9).

When carnivores or diseases are exerting control of age distribution, they are doing work beneficial to their prey populations as well as drawing energy from them. This is not so different from farmers selecting as they manage their cattle. Thus, the energy diagram in fig. 7.4 showed a reinforcing loop from humans in managing the cattle.

Genetic Depreciation

With billions of dollars spent for human medicine each year, society in the late 20th century appeared to be removing the extra duplication, selection, and system reinforcement that used to eliminate genetic errors and encourage evolutionary progress. Brilliant medical achievements are directed toward saving all offspring, thus eliminating the quality selection process. As the genetic errors are retained, medicine steps in with ever-greater programs and power requirements to repair, support, and permit the errors to continue to the next generations. Each generation needs more emergy because of the accumulation of genetic errors and losses of function.

Deaths

Human beings, like other tightly organized, concentrated structural storages, reach a stage of deterioration as they age. In aging, the requirements for repair and health maintenance are greater than those for replacement (chapter 8). At an earlier time, the human death system used parasites and diseases that could neatly and cleanly remove people when general deterioration brought the energy reserves of the body below certain minima.[3] Many bodies had weak structures in hearts or kidneys that failed long before the rest of the body had deteriorated. This older death system removed weakened individuals, and epidemic disease dispersed populations just as carnivores do in ecosystems. The diseases provided an automatic energy test of the strength to continue life with the medical help of society then available.

But in two centuries of high emergy availability, the global religious–medical ethic has become committed to using all resources possible to preserve life. Medicine and science have greatly restricted the disease method of death without providing a substitute that is humane for individuals. As a consequence of high-emergy medicine, the old, too-efficient death system has been removed. People live longer but often last until they die in misery of cancer or after years of incapacitated immobility, with regret.

Emergy and Health Management Alternatives

Emergy evaluation of health maintenance alternatives can help make health management choices now and in times of less emergy ahead. Currently less emergy now goes into reproduction to replace early deaths, and more emergy goes into medical maintenance. The longer people live, the higher the emergy and monetary costs of medical maintenance. Eliminating the disease control system produced a global population explosion. As a result, we are forced to use more emergy to control births.

There are advantages of older adults to society as long as they can contribute. With the increasing complexity of the power-rich networks, the human computer (educated brain) becomes ever more valuable, with greater and greater empower investments in its training and memory storages. Now individuals begin to be truly unique because combinations of knowledge and training allow them to do things no one has ever done before or may ever do again. The value of individuals who have received large educational investments is vast, but perhaps emergy should not be used to sustain people in vegetative states.

We might wonder whether disease and subsequent death served primitive people well, both for individual lives and for the effectiveness of their role in the system. Is one way any more humanitarian than another, or does it have any energetic advantage? When there is retrenchment of the rich energy sources, the old system may return with a vengeance. Do epidemics and hardships for the handicapped follow? Primitive, disease-resistant human stocks may be available from countries where there has been more natural selection in recent years.

ORGANIZATION AND GOVERNANCE

Social relationships make society an organized structure of cooperative functions instead of a crowded mob. Organized social systems displace those that seek to disorganize, such as anarchy. On each scale, from the family to global society, there are groups of people who support and are led by a more important but smaller leadership group operating at the next larger scale (fig. 10.5). A family has children led by the parents. The military organization has designated ratios of soldiers, corporals, sergeants, lieutenants, captains, majors, colonels, and generals in which transformity increases and numbers decrease with rank. Human organization appears to follow principles of energy hierarchy (chapter 4). Other studies have examined political organization as a system and diagrammed pathways of political power (Lasswell and Kaplan 1950). Here they are given their position in the energy hierarchy (figs. 10.5 through 10.8).

Human decisions open and close the main empower flows of the economy and environment. Control networks exist in mental concepts and relationships, learned and accepted by individuals to guide their behavior. Through votes on committees, acceptance of authority, participation in elections, and other control mechanisms, humans ensure continued flows of their own emergy and that of their institutions. Figure 10.5 shows the general form of the loops of reinforcement in hierarchies of government.[4]

Successful management in industry has a cascade of control spreading out for decentralized action. Top management has high transformity and the empower

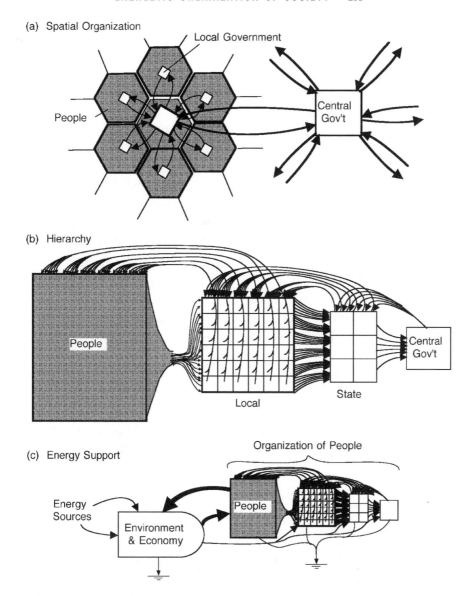

(a) Spatial Organization

Local Government

People

Central Gov't

(b) Hierarchy

People

Local

State

Central Gov't

(c) Energy Support

Organization of People

Energy Sources

Environment & Economy

People

FIGURE 10.5 Superstructure of human social organization that operates environment and economy. **(a)** Spatial convergence and divergence; **(b)** people and units of government arranged left to right in order of transformity; **(c)** energy basis for social organization.

to direct a few large-scale functions, but the main volume of decisions and energy transformation is decentralized in steps down to the leaders of small groups.

Empower concentration is necessary for large-scale actions, but most of the decisions and controls are small scale and are relegated to the smaller units of government and individual decisions. We call the distribution of power control at many levels *democracy*. Thus, emergy-based hierarchical organization resolves the old argument about which society is best, the highly centralized or the entirely decentralized.

Law

As systems become older and complex, relationships can be coded as laws, which are society's means of maintaining a workable network. In systems models, a law is a control pathway, often turning actions on or off.

Using Hohfeld's theory of dynamic law, Hoebel (1961) examined primitive cultures for the rudiments of law. Four types of interpersonal relationships of the Hohfeldian system were used for eight basic relationships. In systems diagrams *rights* are information pathways that control function.[5] The *demand right* of one person to draw a *duty* from another is an energy circuit in which the downstream recipient may elicit some upstream work, but the upstream unit may not interrupt the flow.

The *privilege right* of one person to operate without a *demand* right from the other is an energy flow without connection from the other. It is an insulated relationship only the privileged may change with switching action.

A *power* individual may generate a work flow involving a *liable* individual with or without a contract for return exchange. The liable individual may form a flow only with the switching action of the power individual.

The *immunity* of a unit to any *power* flow from another unit affecting it means that the immune unit has the right to remain insulated, without a network connection.

Whereas smaller groups can be organized with personal relationships, it took the innovation of written law and enforcement to make larger groups stable.

Spread of Innovation

The spread of innovations in society is like biological evolution but much more flexible. Toynbee's ideas (1935–1939) about the sequence of innovation are diagrammed in fig. 10.6. Various relationships are tried in the self-organization process of many groups on a small scale. An innovation that reinforces production and increases empower prevails, and that system expands, capturing more emergy. Then the information regarding that innovation spreads to other areas, and the advantage of the early monopoly is lost. The dominant system loses its supremacy, and the advance becomes general. Innovations sometimes spread gradually and peacefully; at other times, change is resisted, as with civil war.

Roman Empire

Historians often cite the innovative system of laws, slaves, and military legions that sustained organization in the Roman empire. It was the greatest world order ever built on solar energy alone. Figure 10.7 is a highly aggregated diagram of the Roman system. With innovations in organization, military mechanisms, and gov-

(a) Before Innovation

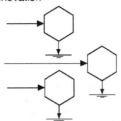

(b) Supremacy Based on Innovation

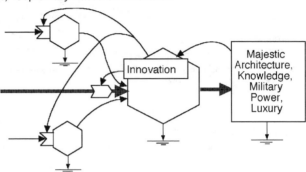

(c) Equality after Spread of Innovation

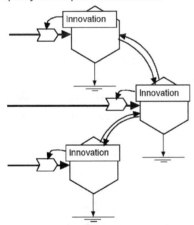

FIGURE 10.6 Systems view of the spread of innovation, after Toynbee (1935–1939). **(a)** Isolated units with low empower; **(b)** monopoly with a new innovation draws high empower and control of other units; **(c)** spread of the innovation equalizes empower support.

ernment service, a reinforcing loop directed enough emergy into organizational work to produce innovative culture, structures, and institutions. Whereas money was used in smaller transactions, grain and slave levies were exchanged on a larger scale and enforced by the Roman legions. The people of Rome were fed with grain levies from provinces such as Africa, whereas other provinces contributed slaves, recruits for the legions, manufactured goods, and literary services. The feedback of organizational services to the provinces generated support for Roman authority. The high-transformity military work of the legions was supported by the conquered

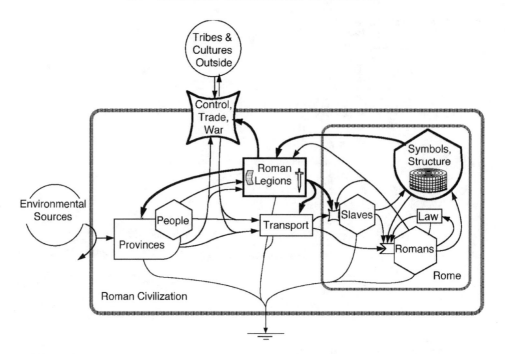

FIGURE 10.7 Systems diagram of the Roman Empire.

lands. The legions prevented the surrounding tribes from draining energies from the main Roman provinces for several centuries.

The fall of the Roman Empire fascinates historians and scientists. Consider how some factors (Gibbon 1962) could have reduced pathway reinforcement on the energy diagram. Strength of discipline in the Roman legions was weakened by loss of a common image of invincible loyalty. The central government was weakened by the competing hierarchy of Christianity, and as Christian rule became absolute, it eliminated polytheism and the information associated with it (the great library at Alexandria). A shift in rainfall belts may have reduced the grain production of North Africa, diminishing the emergy support for Rome, the center of power. A contributing mechanism was the loss of monopoly on innovations (Toynbee's theory in fig. 10.6). If part of the early Roman success was based on their innovations, the spread of this knowledge might have allowed surrounding provinces to organize and begin to take some lateral energy flow formerly directed toward Rome. An emergy evaluation of the invasions of Rome is needed, similar to the one done by Bob Woithe (1994) for the U.S. Civil War.

Organization of Two Units

After interacting to test relationships, two competitive individuals can adopt a plan of work sharing with mutually reinforcing loops, shown in fig. 10.8a. This simple organization eliminates competition, conserves energies, and is able to evolve because the two units help each other and together do more. The design applies

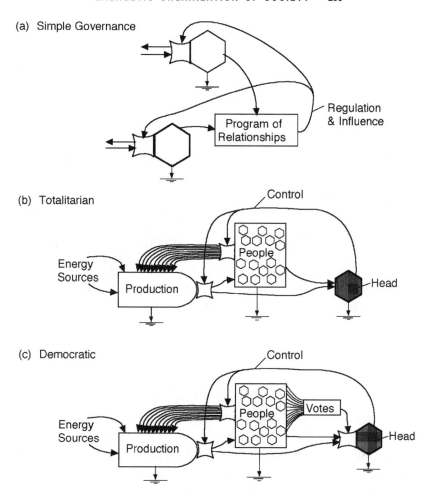

FIGURE 10.8 Diagrams of the flows of energy convergence and control loops for different forms of government.

to many relationships, such as love, marriage, partnership, contract, symbiosis, and business partnership. Primitive examples are the pecking-order relationships among chickens that become organized after a learning period (Allee 1938, 1951). Berne (1964) used a special language of two-way vectors to show types of psychological organization between two individuals.

Types of Governments

Government is a group's system of control and public services. It is a sharing of information that makes a society of people follow its organization pattern. The design in fig. 10.8a for symbiosis of two individuals is a simple government.

Figure 10.8b is a totalitarian design between a head and its supporting group. Lower-level people supply products and services that are controlled by the "head." The products may be diverted from the work and welfare of the people. There is no automatic mechanism to keep the top level efficient in reinforcing the group or its production. Examples are slaves and owners, class control by an aristocracy, government by inherited royalty, theocracy or government by religious authority, and military dictatorships. Without some way to ensure that actions of the head are reinforcing, the system is not self-correcting, and revolutions may occur.

Figure 10.8c demonstrates a more democratic design in which the head is controlled by votes of the majority. Votes can reinforce the controls that work. Empower converges to the head but is distributed out to the group in services. Democracy is an innovation in which control is mutual, passing up and down the hierarchy. Elections and replacement of leaders allow change according to the society's collective image of performance. Systems with individual rights prevail, and the freedom improves the productive performance of the people.

Political "left" and "right" refer to the left and right ends of the political hierarchy (fig. 10.8c). Many people of smaller individual empower on the left support and are controlled by a few of high transformity and influence on the right. The social hierarchy is analogous to the food chain hierarchy of ecosystems and is energetically similar (Odum and Scienceman 1986).

Social Structure and Political Power

The influence of individuals, groups, and political bodies lies in the useful potential energies that flow under their control. Political power does work, gains and manipulates storages of energy, and directs forces. The energetic laws are as much first principles of political science as they are first principles of any other process on Earth. Control is measured by transformity and empower. We can clarify many of the political, military, and international problems of our times by evaluating them in empower units. A government that prevails controls the most empower.

Social animals and humans, through their programs of information transfer and response to each other's information, exert controls and influences on each other's work. Transmissions of opinions are adaptive and produce social structure and actions. In psychological studies, social power and force have long been recognized and described with factors and matrices of different combinations of interaction (Cartwright 1959). Concepts include resistances to forces, opposing forces, and interactions, which have their counterparts in the physical and biological systems. Measures of correlation, frequency, and probability have been used to indicate pathway strength (French 1956). Using energy systems diagrams and emergy evaluation for these pathways unites physical, biological, and social concepts of power.

CLIMAX SOCIETY

Societies of developed nations are becoming mature and diverse. Sequences of human history are like ecological succession. Competition, overgrowth, and simplicity in times of growth are followed by a climax stage of accumulated assets and diversity. Later there is descent. In the mature stage units and functions are complex and interconnected. Figure 10.9 shows similarities between the diversity of species and the occupational diversity of society.

Chapter 8 (fig. 8.10) explains that more energy is required for organizing diversity, but the consequence is also greater efficiency in converting energy inputs into production. Sometimes the greater energy required for organization is said to be an inefficiency and is called Parkinson's law. However, organizational energy is a

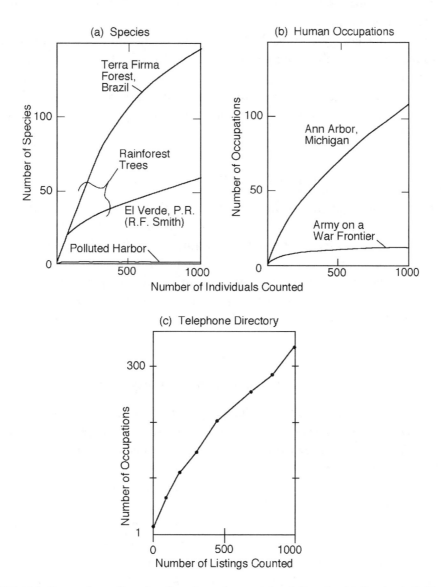

FIGURE 10.9 Comparison of species diversity and occupational diversity. (a) Species–individual curves for ecosystems; (b) occupation–individual curves for human occupations in Ann Arbor, Michigan, in 1964 (Clark et al. 1964); (c) occupational listing curve for human business in a telephone directory in Gainesville, Florida, in 1999 (Odum 1999).

necessity for organizing work that cannot be done by individual parts; we may accept a small power diversion as necessary. The animals in a forest or a sea use energies in their behavioral interactions to perform certain activities in unison. Fishes school, birds divide up a food area by zoning themselves in territories, and insects divide up the feeding times, with some operating at night and some during the day. In forests, the larger hawks and the migrating flocks of animals and birds compose a weak government, integrating the energy flows, cropping excesses wherever they tend to develop, and redistributing the minerals involved in their waste dispositions. In organized society, 20–50% of each citizen's dollar budget goes into supporting the government.

Where little energy is available, there is little energy for complex organization because there is hardly enough for individuals. Centralization of power and individual freedom in using energy were rarely possible before 1800. Humans were able to organize large areas of the world only when large excess fuel inputs became available for integrating systems of communication, government organization, and military outlays.

Images of Public Belief and Motivation

Part of the social nature of human beings as they have evolved so far is the exchange of information by which a consensus of attitudes toward their own collective society is taught, shared, and dramatized with such symbols as flags. Figure 10.10 summarizes the emergy-based social hierarchy supporting Asian societies. The dragon symbolizes the unity and power of society that is celebrated with fireworks at festival time. In some countries there are military parades; in others there are parades of commerce and customers celebrating an image of consumer power.

Competition and Sports Extravaganzas

Sports are a part of most cultures, with many useful functions: maintaining healthy physiques, teaching how to improve performance through competition, and teaching social cooperation toward group goals. These benefits are important in the education of the young.

As systems of society pass through stages of succession into complex mature stages, sports achieve a central importance to the people who share and revel in public games as spectators. Sports arenas are found in the Mayan ruins, in the stadiums of ancient Rome, and increasingly in the priorities of budgets and time of developed nations. Sports winners become national heroes and highly paid celebrities. Sports events and images become a way for people of greatly different cultures and backgrounds to share common discourse. Integration of sports in the southern United States was a major force accelerating racial integration after World War II.

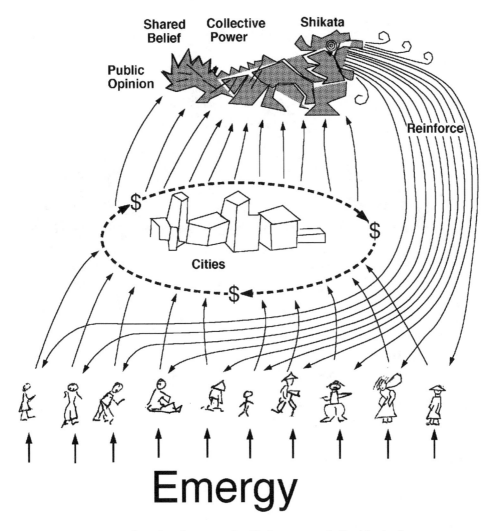

FIGURE 10.10 Society and its shared images of unified power, symbolized by the dragon.

The nature of the sports system in the mature society is suggested in fig. 10.11. The sketch shows the spatial hierarchy of converging input and diverging aftereffects as people assemble and disperse. It is an information-sharing phenomenon that appears to reinforce social cohesion. It provides a fairly harmless outlet for some of the competitive behavior that is not needed when society is in its complex mature stage. Championships, Olympic games, and Superbowls are high-level storms of society that absorb the group focus and fanatic attention that might otherwise go into destructive competition and war. In the highly energized society of the developed world, sports generate social eddies analogous to the turbulent swirl of fluids in stormy seas. Repeated pulses of excess energy may generate a natural selection of designs that improve empower on the average (chapter 4).

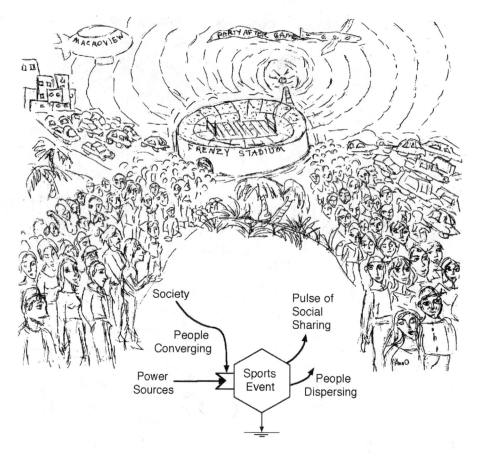

FIGURE 10.11 The system of high-emergy sports extravaganza, a pulsing surge of social sharing (Ann Odum).

Within our present consumer society organized by a market economy, commercial advertising has been coupled to the sports system, causing extraordinary money to go to television sports advertising and celebrity salaries. Clearly there is pathology when entertainment enterprises take more money and emergy than necessary away from the more essential tasks of making everyone adequately educated and employed to contribute to basic production. Sports heroes are a new kind of aristocracy. Some of these celebrities are beginning to feed their wealth back to society as a form of noblesse oblige.

Emergy Diversion and Destruction

When work is done by one process on another in productive interactions, large amplification factors are important if the first is to control the second. When the work operation is one of disordering the other system, it is operating in the direction of the natural tendency for disordering and dispersion of the second energy law. Less potential energy is required to disorder than to order. An illustration of this is the role of light energy in plants in two processes: photosynthesis, in which light builds up structure, and photorespiration, in which light captured by rhodopsin is harnessed to help the disordering of respiration.[6] Kowallik (1967) found much less energy required for disordering than for reordering.

In the high-emergy developed world, resources are easily diverted from the useful pathways into endeavors that appear to be destructive. Diversion of emergy from useful pathways by crime is analogous to the short circuits of electrical systems. To prevent losses, insulation is used in electric systems and hot air ducts to keep energy in useful pathways. In society, police institutions and crime prevention are analogous to preventive insulation and are observed to increase with energy levels. As the emergy within societies increases, it takes more emergy for police to counter the destruction by criminals and terrorists.

High-transformity pulses have great destructive power when applied to disorder high-emergy structural storages. Helping available energies to disperse more rapidly creates momentary and destructive autocatalytic storms. However, destruction often is used by one system to replace another. Whenever disordering is adaptive, it is cheap. For example, explosives have great effect in removing city buildings during urban renewal before reconstruction. Civil revolutions are examples on a larger scale. By using destructive means, terrorism allows a few people with high-transformity weapons and motivation to have major effects on society as they set off secondary physical and social eddies. Whether destruction is useful depends on its effect on the social empower examined on the larger scale of time and space.

Uncoupled Energy and Social Eddies

Like murder, so the expression goes, energy will out. A flow of energy either goes into storage or drives processes. If it goes into storage, it builds up its ability to drive more processes in that place. If a flow of energy does not do useful work and does not have feedback loop controls, that energy flow may do work that is detrimental and explosive. Laws of energetics limit not only the nature of useful work but also our ability to discard potential energy quietly.

In electrical systems, short circuits are flows of energy not harnessed to useful work. Instead, the rapid dispersion of energy into heat forms heat gradients and does detrimental work such as melting the insulation of wires. There is deafening noise, clouds of steam, and managerial confusion when an electric power plant is suddenly disconnected from the load of its powerlines. In forest fires, sudden release of heat develops strong thermal winds, a pattern that is similar to river systems in which a waterfall throws its potential energies into geological work on the riverbed. In all of these examples, the proffering of potential energy leads to dispersive processes that also do work as they shunt away energy. Available energy cannot be released without doing work. Hence, any design or energy network must provide a controlled and useful channel for the energies supplied or the system will find its own circuit, one that may be inimical to society.

In social systems in the 1980s, money was given for welfare support of idle youth, for the unemployed, and in other arrangements, without coupling the people aided to useful work. Then, as in the short circuit or the forest fire, the energy emerged in useless social eddies, additional reproduction, group agitation, disorganized play, theft, and paramilitary adventures.

In natural pine forests, fire is part of the stable system as long as it is regular, helps release minerals, and restarts the succession of tender plants. Social eddies can become part of a stable system if they are regular and serve some useful feedback-reinforcing functions. Simple war and energy-using defense games may have shunted excess energy in the past (e.g., fig. 7.1). If means for the regular use of excess energy storages are not provided, they may accumulate until fire, explosion, or social disorder is so great that it is destructive to the whole system. The elimination of fire from pinewoods sets the stage for disastrous fires 20 years later. Eliminating energy outlets from energy-rich economies may cause the buildup of a potential for greater damage at a later time.

Learning Pathology and Addiction

The spread of human society as the world's information processor evolved with the great learning capacity of the human brain. Patterns of information processing that generate useful functions are rewarded with neurochemical reinforcement that establishes pathways in memory for frequent reuse. Thus people store information as they learn useful functions, including those within the body and the behavioral interactions that are reinforced by society.[7] Addictive drugs are chemicals that steal and divert learning mechanisms and attention by reinforcing pleasure pathways and priorities for more drug use and access. The addictive drugs are another case of energized destruction.

Use of illicit drugs has reached extremes in developed society, diverting emergy, corrupting governments, overloading the prison system, and fostering disrespect for government. Most people acknowledge that such habit-forming drugs as tobacco, alcohol, marijuana, heroin, cocaine, and many others are destructive. These are high-transformity substances with great amplifier impact on neurobiological learning systems. Most agree that becoming addicted often ruins health, causes financial ruin, destroys social relationships, and stimulates crime. It may be that as society moves from the simple focus on growth to the complexity of climax, many people lose motivation, making them drug susceptible.

Society has vacillated in its approach to drug use. With addictive alcohol, society has sometimes prohibited sales, sometimes legalized alcohol, and at other times limited the times and places of use. Legalizing drugs may make it easier for the young to become addicted, but money can go for education and treatment. Making the drugs illegal leads to huge black markets, with international smuggling and incomes as great as the budgets of governments. The underground economy and the economy of drug enforcement have become major parts of the economy of many nations. Production, distribution, and consumption of addictive drugs have increased with energy use, global transportation, and excess wealth in developed countries. Addictive drug use may be expected to diminish in a lower-emergy world.

Whenever something destructive becomes important, one needs to look to the larger scale to see what it may be reinforcing. The flow of drug money from developed to undeveloped areas may be one of the ways in which global self-organization is equalizing emergy exchange. For example, the emergy that can be purchased with drug money returning to Colombia may exceed the emergy in the cocaine transfer to the United States.

INTERNATIONAL EMPOWER

Society on the global scale is a family of self-organizing states and nations. Each society is controlled by the global earth processes, the international circulation of money, the shared genetics of humanity, and the ideas and ethics of international behavior learned and globally shared. By the end of the second millennium the explosive wave of development of the Industrial Revolution culminating in the information revolution was uneven, with great differences in population, empower, and emergy storage. Table 10.1 compares indices of the nations.

Unequal Exchange

As explained with fig. 9.15, each country has a different emergy/money ratio, which determines how much of its real wealth goes abroad with trade and fiscal exchanges. Included in table 10.1 is the ratio of emergy received to emergy given in international trade. In the recent past, populations reproduced faster in the undeveloped countries and migrated to the developed nations, representing an emergy drain from the country left behind. In addition, resources from the undeveloped country often were exported to the developed nations, where they stimulated the economy. The combination of the "brain drain" and exports of resources often left the undeveloped country impoverished.

The accumulation of high emergy in developed countries allowed the capital that was accumulated there to be directed in ways that autocatalytically drained the other countries. This was done with the help of colonial military invasions in the 19th century and with economic imperialism in the 20th century. It was fostered by the belief that price measures real value and that loaning money is a benefit to less developed areas.

Organizations at great distance from their source of control develop tensions if the reinforcement loop between the people and distant governance fails. For example, in the early American colonies, self-organization on the frontier reinforced local needs, whereas the English Crown tried to sustain the inequitable exchange

TABLE 10.1 National Emergy Indices

Nation	Empower (E14 secal/ person/yr)	Import/Export Ratio	Emergy/Money (E11 secal/$)
Liberia	62	0.15	82.0
Dominica	31	0.84	36.0
Ecuador	26	0.12	21.0
China	17	0.28	21.0
Brazil	36	0.98	20.0
India	2	1.5	15.0
Australia	141	0.39	15.0
Poland	24	0.65	14.0
Soviet Union	38	0.23	8.1
United States	69	2.2	6.2
New Zealand	62	0.76	6.0
Germany	67	4.2	6.0
Netherlands	62	4.3	5.2
World	10	—	4.7
Spain	14	2.3	3.8
Switzerland	29	3.2	1.7
Japan	29	4.2	1.2

Note: Based on tables 10.7, 10.8, and 11.1 in *Environmental Accounting* (Odum 1996). The indices are from national evaluations from different years between 1986 and 2000.

of empower favoring British needs in its trade with the colonies. Thus two feedback loops were competing for control, one for local benefit and one by and for the governing parent. Protests such as the Boston tea party started the American Revolution. In the subsequent empower testing of war, there was enough empower in the 13 colonies, supplemented with empower aid from the French, to overcome the empower that Britain could transfer easily across the ocean. In 1864, during the American Civil War, the empower of the northern states eventually exceeded that of the south (Woithe 1994).

Relationships of Nations

At the global scale, nations have some of the properties of businesses competing in an economy or species in an ecosystem. Minimodels of some parallel relationships are shown in fig. 10.12. Simple competition for resources (fig. 10.12a) and active competition with war (fig. 10.12b) prevail when energy use is expanding. Competitive exclusion occurs because overgrowth of the more rapid colonizers maximizes empower at that stage. Later, when few unused resources remain, trade and specialization develop (fig. 10.12c), which makes the two units cooperatively coexistent.

Military Controls and War

In war, disordering disrupts the military systems, and much damage is accomplished with high-transformity weapons. Perhaps the prevalence of war in some earlier systems for such purposes as territorial organization and renewal was useful. Now, however, with vast new quantities of energy injected into war programs, the system has shifted to one of waste. A war zone develops a high emergy flow because it receives the emergy from both systems for competition and for control. Emergy evaluation was made of several wars.[8]

A forest edge is another high-emergy boundary and is a zone of conflict. It has more birds that enforce territories and use the emergy of both forests and fields. Some complex ecological systems that are adaptable and resilient have *ecosystem soldiers*. These are specialists whose biological behavior is programmed to destroy the outside agents that are dangerous to the parent ecosystem. The kelp beds of the sea raise kelp bass, which are programmed to concentrate on urchins that would otherwise cut the kelp and float it away. McClanahan (1990) showed that East African coral reefs had triggerfish regulating the urchins. After the triggerfish were overfished, urchins destroyed reef complexity. Similarly, the fishes of tropical reefs graze back turtle grass in the vicinity of corals, thereby preserving the reef.

Behavior is a mechanism for self-organizing. Some human behavior makes sense only at the system level. Ardrey (1966) recounted the behavior programs of animals and people distributing themselves over the land. Before the fossil fuel era, "the territorial imperative," as Ardrey called it, distributed human populations evenly in relation to the energy incoming from the sun, which was roughly proportional to area. Now, however, the new extra energies of fossil fuel are not coming in on an area basis but are available at any place that can receive a tanker.

The ability of nations to incorporate new power and transform that power into military power has grown along with other structuring skills. In earlier times of low energy, wars helped expand organizations. For tribes engaged in border conflicts, the wars helped maintain territories, achieving a division of the resources, which was a form of spatial organization. When these wars were decisive so that one group came under the control of another, an organizational structure of greater dimension was imposed. War at this stage channeled emergy into organization.

In the late 20th century, international agencies and peace forces emerged, using police and peacekeeping actions and education to control conflict. Emergy evaluation is needed to determine which system results in higher empower and which takes more emergy to support: a world of decentralized power politics with boundaries established by defense or a world with boundaries maintained by unified international enforcement.

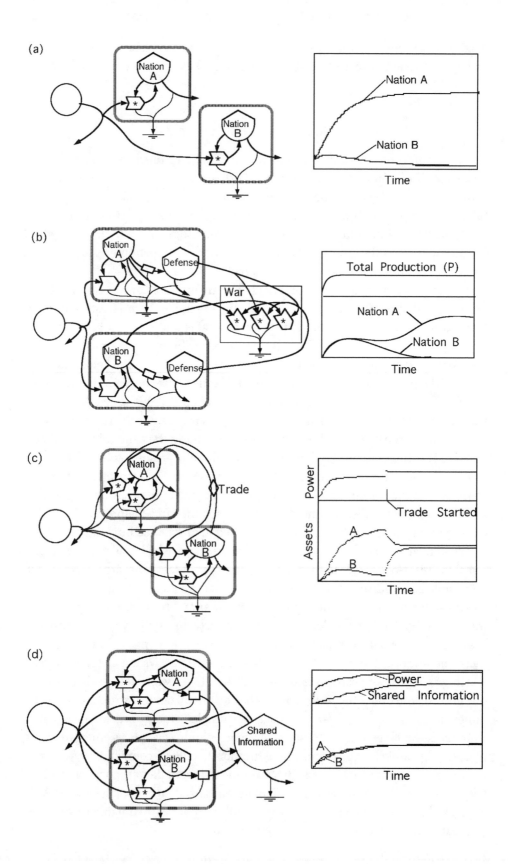

(a)

Nation A

Nation B

Nation A

Nation B

Time

(b)

Nation A

Defense

War

Nation B

Defense

Total Production (P)

Nation A

Nation B

Time

(c)

Nation A

Trade

Nation B

Power

Trade Started

Assets

A

B

Time

(d)

Nation A

Nation B

Shared Information

Power

Shared Information

A

B

Time

An Emergy Basis for International Organization

When the emergy flow to human groups was small, boundaries resulted from testing of the power to defend. Developing territories according to the emergy available is analogous to birds defending territories. These mechanisms kept populations distributed in proportion to resources. Now, however, countries have enormous empower budgets. Even ritualistic empower testing by military defenses at boundaries is too wasteful and could be replaced by agreements to use empower evaluation.

FIGURE 10.12 (*opposite page*) Energy systems minimodels of national relationships and typical computer simulations. **(a)** Simple competition for resources; **(b)** active competition of war; **(c)** action of trade in making competitors cooperative; **(d)** national cooperation through the sharing of information. Equations for the simulation of national relationships are as follows:

(a) Nations with simple competition for resources:

Available energy: $R = I - (k_1 * R * A) - (k_2 * R * B)$;

$R = I/[1 + (k_1 * A) + (k_2 * B)]$.

Population Q_1: $dA/dt = (k_5 * R * A) - (k_3 * A)$.

Population Q_2: $dB/dt = (k_6 * R * B) - (k_4 * B)$.

(b) Nations at war:

Resources: $R = I - (k_0 * R * A) - (k_1 * R * B)$;

$R = I/[1 + (k_0 * A) + (k_1 * B)]$.

Assets: $dA/dt = (k_2 * R * A) - (k_3 * A) - (k_6 * A) - (k_{13} * A * Mb)$;

$dB/dt = (k_4 * R * B) - (k_5 * B) - (k_9 * B) - (k_{12} * B * Ma)$.

Defense: $dMa/dt = (k_7 * A) - (k_8 * Ma) - (k_{16} * Ma * B) - (k_{15} * Ma * Mb)$;

$dMb/dt = (k_{10} * B) - (k_{11} * Mb) - (k_{14} * Mb * A) - (k_{17} * Ma * Mb)$.

(c) Trading relationship:

Resources: $R = I - (k_1 * R * B) - (k_{18} * R * B) - (k_{19} * R * A) - (k_0 * R * A)$;

$R = I/[1 + (k_1 * B) + (k_{18} * B) + (k_{19} * A) + (k_0 * R)]$.

Assets A: $dA/dt = (k_2 * R * A) + (k_{20} * R * B) - (k_3 * A)$;

$dB/dt = (k_4 * R * B) + (k_{21} * R * A) - (k_5 * B)$.

(d) Nations sharing information:

Resources available: $R = I - (k_0 * R * A) - (k_1 * R * B) - (k_{13} * R * S) - (k_{14} * R * S)$;

$R = I/[1 + (k_0 * A) + (k_1 * B) + (k_{13} * S) + (k_{14} * S)]$.

Assets, A and B: $dA/dt = (k_{11} * R * S) + (k_2 * R * A) - (k_3 * A)$;

$dB/dt = (k_4 * R * B) + (k_{12} * R * S) - (k_5 * B)$.

Shared information S: $dS/dt = (k_6 * A) + (k_7 * B) - (k_9 * R * S) - (k_{10} * R * S) - (k_8 * S)$.

Total power P: $P = (k_2 * R * A) + (k_{11} * R * S) + (k_4 * R * B) + (k_{12} * R * S)$.

The world evolved international organizations to maintain peace, but organizations such as the League of Nations after World War I were soon bypassed when their real empower did not correspond to their votes. Currently in the United Nations General Assembly the votes by countries with small energy budgets are equal to the votes of countries with vast energy budgets. As a result, some international disputes have moved outside the United Nations. This degrades the organization's potential might to that of a paper tiger. In order for votes to anticipate and make unnecessary the wielding of economic and military power, voting could be adjusted to correspond to annual empower.

The power budget of a country is readily computed (table 10.1) and could be determined annually. Instead of the yearly assignment of votes in proportion to the number of nations, they could be assigned in proportion to each country's emergy budget of the previous year. Thus, the voting power would always correspond to the country's real ability to influence the world. Because economic power and military power stem from empower, these influences would then correspond to the vote in the world organization. A voting demonstration of power could substitute for the vast expense of a direct economic or military test. Any majority vote would be backed automatically by an empower excess, which might be a deterrent.

The use of empower as the means of allocating votes has more flexibility than votes fixed by an organizational charter. An undeveloped country that increases its power budget with industrialization gains votes in proportion. The empower budget includes all kinds of energy. The process of annually computing votes from power budgets for everyone to see also eliminates posturing, apprehensions, propaganda, and saber rattling about strength. True empower becomes known as it develops without military and economic demonstrations. Perhaps empower voting can eliminate wars of miscalculation.

Organization by Global Information Sharing

What is most exciting is the global sharing of information that was already predominant in the year 2002. Although agents of capitalism seek to restrict information as a sales commodity, large-scale shared information, such as common genetic varieties for agriculture and common attitudes toward peace and global pluralism, reinforce all countries, making them coexist. The model in fig. 10.12d generates coexistence and greater global empower. The important question is whether enough global attitude sharing can exist in times of lower energy to preserve a system of peaceful cooperation. Or will the system of defended boundaries return?

Preservation of Cultures

In low-energy times many cultures provided a stable existence for their people, but over the past two centuries, low-energy systems have been gradually invaded with the ways of the empower rich. Low-emergy budget systems ultimately are displaced by systems that involve more empower. However, an energy-rich system can preserve some of the simpler culture. Many countries set aside native reservations, but most of the youth on the reservations seek higher-emergy ways. But some proud people of older cultures are preserving language, rituals, and symbols as an avocation, for visitors, for group cohesion, and for historical education. Some older ways, persisting in cultural memory, may be needed again when times of lower energy return.

Pluralism and the Global Organization of Society

Just as species of large territory and influence tend to stabilize global ecosystems, units of society are evolving with global territory and influence. Religious differences decrease as migration causes all countries to contain some of all religions. Some global corporations may be evolving away from emphasis on profit for growth toward a mode of service for survival. Many nongovernment organizations try to protect diversity of life and learning. The United Nations has gradually expanded its role from health, welfare, and information to forceful peacekeeping. Religious institutions may be evolving a new role (see chapter 11).

As said before, the higher information resides in the energy hierarchy, the more emergy it requires. Shared ideals on a global scale have great impact, but the quantity of information that can be transformed, distributed, and absorbed on that scale is emergy limited. In other words, it is not desirable, nor is there enough emergy to homogenize information sharing to make one global culture. Besides, diversity of culture, like the diversity of occupations and species, helps maximize function. Thus energy hierarchy principles support global *pluralism,* in which there is diversity of lifestyles and functions on smaller scales while at the global level a unified concept of respect for differences, international equity, and unity with nature is shared.

SUMMARY

Energy principles explain the growth of world populations and the designs by which people are self-organized into social relationships and institutions. In their lifetimes, human beings pass through an energy hierarchy of increasing transformity and influence. Human societies alternate between a stage of overgrowth when there are excess resources that cause simple conformity and a complex stage of high emergy, diversity, and information. Among nations, competition and war prevail when there is growth in emergy use. When resources are not in excess, nations develop mechanisms of cooperation and peace, including trade and information sharing. The climax states with excess emergy have high-transformity symbols such as dragons, sports, military shows, and social causes to focus national attention on the system. When group visions for a better system go unrealized, these high-empower pathways can short circuit through destructive social eddies. The emergy of organization and diversity increases with the numbers of types and their connections. The successful organization of people into a society has positive empower reinforcement loops, which keep the governing controls adapted for maximum performance.

BIBLIOGRAPHY

Note: An asterisk denotes additional reading.

Allee, W. C. 1938, 1951. *Cooperation Among Animals.* New York: Henry-Schuman.

Altman, D. E. and L. Levitt. (2002, January 23). The sad history of health care cost containment as told in one chart. *Health Affairs,* Web exclusive. Retrieved April 1, 2002, from http://healthaffairs.org/1110_web_exclusives.php.

Ardrey, R. 1966. *The Territorial Imperative.* New York: Atheneum.

Berne, E. 1964. *Games People Play.* New York: Grove Press.

Brown, M. T. 1977. Ordering and disordering in South Vietnam by energy calculations. In H. T. Odum, M. Sell, M. Brown, J. Zucchetto, C. Swallows, J. Browder, T. Ahlstrom, and L. Peterson, *The Effects of Herbicides in South Vietnam, Part B: Working Papers. Models of Herbicide, Mangroves and War in Vietnam,* 165–191. Washington, DC: National Academy of Sciences.

Cartwright, D. 1959. *Studies in Social Power.* Ann Arbor: University of Michigan.

Clark, P. J., P. T. Eckstrom, and L. C. Linden. 1964. On the number of individuals per occupation in a human society. *Ecology* 45: 367–372.

Easton, D. 1965. *A Systems Analysis of Political Life.* New York: Wiley.

Edmondson, W. T. 1965. Reproductive rate of planktonic rotifers as related to food and temperature in nature. *Ecological Monographs,* 35: 61–111.

French, J. R. P. 1956. A formal theory of social power. *Psychological Review,* 63: 181–194.

Gibbon, E. 1962. *Barbarism and the Fall of Rome. Vol. II of the History of the Decline and Fall of the Roman Empire, 1776, 1778,* abridged and edited by J. Sloan. New York: Collier.

Henderson, K. 1977. *Senility and mortality.* Master's paper, University of Florida, Gainesville.

Hoebel, E. A. 1961. *The Law of Primitive Man.* Cambridge, MA: Harvard University Press.

Kowallik, W. 1967. Chlorophyll-independent photochemistry in algae. In *Energy Conversion by the Photosynthetic Apparatus,* 467–477. Brookhaven Symposia in Biology, No. 19. Upton, NY: Brookhaven National Laboratory.

*Kroeber, A. L. 1963. *An Anthropologist Looks at History.* Berkeley: University of California Press.

*Lack, D. 1954. *The Natural Regulation of Animal Numbers.* Oxford, UK: Clarendon.

Lasswell, H. D. and A. Kaplan. 1950. *Power and Society.* New Haven, CT: Yale University Press.

Logan, M. O. 1998. *A Simulation Model of the System of Health Care in the United States.* Ph.D. dissertation, George Mason University, Fairfax, VA.

McClanahan, T. R. 1990. *Hierarchical Control of Coral Reef Ecosystems.* Ph.D. dissertation, University of Florida, Gainesville.

Nixon, S. 1969. *Characteristics of Some Hypersaline Ecosystems.* Ph.D. dissertation, University of North Carolina, Chapel Hill.

Odum, H. T. 1983. *Systems Ecology.* New York: Wiley.

———. 1996. *Environmental Accounting, Emergy and Decision Making.* New York: Wiley.

———. 1999. Limits of information and biodiversity. In H. Loeffler and E. W. Streissler, eds., *Sozialpolitik und Okologieprobleme der Zukunft,* 229–269. Vienna: Austrian Academy of Science.

Odum, H. T., S. Nixon, and L. DiSalvo. 1971. Characteristics of photoregenerate systems. In J. Cairns, ed., *The Structure and Function of Freshwater Microbial Communities,* 1–29. Blacksburg: American Microbiology Society, Virginia Polytechnic University.

Odum, H. T. and D. M. Scienceman. 1986. Commonalities between hierarchies of ecosystems and political institutions. *Yearbook of the International Society for the Systems Sciences,* 29: 23–32.

Slobodkin, L. B. and S. Richman. 1965. The effect of removal of fixed percentages of the new born on size and variability in populations of *Daphnia pulicaria* (Forbes). *Limnology and Oceanography,* 1(3): 209–237.

Stoekenius, W. 1978. Bioenergetic mechanisms in Halobacteria. In S. R. Caplan and M. Ginsburg, eds., *Energetics and Structure of Halophile Microorganisms,* 185–198. North Holland: Elsevier.

Strehler, B. L. and A. S. Mildvan. 1960. General theory of mortality and aging: A stochastic model relates observations on aging, physiologic decline, mortality, and radiation. *Science,* 132: 14–21.

Sundberg, U., J. Lindegren, H. T. Odum, and S. Doherty. 1994. Forest emergy basis for Swedish power in the 17th century. *Scandinavian Journal of Forest Research,* Supplement No. 1.

Toynbee, A. 1935–1939. *A Study in History,* Vols. 1–6. London: Oxford University Press.

Von Foerster, H., P. M. Mora, and L. W. Amiot. 1960. Doomsday: Friday 13 Nov. A.D. 2026. *Science,* 132: 1291–1295.

Woithe, R. 1994. *Emergy Evaluation of the United States Civil War.* Ph.D. dissertation, University of Florida, Gainesville.

Wynne-Edwards, V. C. 1962. *Animal Dispersion in Relation to Social Behavior.* New York: Hafner.

Notes

1. Malthus was a demographer and political economist in England who wrote an important essay in 1798 on population growth titled "An Essay on the Principle of Population, as It Affects the Future Improvement of Society with Remarks on the Speculations of Mr. Godwin, M. Condorcet, and Other Writers." Malthus's pessimistic views on the human struggle for existence influenced Charles Darwin and his theory of evolution.

2. This section was written by Mary Odum Logan and gives the results of an energy systems analysis of health care in the United States, which she performed with advice from her father.

3. Strehler and Mildvan (1960) used the biomathematics of death curves with time to show that human death often may be caused by the chance that the energy demands of random stresses exceed reserves as the efficiency of functions decline with senescence. They showed how to derive the Gompertz equation from this premise. Human death curves follow the Gompertz graph from age 30 to 75. A Gompertz curve is straight on a log–log graph with time. Kevin Henderson (1977) improved the fit of the Gompertz function to human survivorship curves by modifying the relationship with a negative exponential term to account for infant mortality and demonstrating the new mortality function with analog computer simulation (Odum 1983).

4. The form of feedbacks is that of political institutions given by Easton (1965) but here recognized as pathways of high emergy and transformity.

5. A diagram with these pathways was given in *Environment, Power, and Society*.

6. Ecological studies of diurnal changes in oxygen in brine microcosms by Scott Nixon (1969) (see fig. 12.9) showed the role of rhodopsin before its mechanisms were worked out physiologically (Odum et al. 1971; Stoekenius 1978).

7. A mechanism to represent the essence of learning with an energy systems diagram includes a sensor from an operating pathway storing information in a tank symbol that auto-catalytically reinforces further flow in that pathway.

8. Emergy evaluations were made of the Vietnam War (Brown 1977), the U.S. Civil War (Woithe 1994), and the Swedish wars of the 17th century (Sundberg et al. 1994). Using numbers from newspaper clippings, the author presented an emergy evaluation of the Persian Gulf War and its oil well fires to the American Association for the Advancement of Science meeting in Chicago (1998). The emergy used and destroyed in the war exceeded the Near East oil budgets for that year but was much less than the Near East oil reserves.

CHAPTER 11

ENERGETIC BASIS FOR RELIGION

Apparently, all societies develop religious institutions that give human individuals learned programs of dedicated behavior. Cultures prevail that motivate people to contribute to the maximum empower of society,[1] but poorly adapted religions interfere with optimum functions. With the expanding role of society on Earth, the ethics of human behavior requires morality on a larger scale not much covered by earlier religious teaching. This chapter shows the systems nature of religion, its place in the energy hierarchy, relationships of religion and science, global pluralism, adaptive God, alternative views of the cosmos, and the kind of religion needed for times ahead.

Functions of Religion

Religion consists of programs of learned human behavior shared with other people and taught in religious institutions controlled by religious leaders. Religion guides the participation of human individuals in the energy system of society and environment. Religions usually motivate people to support their society by teaching them to have faith in principles set out by a larger authority. Individuals are reinforced and meet their inherited need to share purpose and feel secure. Usually there is a book of principles, which has ideas or examples that seem to fit the many stages of society including growth, climax, and descent. Leading religions and their books, symbols, and architecture are at least 1,000 years old, have millions of followers, and thus have very high empower and transformities.

Nature Versus Nurture and Genetic Codes

People, like other organisms, have their operational programs coded in genes of their chromosomes. With the genetic code in every cell of the body, the information is always at hand for whatever cell function the body system calls into operation. Religious learning, as taught at an early age to children, is to society what genetic codes are to the biological system. In religious society, copies of shared operational programs reside in each person so that individuals can all operate together for a common purpose.

However, humans are much more adaptable than other organisms. People have covered the world because half of their operating information is learned. Learned information adapts the genetic information so that humans can fit the conditions of environment and economy. The old argument about which is more important, nature or nurture, is silly because both are required and either can be limiting.

Each person has a set of programmed functions, which they contribute to society by *plugging in*. Individuals will plug into a network of functions that is consistent with their ethical beliefs. The property of an individual as a plug-in unit is illustrated in fig. 11.1, showing the required material and energy connections. The need for single humans to connect their behavioral inputs and outputs with others is a well-known requirement for good mental health. Society is cemented by the propensity for individuals to join.

Social Switching and the Energy Hierarchy

Societies with strong programs of morality, religion, and ethics focus and unite the dispersed power resources of individuals as needed for group protection and unified actions. Especially when human societies must exist with famine, war, rapid change,

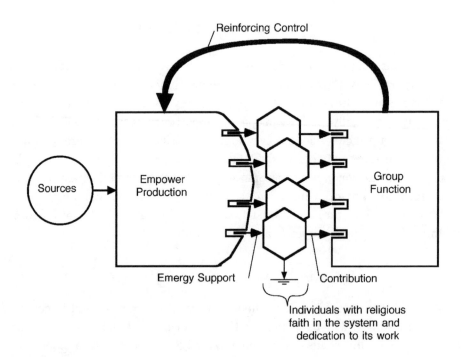

FIGURE 11.1 Analogy of human beings as plug-in units, each containing a behavioral program of use and contribution.

and disrupted central governments, a strong religion provides a flexible focus of power, used at times for individual works and at other times for group action.

Social animals have an analogous switching of individuals from individual efforts to group efforts, but their neural systems are simple, and more of the programming is genetic and automatic. In the tropical rainforest, swarms of army ants alternately spread through the forest as individuals, then switch in behavior and converge their energies into common functions, exerting work flows vastly greater than those they might achieve separately. They build houses out of their own bodies or devastate a food mass. Such a network has great flexibility because of the self-switching action of the individuals. Figure 11.2 has each individual in

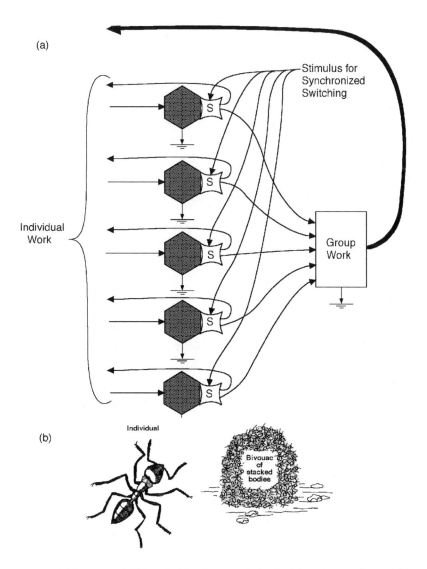

FIGURE 11.2 Social insects and their switching between individual functions and group functions. **(a)** Systems summary; **(b)** army ant example.

two modules, separate operation and group operation.Synchronized switches shift behavior from one to the other.

In a vastly more complex way people also distribute their work energies in dispersed activities until some common need causes their behavioral programs to unite, focusing their work to meet emergencies, achieve unity, and perform group functions swiftly (fig. 11.2). Although each person in a group may have the same religious teachings, they usually have differences in genetics and learning. When something concerns the group for which the programming is the same, the group can act with shared functions and great power. For other functions for which the individual behavior programs are different, the group profits from individual creativity and diversity.

Scales of Religious Power

In social insects such as bees, ants, and termites, the separate individuals are small on the scale of turnover and transformity. When they converge into a group in a central place (an animal city), however, they share a larger scale and transformity. The shared program of social information has a large territory of operation and longer turnover time than a single individual (higher transformity is to the right in fig. 11.1). Humans also create higher-level functions in the energy hierarchy when they respond in zealous group action. Their unity has high transformity and great effect. Whether the action taken leads to system survival (good) or system collapse (evil) depends on the larger system.

Reinforcing Structure of Religion

Figure 11.3 shows the role of decentralized religious programs, religious leaders, and their institutions in a society. Each level has the loops of reinforcing amplification that can adapt both morality and religious institutions to meet external challenges. Individuals have faith in the social system and therefore contribute their efforts. Of the many variable flows and actions resulting, those that develop reinforcement loops are retained for the future. Faith in a religion that supports the chosen way of life is necessary to the group's effective energy use and survival.

It takes a leader to recognize group needs and put them in the symbolic and vernacular language for the times, in effect interpreting the network. It takes a religious institution to train and indoctrinate people, especially the young, to learn the appropriate behavior, adopt the faith, and accept the symbols of the religion's conception of what is *good*. The loop through the religious leader can keep religion adapted to changing times. If there is no mechanism for change, one religion may be replaced by another. What has been energetically reinforced is taught as religious truth to strengthen faith, cause further reinforcement, and help the system continue.

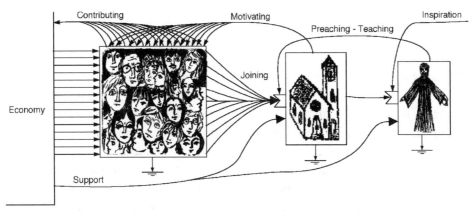

FIGURE 11.3 Religious system of emergy management (Ann Odum).

Faith to Motivate

An organization survives if its members are motivated to contribute. An inner satisfaction from making a contribution reinforces the motivation to contribute more. Satisfaction with useful work may be part of inherited human nature genetically established by natural selection in the past.

The flexible minds of humans learn programs of behavior and develop faith in what they learn. As children develop, what they learn is first tied to faith in their parents who provide the means for life. Later their faith may be transferred to personified images of infallible higher authorities out of sight, often called God. Some people transfer their faith to their society, to the earth, or to the universe, which may or may not be labeled God. For the humans to be contributors to the systems of which they are a part, there has to be faith that the larger systems justify their purpose. Old organized religions have long been successful in providing ready-made programs of faith to motivate useful lives. These messages are helped by their mystic flavor.

When a person is within a system, that person may not easily see his or her part in the whole. A man in an auto assembly line who never views the whole plant cannot understand what his work accomplishes. He might not have any inkling of the complex processes that keep a continuous production of cars adapted to inputs, adapted to outside demands, and stabilized in the face of fluctuations. The situation is similar for people in the bewildering complexity of the modern network of society and environment.[2] Each person may understand her or his local interactions but not how they relate to the larger realms of earth and beyond. But having faith in the worth of their systems simplifies thinking and motivates their efforts.

RELIGIOUS PATHOLOGY

At times societies can have too little religion. If there is not some common basic program of shared useful behavior, there is not enough group cohesion. Individuals are motivated in different directions or not motivated enough, and as a result emergy disperses. Functions of the society fail, the economy does not compete, and that social system may be displaced.

At other times the religious programs and institutions gain too much of society's power, building more religious structure than needed and taking too much time from individuals, thereby preventing other functions, eliminating information, and choking off the necessary creativity of trials and errors. Again, the society loses vital functions, its economy declines, and it is displaced. The Spanish inquisition, witch burning in New England, and lynch mobs often are cited as examples. Clearly there is an optimum energy allocation to religion for maximum empower and survival of society. Religious action can be perverted when it engages society against the realities of energy principles. Religion can become pathological when religious leaders learn only dogma without understanding how it should relate to individuals and society.

Holy War

World history records religious zeal as the cause or a contributing factor to many wars of territorial power. Spreading of the Christian religion among primitive cultures was used as the excuse to enlist group support for the military conquest and displacement of less developed societies by nations of Europe in the colonial era from 1500 to 1750. In recent years, extreme religious zeal has resulted in intermittent, devastating war between Israel and Palestine; however, the wasted emergy may have had one benefit by creating globally shared images of what not to do.

Religious Conflagration and Chain Reactions

Everyone knows that chain reactions can result in explosions; for example, atom bombs work by chain reactions. A chemical chain reaction is diagrammed in fig. 11.4a. Note the closed loop in which the interactions (pointed block symbols) stimulate each other. If sufficient energy is available, the system explodes. If the autocatalytic reinforcing is diluted by distance or an absorbing medium, the reaction is quenched and dies away.

Religious autocatalysis has a similar design (fig. 11.4b) and can go into a chain reaction, producing an explosive social revolution. In the late 1970s, the people of Iran, in the absence of other information input, circulated tape recordings from Ayatollahs enlisting their uniform religious code in a revolution. At the same time, there was a huge expansion in the global available energy based on oil. Yet the ex-

(a) Explosive Chain Reaction of Two Chemicals, A & B

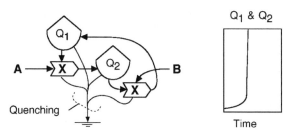

(b) Zealous Religious Revolution

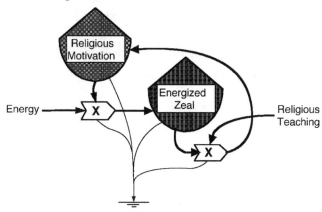

FIGURE 11.4 System design of chain reactions that explode. (a) Chemical reactions; (b) unholy combination of religious dogma and excess oil.

plosive change converted the nation into a fundamentalist theocracy using beliefs from the 7th century, which destroyed their reservoir of knowledge and technology and wrecked their economy. By not maximizing their potential empower, the religious excess interrupted their progress.

Sins of Energy Dissipation, Autopower

Love of power, like hunger and sexual drives, can have important uses when properly balanced in the whole pattern of adaptive behavior. Like these other packaged drives, lust for excess power destroys. The old religions warn that excess energy dissipation is sinful. They preach to avoid false gods that lead cultures into waste and individuals into destruction. Surviving cultures channelize energy to maximize empower.

Cheap high-grade energy was not generally available in earlier times to inflame society. But in the 20th century high-powered, finely tuned fossil fuel burners, the automobiles, dispersed emergy everywhere. The old religions had nothing to say about these new forms of excess and were blind to the new sin: a worship of cars and trucks and the mob demand for endless strips of asphalt and concrete, standing idols to the false God of motor power. In the 20th century the sins of excess power dissipation began displacing older weaknesses as primary evils.

The lust for speed and acceleration easily takes hold of all of us, drawing the first buying power of the adolescent, dominating the thinking of his elders in the budgeting of funds, and pumping more and more fuel into acceleration and speed. Luxury automobiles and expressways are surely the dominant part of our culture that is unnecessary, destroying human cultural coalescence, promoting fatal crashes, keeping people on the move, and separating society.[3] Urban people speeding around in circles are fossil fuel storms. Urban culture will be reformed when the fuels dwindle and a new culture uses energy better. Church doctrines have been slow to include these new sins because the written testaments from earlier times did not mention automobiles, highway folly, or giant airplanes. To succeed, religions today must find images and metaphors to help people fit reality.

High-Energy Idols

When anthropologists look at the remnants of past civilizations, they find large structures such as the pyramids, apparently important to those civilizations. They were built when there was excess energy. Future cultures looking at the remnants of our culture may find giant, empty expressways. The concrete network may even develop religious significance to less energized cultures, in the same way that the remnants of Roman culture achieved religious significance in the Middle Ages. Symbols of ancient high-energy eras imply the past magnificence of God to those living with low energy. Perhaps civilizations are marked most by the false gods of their times of excess energy.

Celebrity Worship

In the rapidly changing world of the new millennium, huge energies flow into sports, music, movies, politics advertising, journalistic entertainment, and especially advertising. The information system competes for people's attention by generating celebrities. Many young people copy celebrities' behavior and pronouncements, displacing their normal morality, especially if they have not received traditional family and religious teaching. For example, good sportsmanship, humility, hard work, dedication to others, and honesty get lost in celebrity pathology.

Religious Ignorance

To reinforce society well, a religion motivates people without fostering ignorance or hiding truth. A recurring religious pathology is teaching untruths in conflict with knowledge well established by scientific study. *In each person's mind truth is the absence of contradiction.* Humans can be overzealous about what they are taught as children, even when it is wrong, unless they are exposed to contradictory information while they are receptive. When religious leaders only learn dogma without being educated in science, their minds have no contradiction, and they can teach erroneous belief. It is pathological to attack facts as evil. A major challenge is how to correct factual error taught in religion without undermining motivation.

Perception of the Universe

Humans looking out into space at the starry heavens cannot ignore the universe. Some concept of the cosmos is included in their basic view of their existence. Religious documents on the origin of the universe, such as the Book of Genesis, are not factual but were written by inspired people without scientific knowledge. Yet even with rapid advancement in the science of astronomy, there are many alternative theories. The big bang concept sees the universe as starting from one point expanding outward for several billion years, developing galaxies and stars as it goes. According to one idea, the universe will collapse again, followed by a new exploding expansion so that there is a sequence of successive pulses, a kind of steady state. It may be that the scale of some of these present concepts is too small.

The energy hierarchy concept suggests a larger-scale view of the universe. Instead of one big bang, there is a universe of many centers, each pulsing at different times. An energy hierarchy view of the universe is summarized in fig. 11.5.[4] Space is occupied by scattered low-temperature radiation (3°K) and scattered dark matter. These are shown on the left, where they constitute a large quantity of units of small scale.

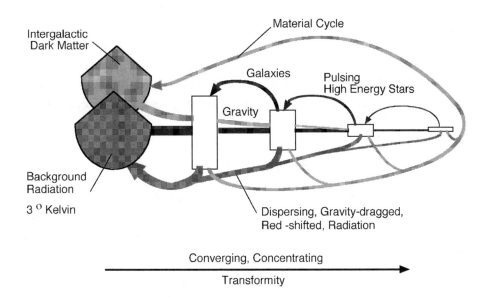

FIGURE 11.5 Energy hierarchy of the universe representing the ecoenergetic concept of a many-centered, multipulsing, steady state cosmos.

Matter falling together concentrates energy and generates galaxies of planetary bodies. By the natural selection described in fig. 4.13, units develop that have the most autocatalytic capture of matter and conversion of kinetic energy into heat (fig. 4.16). There is a food chain of units consuming other units as gravity attracts their masses to fall together. Energy is transformed and concentrated in successive stages to high-transformity centers, the stars, which have enough energy to set off nuclear reactions and conversion of matter to energy. Matter and energy recycle and disperse back out to space with pulsing explosions of the highest-energy stars and black holes. The high-intensity nuclear energy transformations in the stars generate the chemical elements with a hierarchical distribution, having a large quantity of light elements and a small quantity of heavy elements with higher transformities (fig. 4.18).

One evidence cited for the big bang is the shift in the wavelength of light to longer values (and lower energy per photon). The red shift is assumed to be the Doppler effect of the light coming from objects traveling away. Some of the red shift in the light may be caused by the removal of energy from light by gravity fields. The hypothesis in fig. 4.19 invites evaluation of emergy and transformity of all the components of the heavens in the same way as is done for the species of ecosystems. Instead of using the earth's solar insolation emergy as the base for calculating transformities, as in this book, one can define a universal transformity as the emcalories of 3°K space radiation required to make hierarchical structures.

Angels, Devils, and the Energy Laws

Personifications and symbols of religious teachings can reach children in their formative stages, imprinting them for a lifetime. Figure 11.6 is an analogy between the images of religion and system thermodynamics. The building of structural storages and useful information (chapter 8) is shown at the top as angelic, whereas the dispersal and losses that generate and disperse heat are shown as the realm of devils, hot and down below.[5] The selection of souls at the pearly gates may be compared with the reinforcement of functional information by energy networks. Increasing the available energy increases the evil of disordering heat but also the evolution of even greater good.

Human Value and Immortality

As explained in fig. 10.3, when people pass from the fertilized egg through birth, childhood, schooling, and early job experiences, their biological network develops greater structure, more memory storages, and more of the abilities that serve the system well at the individual, family, local, and worldwide levels of human participation. The emergy required for all this improvement is the work spent by and for the individual, which causes gradually increasing transformity. The core set of information that is stored in human minds and operates with very tiny mental energies has the highest transformity of all those in the body.

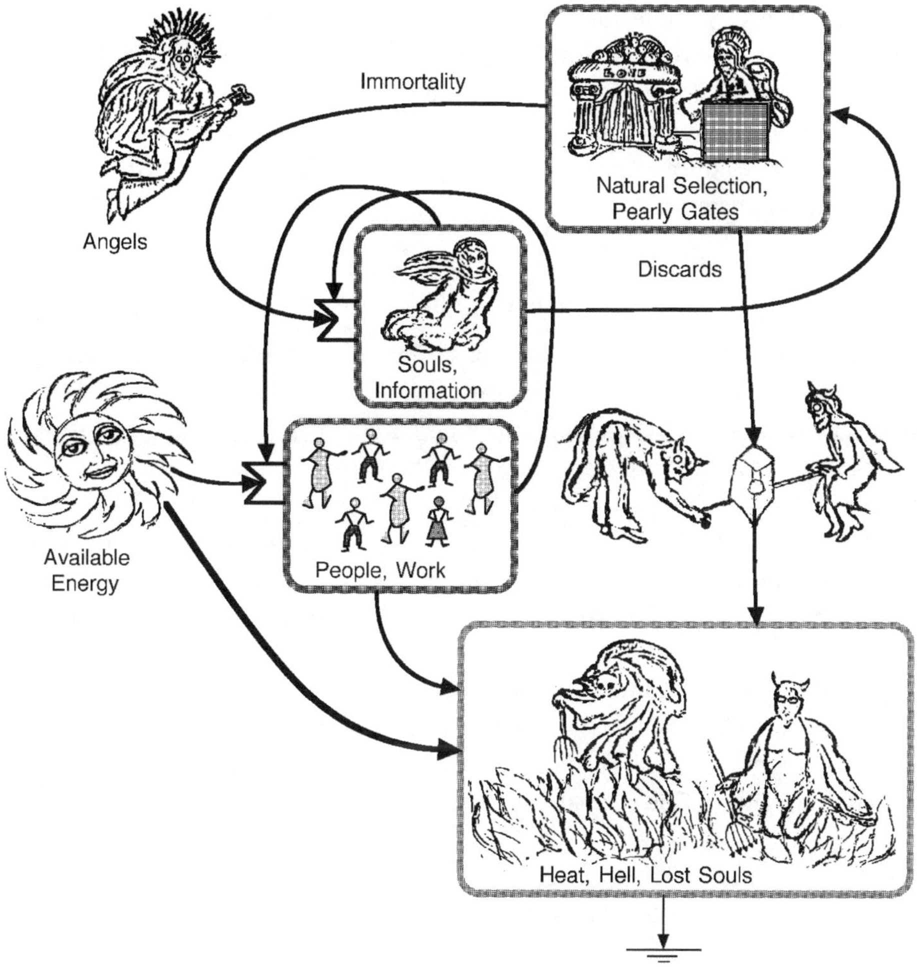

FIGURE 11.6 Energy systems diagram that represents thermodynamic concepts and analogous religious concepts (Ann Odum). Souls are connected to work with chain reaction design (fig. 11.4).

Some religions teach that the core essence of a human being is its soul, a nonmaterial entity that exits the body at death and is immortal. When an individual makes contributions to society, some of the core information is transferred to relatives, students, libraries, and other legacies. In a limited sense, legacies give that unique information a longer life and a limited immortality.

Infinite Transformity Information

There is an intriguing idea about energy and souls. Along the energy transformation series emphasized in this book, energy flow decreases through each transformation, and the transformity increases (to the right in our diagrams). If transformations continue successively, the energy flows approach zero, becoming unmeasurable, as the transformity (emergy/energy) becomes nearly infinite. Are there information flows on Earth or in the universe that are so far undetectable that have nearly unlimited ability to transport and amplify? The reader may recall the controversies over extrasensory perception and the difficulty with proving or disproving its existence.

God and the System

Religions are based on the individuals of society sharing faith in God or gods. Gods sometimes are thought of as supernatural beings to be represented with images by artists. Or God is said to be a social invention, undefined, intangible, or imaginary. But the images of God are global, are high in emergy, and have the greatest effect of all energy-controlling information.

After learning how to model systems and use our symbols to diagram aspects of our world, students in some classes were asked to include their concept of God. Some drew God on the left side as the source supporting humanity from the heavenly cosmos. Others put in God as the item of highest transformity to the far right of global systems diagrams with loops back to control humans in the center. For some God was identified as the unknown but somehow connected to everything. Still others regarded God as the network, the whole of everything. Apparently, in our culture, faith in God has various images of system.

LARGE-SCALE RELIGION FOR THE FUTURE

New kinds of religions follow the emergence of new kinds of societies. Christ arose in Roman times, Joseph Smith on the American frontier, and the communist ideology in the Bolshevik revolution. Religious learning used to prepare people with ethics for the small scale of family, friends, occupation, and local society. But doing what seems right on a small scale often is wrong in a larger view. As societies and cultures of humans have grown and evolved, humanity has moved its role to larger and larger scale, now operating the whole earth. People share in decisions about international relations and the global environment that the economy now affects.

Environmental Imperative

The need to include nature and the earth's energy systems in our religious ethic was named the *Land Ethic* by Aldo Leopold (1949) and well stated by White (1967). Many tribal religions pay allegiance to nature, on which they depend. In the Tlingit Indian tribes of the northwest coast of North America, the system of nature is personified by the raven, the salmon, and the wolf (fig. 11.7). They believed that the virtuous return in the afterlife as respected animals.

FIGURE 11.7 Religious tribal art of the Tlingit tribe of southern Alaska: the salmon, a basis of life, and the raven and the wolf, around which clans and marriages are arranged. Art by Kashudoha Wanda Culp and Michael L. Beasley, Juneau, Alaska, reprinted with permission.

In the early Industrial Revolution based on fossil fuels, humans were less dependent on the environment and were protected by the large capacity of the earth's life support system. Urban society swarmed over the earth, forgetting the dependence on environment. Few regarded the loss of soils or the draining of swamps as evil, but preserving life support requires the ethic "do unto nature as to oneself" (because nature is the basis for self).

Although there are already initiatives in the established religions to include larger-scale ethics, great inertia makes them slow to change. Orin Gelderloos (1992) reviewed environmental stewardship and scripture in the Judeo-Christian religion, ways to include environment in worship of creation, declarations of what is environmentally right, and institutional measures such as ordaining earth keepers.

Figure 11.8 illustrates a simple mission for people in nature as the earth's information processing specialists. The closed loop design shows that society will receive resource support only as long as people use their high-transformity work to reinforce nature. Voices of conservation are too timid in merely asking religions to teach stewardship. Although most people don't know it yet, society has to fit earth's ecosystems to survive. Humans cutting mountain forests for private profit in Honduras devastated its economy. And in El Salvador, forest cutting caused the land to slip in an earthquake and bury a portion of the village of Santa Tecla, near the capital city, San Salvador. High energy has caused many human settlements to be poorly located, causing environmental disasters to increase.

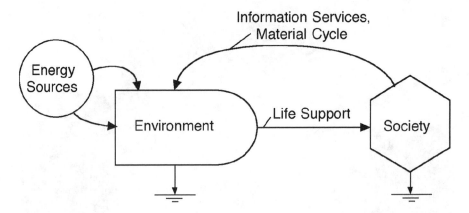

FIGURE 11.8 A religious imperative, the symbiosis of human service for the earth.

Watershed and Religion on Bali

On the island of Bali in Indonesia, the Hindu culture has a close union of religion and environment. Water from rains at the top of a volcanic island is passed downhill to the network of rice fields controlled by religious authorities with temples along canals at points of control. Kremer and Lansing (1995) developed computer programs that show how the evolution of fertile agriculture was reinforced with religion. The people also developed an extraordinary religious emphasis on environmental respect in their painting and wood carving. Resources converged from other islands—the wood came from Borneo. The special atmosphere of art and religion attracted high-intensity tourism.

Systems Faith

When people were a tiny part of the stable complex forest, their faith was in the energy system umbrella, with God identified as the intelligence of forest control. For example, the early Druids of Europe had religious faith in the forest network. A stable forest actually has a system of compartments, connections, flows, and logic circuits that is a kind of intelligence beyond that of its individual organisms.

When people entered the sharply pulsing climates, the erratic alternation of favorable energy flows and famine was capricious. The faith that developed entertained miracles, infinite patience with catastrophe, and respect for outside energy surges. Story (1964) found the Bushmen of southern Africa with intimate knowledge of its natural dry ecosystem, which they used to fit their culture to the earth.

Most people in modern society have faith in their system, whether that faith is learned from family, school, or church. Now old faiths are disturbed by the spread of the information society, especially when people see the old patterns disintegrate. But a new faith is emerging for those who receive enough education to participate.

The respect of modern people for the magic of their economic system is not so different from the faith of the primitives in their forest network.

Supernetworks and Adaptive God

Networks in which individuals are mere parts have emergent properties, superintelligence, mechanisms of protection, means of arriving at the good, and energetic functions justifying faith. Religious spokespeople such as de Chardin (1961:290) refer to "system psyche, consciousness of the system, and love, as measurable by energy, noogenesis against the entropy stream." The supernetwork of humanity, machines, and nature has its own controls, with patterns for inheritance and transmission to the future. The flow of information is a kind of thinking on a larger scale than that of the individuals. Its operational rules are a kind of morality of the supernetwork not yet visible to individuals. But do the supernetworks have self-consciousness?

Sometimes called Gödel's theorem (1931) is the principle that no system can understand itself because understanding requires more structure than operation. A complex global system cannot understand itself, but it can evolve humans to make simplified overviews. Society has become the system's brain, which uses its information for system benefit.[6] Humans make simplified models to portray what goes on. Religious teachings speak of humans in the image of God. More apt is the image of God from the models of humans.

Religion for the New Millennium

In the years ahead, international information sharing may be able to sustain the scale of global unity and peaceful organization (chapter 10). The most important sharing could be a code of earthwide attitudes that includes pluralistic respect for other faiths. Either new religions will develop, or changes will occur in the older religions. The religious belief that other religions are evil causes conflict and war. At the beginning of the 21st century, the restless, energized movements of populations everywhere were distributing people of the large religions across many countries, helping to establish pluralistic respect.

New religions are developing, some with new music. For example, Rastafarian religion blends Judaism and Christianity in some aspects. Kwanzaa was formed in 1966 with tenets derived from earlier cultures of Swahili-speaking peoples of east Africa and had an estimated 26 million followers in the year 2001.[7] In all cases, new religions use faith to motivate people to contribute together.

In the modern world, faith in our systems, science, and big institutions is receiving reinforcement, although these faiths are not yet considered religious. Older faiths still serve for those who are not receiving much from the new systems or for those who are conservative about redesign. Some allegiance is dedicated to the schools through which new ethics are taught. Because systems are changing so fast, there is not time for one pattern to develop to the status of a religion before the pattern changes again. Many young people, seeking a workable faith for the new patterns they are experiencing, do not find it among the religious institutions where they search, for it is evolving elsewhere and is still hidden. The time may be right for new prophets.

Emergy Test of Morality

Because energy laws explain the self-organization of society, these principles define what is moral. Emergy waste is immoral. For example, excesses of individual sexual behavior were seriously detrimental to the scarce energies of a primitive culture and regarded as immoral. In the energy-rich culture of today, sex and other drives have been stimulated but partly separated from reproduction and diverted to other uses. Sexual images are promoted to compete for television attention and stimulate the economy with advertising. Efforts to use old-time morality regarding sex to remove a president of the United States failed in 1999 because the majority believed that sexual conformity was not that important. Greater sins are the enormous wastes of calories in destruction of environmental life support, useless worship of automobile power, gun purchases without purpose, and the greed of stealing resources from future generations.

The emergy ethic leads us to different conclusions about trends of our times and the needs of our youth from those of evangelists, whose personifications and language sometimes lose them in the new wilderness. In energy-rich times, which religion is devout: one that worships a past network with ancient language and persecutes those who change or one that accepts the new information network erupting miraculously from earth fuels as an expansion of God?

Religion with Declining Energy

In times of declining energy ahead, the economy will decrease (chapter 13). As soon as there is an economic downturn and public consternation, some people seek security in religion of the past. Other people believe that the new conditions require revised ethics. Some of the conflict between liberal and conservative attitudes in politics concerns the conflict between faith in the emergent new network of society and the old-time religions for simpler times of growth. New religions could help unify people around the pathway down.

In medieval centuries, the information of earlier civilizations was preserved by religious institutions. It was appropriate to serve God by entering a monastic refuge to aid the retention of knowledge about the former network. Faith was directed to sustain knowledge, which developed religious value. This information was used later, in the Renaissance, to start the next pulse of civilization.

In the complex world of the 21st century, thousands of organizations and controls are operated by the systems of economics, law, and information. All have loop reinforcement designs (chapters 7–10). Energy-rich systems support many kinds of controls at larger scales, overriding religious influence at the level of the individual. In simpler times, religion was the main control integrating society. When less energy is available for social controls, the religious mechanisms may be the more efficient choice, especially if large-scale connections are lost. Systems with strong religion may not need so many laws.

Summary

Surviving systems of humans and nature use learned programs of religious teaching to motivate individuals to follow the ethics of energy laws. Religious faith aids the classic struggle between structure and dispersal, useful work and dissipation, angels and devils. Although programs of religious control develop independently, they seem to have similar principles derived from their energy basis. There is an optimum intensity of religion for maximum empower, and religious pathology results when religion becomes too weak or too fanatic. Energy commandments guide survival, and system survival makes moral right. Table 11.1 summarizes energy-based ethics.[8] Whereas the earlier tenets of religions were developed for the small scale, the high emergy of the human place in the larger scale of environment and global sharing needs large-scale morality. In the schools, we can learn energy truth through general science, and in the changing churches we can teach the love of system and the mission of each to contribute.

TABLE 11.1   Energy System Ethics for All Scales

Seek satisfaction in useful contribution.
Help maximize real wealth (empower).
Reinforce environmental sources.
Treasure genetic and cultural diversity.
Adapt to natural hierarchy.
Minimize luxury.
Minimize waste.
Adapt to system rhythm.
Share information.
Optimize efficiency.
Circulate materials.
Circulate money.
Fit the earth.
Reproduce only as needed.
Have faith in self-organization.

BIBLIOGRAPHY

Note: An asterisk denotes additional reading.

*Curzon, F. I. and B. Ahlborn. 1975. Efficiency of a Carnot engine at maximum power output. *American Journal of Physics,* 43: 22–24.

de Chardin, T. 1961. *The Phenomena of Man.* New York: Harper & Row.

Gelderloos, O. 1992. *Eco-Theology, The Judeo-Christian Tradition and the Politics of Ecological Decision Making.* Glasgow, Scotland: Wild Goose Publications, Pearce Institute.

Gödel, K. 1931. Über formal unentscheidbare Sätze der *Principia Mathematica* und verwandter Systeme I. *Monatshefte für Mathematik und Physik,* 38: 173–198.

Hall, C. A. S. 1995. *Maximum Power.* New York: Wiley.

Keith, A. 1948. *A New Theory of Human Evolution.* London: Watts.

Kremer, J. N. and J. S. Lansing. 1995. Modeling water temples and rice irrigation in Bali: A lesson in socio-ecological communication. In C. A. S. Hall, ed., *Maximum Power,* 100–108. New York: Wiley.

Leopold, A. 1949. *A Sand County Almanac, and Sketches Here and There.* New York: Oxford University Press.

Northrop, F. S. C. 1962. *Man, Nature and God.* New York: Simon & Schuster.

Odum, H. T. 1977. The ecosystem, energy, and human values. *Zygon,* 12(2): 109–133.

——. 1995. Self organization and maximum empower. In C. A. S. Hall, ed., *Maximum Power,* 311–329. Boulder: Colorado Press.

*Odum, H. T. and R. C. Pinkerton. 1955. Time's speed regulator: The optimum efficiency for maximum power output in physical and biological systems. *American Scientist* 43: 331–343.

Story, R. 1964. Plant lore of the Bushmen. In D. H. S. Davish, ed., *Ecological Studies in Southern Africa,* 87–99. The Hague: Junk.

White, L. 1967. The historical roots of our ecologic crisis. *Science,* 155: 1203–1207.

NOTES

1. This chapter extends the author's earlier article (Odum 1977). Among those after Darwin who have developed the theme of natural selection at the group level of human society is Keith (1948). Hall (1995) assembled chapters concerned with selection for maximum power.

2. The Vernadsky noosphere is defined as realms of phenomena so dominated by the activities of people that they determine energy flow. A noosphere is possible only where and when the power flows of people displace those of nature. Industrialized areas that displace nature survive only because of the purifying stability of the greater areas of the globe not yet stressed.

3. Because the kinetic energy of a moving object goes up as the square of the velocity, acceleration followed by braking leads to much greater energy losses in transportation than a slow speed against a steady friction. For any process in which movement is needed, there is a speed

that is too slow, so that competing activities will displace it. There is also a speed that is too fast, dissipating so much energy unnecessarily that the unit loses out in competition because waste prevents it from maximum contribution. There is an optimum speed for processes to survive (chapters 1, 3, 9).

4. The author presented the concepts at a symposium of the Swedish Royal Society in Stockholm in 1989, and a more complete diagram was given on page 328 of C. A. S. Hall's book *Maximum Power* (1995; Odum 1995).

5. A similar diagram was published in *Environment, Power, and Society* without the pictures of angels and devils, which were deleted by one of the publisher's executives.

6. Possibly consistent with emergent properties of networks as the essence of adaptive God is the thesis of Northrop (1962). In his chapter 1 he notes the creativity of human neural networks and also finds nature creative, the source being called "evolution, God, Allah, Yahweh, Brahman, Nirvana, or Tao" (1962:44).

7. An article in the *Gainesville Sun*, Dec. 26, 2000, described seven principles for Kwanzaa given by Professor Maulana Karenga of California State University in Long Beach: unity, self-determination, collective work and responsibility, cooperative economics, purpose, creativity, and faith.

8. The table in *Environment, Power, and Society* was labeled "Ten Commandments of the Energy Ethic for Survival of Man in Nature."

S ELF-ORGANIZATION is rapidly adapting our fuel-driven, urban economy to the environmental systems of atmosphere, oceans, and landscapes. As our fuel culture waxes and wanes, *environmental fit* is likely to become society's next concern. With survival at stake, humanity will need knowledge and faith to refit society to renewable resources. New kinds of systems will interface the human culture and the environment of the future. Innovative combinations will emerge as the global diversity of species interacts with the diversity of human technology. In this chapter we consider principles for restoring humanity's partnership with nature.

ENVIRONMENTAL INTERFACES AND MANAGEMENT

When the energy under human control was but a tiny percentage of the environment, people could rarely manage much of nature. For survival, humans relied on small domestic ecosystems to supply food, clothing, and fiber and on nature to provide clean air and waters, minerals, and aesthetic incentives. Now, with rich but temporary fuels, humans control and manage enormous quantities of the world's emergy, affecting the whole biosphere and displacing ecosystems. The empower of human operations is as great as that of the natural processes that maintain the stability of the air and ocean. Concentrations of carbon dioxide rise, other wastes increase, and the productive buffer of the natural systems becomes less and less protective. Human impacts become more and more important to the design of ecosystems. As they change, human culture has to change to fit. In *Governing Nature,* Murphy (1967) traced "the changing currents of public opinion and legal practices which have moved away from the assumption of nature as a free good."

Figure 12.1 shows economic society having three kinds of interfaces with the environment: protected wild ecosystems used for watershed control, life support, and tourism; yield systems of agriculture, forestry, mining, aquaculture, and fisheries that provide products to the economy; and new *interface ecosystems* self-organizing with society.

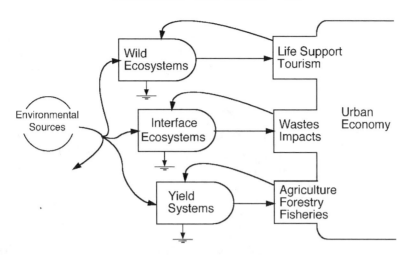

FIGURE 12.1 Types of environmental sectors supporting the human economy.

Adaptive Self-organization

Left alone, nature uses its energy sources and species to develop ecosystems adapted to whatever conditions exist (chapter 6), a consequence of the maximum empower principle (chapter 3). Genetic information is supplied from the earth's pool of species, which are moved around the world by earth processes and people to seed new situations. Figure 12.2 shows the species seeding and economic inputs that lead interface ecosystems into new designs.

People protect wild ecosystems in parks, minimizing impacts to allow nature to organize and supply life support and aesthetics. People plant simple *yield systems* (e.g., agriculture, forestry, aquaculture) and apply enough control energy to prevent nature from reorganizing. But more of the environment is forming new interface ecosystems. The light management of nature's adaptive self-organization is the field of *ecological engineering*, an endeavor that helps the environmental part of the interface to *self-design*. For example, runoff waters are allowed to recharge groundwaters, a filtering process that restores water quality. Ecological engineering also guides the outputs of society and its technology to help environmental self-design.[1] For example, no chemical substance or concentration should be released for which ecosystems cannot self-organize a symbiotic interface.

A Network Nightmare

To understand the difficulties of managing humanity and nature, let us first imagine the nightmare of an electronics technician. After a week of exhausting tedium, soldering circuits and completing a large network of wires connecting thousands of transformers and transistors, he or she goes to bed with the feeling of a system completed. Then, with the veil of the dream, the parts begin to breathe. Next they

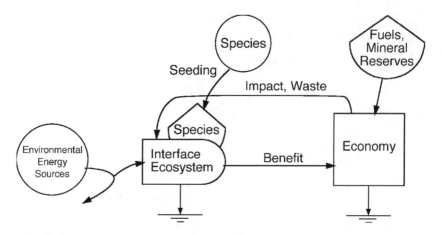

FIGURE 12.2 Interface ecosystems and the species seeding that helps make self-organizing adaptation maximize empower.

grow and divide, making new parts. Then the wires become invisible. The new and old parts disconnect themselves and move into new patterns, reconnecting their inputs and outputs, replacing worn members, and together generating functions and forms not known before. What was neat and known becomes unknown. Soon the new system with its vast capabilities is growing, self-reproducing, and self-sustaining, drawing all the available electric power. Our hero turns off the energy source and awakens. To some visionary engineers the nightmare may seem a preview of a machine world. However, the ecosystem is already this fantasy.

Billions of breathing parts are already covering the earth, each with a program of functions. These parts divide, replace each other as they wear out, and connect themselves to inputs and outputs. Like the flexible circuits in the dream, they develop immense complexities that use all available power sources. These parts are the living populations, and the circuits are the routes of flow of materials and energy.

The people trying to understand the patterns, to predict and design new combinations of the parts and manage the great systems of humanity and nature, experience another nightmare, one of uncertainty. As in the electronics technician's dream, the student of the environmental system is given no circuit diagram, and the pathways are almost all invisible and temporary.

Spontaneous Ecological Engineering

The nature reserve that developed spontaneously around treated sewage wastewaters discharged on bare sand flats at Port Aransas, Texas, starting in the 1950s is an example of adaptive self-organization of an interface ecosystem. As the town's population went from 500 to 5,000, a fertile freshwater lagoon formed, surrounded by freshwater marsh, and around that a saltwater marsh interfacing with seawater (fig. 12.3a). Bird-watching groups put up towers and adopted the ecosystem as a conservation and educational reserve, which was full of waterbirds, turtles, and alligators. Sewage treatment plants in southwestern Florida, near Everglades City (fig. 12.3b) and Naples (fig. 12.3c), discharged treated effluent to mangrove and estuarine ecosystems for many years. Sell (1977) studied the spontaneously developing ecosystems and found effective treatment of the effluents.

Multiple Seeding

The millions of species of plants, animals, and microorganisms are the raw materials of self-design. A species that has evolved to play one role may play a different role in another kind of network. The design of new ecological networks is still in its infancy, but detailed knowledge is not needed. After seeding has introduced new combinations, organisms self-organize their relationships. Multiple seeding and the ensuing self-design provides new ecosystems for new conditions.

We are accustomed to species occurring together in particular associations, but when conditions change, species may group quite differently to effect a workable metabolism. For example, when we constructed saltwater ponds in North Carolina (Odum 1985, 1989), we found freshwater and marine organisms together in new combinations of rotifers with barnacles and water bugs with marine algae. The water bugs turned blue-green mats (fig. 3.4) into blue-green balls. In the inland Salton Sea in California (fig. 12.3d), combinations of marine organisms transported from various places have become established in a new kind of network for a new environment. Walking along the shore, people tread on the shell heaps of the blue barnacle instead of the more familiar shells of present ocean beaches.

Interdependence of Ecosystem Change and Evolution

Hand in hand, evolution controls species, the available species determine the nature of the ecosystems, and the ecosystems then control the next evolution of species. Different species were in existence in past eras, but the variation in the dominants of past seas was caused as much by different conditions eliciting adaptive ecosystem designs as the present adaptation of Earth's ecosystems to fit the changing conditions associated with human activities (see chapter 8). Evolution cannot be explained without an understanding of the sequence of ecosystems. Human society depends on this adaptive process of transmitting and selecting information.

The strange and different associations in the fossil beds of the past are examples of nature's perpetual redesign. We are accustomed to thinking of these differences as the consequence of species evolution. Yet equally strange associations are found today in isolated parts of the world: the Galápagos Islands, the tops of mountains, the bottoms of the deep sea, the outflows of hot springs, and the undersurfaces of polar ice. Redesign has occurred wherever conditions are different or the seeding of species has been different.

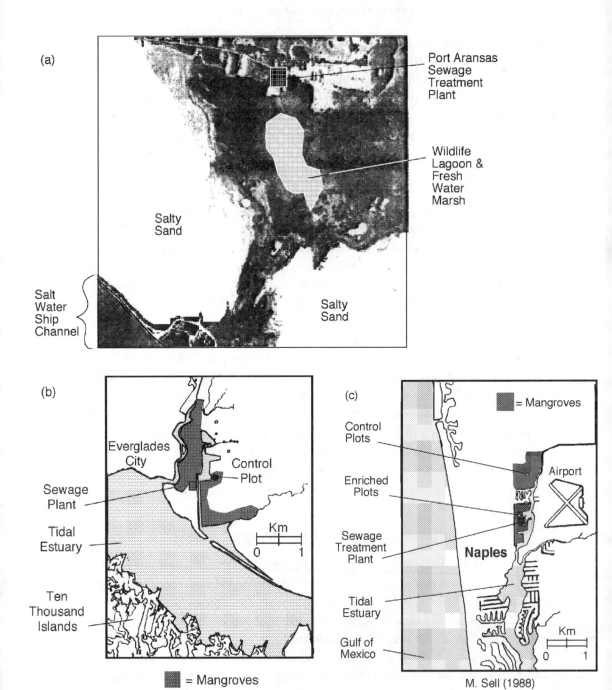

(a)

Port Aransas Sewage Treatment Plant

Wildlife Lagoon & Fresh Water Marsh

Salty Sand

Salt Water Ship Channel

Salty Sand

(b)

Everglades City

Control Plot

Sewage Plant

Tidal Estuary

Ten Thousand Islands

Km
0   1

= Mangroves

(c)

= Mangroves

Control Plots

Enriched Plots

Sewage Treatment Plant

Tidal Estuary

Gulf of Mexico

Airport

Naples

Km
0   1

M. Sell (1988)

(d)

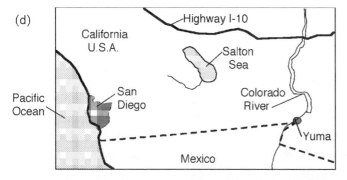

Highway I-10

California
U.S.A.

Salton
Sea

Pacific
Ocean

San
Diego

Colorado
River

Yuma

Mexico

FIGURE 12.3 Examples of adapting interface ecosystems. **(a)** Marshes created by treated sewage outfall at Port Aransas, Texas; **(b)** mangroves enriched by treated sewage in estuary at Everglades City, Florida; **(c)** mangroves enriched by treated sewage in the Gordon River estuary at Naples, Florida; **(d)** new aquatic ecosystems formed in landlocked Salton Sea, California; **(e)** *Spartina anglica*, exotic spread in New Zealand.

(e)

Patches of
*Spartina
Anglica*
Colonizing
Mud Flats;
Developing
Tidal
Channels

River Inflow

Concern with Exotic Invasions

Transplanting species is controversial. From some bad experiences with invaders in the past, people fear that introductions will upset previously stable environments. Species removed from normal roles no longer find the cues by which their behavior was adapted to fit their last system. They may find new energy sources and, at first, reproduce in great numbers, as rabbits did when first introduced into Australia, or as *Spartina* (smooth cordgrass) did when first planted on the coastal mud flats of New Zealand.

Species removed from their habitats may have no controls, at first, other than those from society. If they grow too fast, draining some aspect of the host system, they will eventually achieve their own control because that behavior will not be reinforced. A new parasite or predator may lower the output of its host until both are reduced in role.

When an existing system is disturbed and new species are added, there may be great fluctuations and pushes during the self-designing process. Some of the invasions of exotic organisms such as the gypsy moth, elm disease, and chestnut blight were extremely detrimental in the short term (although they helped restore diversity in the long run). Other invasions and transplantations have enriched the receiving systems with fast-growing trees and birds that are able to live in cities. After hundreds of years, a dozen bird species are adapted to the cities in Europe. Three of these, the pigeon, the English sparrow, and the starling, were transplanted to America, where they became the first birds adapted to its new cities. An invader may be filling a void in a new situation created by development, although humans are tempted to blame the invader for causing the disruptions. In our world of disturbed nature we may need some of these invasions. Sometimes a rare native species will emerge to serve a new situation. Fox squirrels became abundant on golf courses in Naples, Florida, where remnants of small cypress swamps were used as roughs.

When humans have already disturbed and displaced the natural system, conditions are different, and new species may be required to form a new network. The rapid success of some invaders is possible partly because they are able to use portions of the disturbed system that have fallen into disuse.[2] Intensive agriculture has low diversity simplified by insecticides and weeding. Government agencies limit species imports to protect agriculture. Unfavorable history with invaders displacing agricultural species has conditioned many scientists to oppose transplanting as a conservative play-safe policy. Tree species that reproduce with millions of windblown seeds obviously are more unmanageable than trees that reproduce with a few large seeds.

Perhaps conservatism was justified as long as natural systems were little disturbed and knowledge was lacking. Now, however, the natural systems are mostly disrupted, and the economy is integrating small separate systems into one larger system. Perhaps, now, more multiple transplantation is needed to develop new designs. There are likely to be temporary imbalances, epidemic-like population growths, and depletions from overgrazing, predation, and parasitism during the adaptation of a new design, but the final product may very well be a worthwhile new pattern of stability. This kind of self-design will take place anyway, but humans can help by providing multiple-seeding tests in selected situations.

Diversity Protection Against Epidemics

In high-diversity ecosystems there are enough herbivores, carnivores, and diseases to keep the levels of each species regulated so that not too many of any one species are allowed to develop. Local epidemics decimate clusters of single species, which stabilizes the whole ecosystem. Attempts to form plantations with a native species are likely to invoke the mechanisms that evolved earlier that keep all species regulated at safe low levels. For example, expanding the Brazilian rubber industry from trees scattered in a complex forest to plantation monocultures in the 1880s caused disease epidemics and financial disaster.

Vulnerable Species on Islands

Because of their small size, populations of plants and animals on small islands are easily displaced by other species or go extinct through population fluctuation. (See the information maintenance principles in chapter 8.) Remote, isolated islands such as the Galápagos Islands originally were seeded with fewer species than the number of existing opportunities (niches). Specialized endemic species varieties evolved that were locally adapted, including the finches made famous by Darwin (1859/1996). In Hawaii species developed that were wide-ranging generalists adapted to many local conditions but not very efficient in any one. More specialized species, introduced later, displaced the natives from environments where the exotics were more efficient. Preserving an ill-adapted native complex requires a park with a budget for energetic defense, perhaps sustained with tourist fees. In contrast, Puerto Rico, which is less isolated, had enough diversity in its forests to be efficient and stable. Dozens of introduced tree species added to the pool of biodiversity without displacing other species (Wadsworth 1997; Lugo 1997).

Adaptation to New Conditions with Exotics

Even on mainlands, exotics become dominant where society causes new conditions for which the native species are less adapted. Maintaining lower groundwater tables in south Florida caused the Australian *Melaleuca* to take over wetland areas. A strategy of drying out with excessive transpiration was then better adapted than the pond cypress strategy of reflecting infrared sunlight to minimize transpiration and keep wetlands wet (Myers 1975, 1983).

Other exotics were better adapted to south Florida soils after they were changed by agriculture. Natives were restored by first removing the changed soils. Civilization causes cultural eutrophy with increased runoff of nutrients (e.g., phosphorus, usable nitrogen, potassium). As explained with fig. 3.8, overgrowth specialists prevail as long as there is an excess of unused nutrients and energy. The pool of species adapted to the high-nutrient overgrowth stage includes many exotics.

The definition of an exotic depends on your choice of time needed to become native. Most species were exotics whenever they spread from their place of evolution. Cypress moved into Florida 4,000 years ago. Perhaps any species that has become a contributing part of dominant ecosystems, helping maximize empower for many years, should not be called exotic.

Adapting to Toxic Chemicals

Systems may be developed to regenerate heavily polluted waters. For example, many plant varieties are tolerant to toxic heavy metals such as lead and copper (Bradshaw et al. 1965; Odum et al. 2000b). Although energy is required for special physiology, specialization allows such species to predominate in toxic sites. For example, many species are adapted to serpentine soils and their heavy metal concentrations. Some species concentrate the metals (Brock 1978; Brooks 1998) and are proposed as removal agents (phytoremediation). Microorganisms evolve genetic varieties with toxic resistance rapidly. For example, houseflies developed resistance to DDT in a short time (O'Brien 1967).

Adapting to Pulses

Where there is a sharp season, animals may adapt by migrating. For example, in Africa during the wet seasons many animals disperse over savannas and marshes but concentrate at the water holes in dry season (Owen 1966). Fish and shrimp migrate into rivers and estuaries at the time of their greatest productivity. Salmon time their use of stream headwaters and the sea to periods when these habitats can provide appropriate emergy support (fig. 8.5). The life cycles of many animals are adapted so that their maximum energy needs coincide with the energy pulses of their habitats.

SELF-ORGANIZATION IN MICROCOSMS

Much of what is known about ecosystem self-organization came from study of ecological microcosms. *Microcosms* are ecosystems in containers that are small enough to be manipulated in scientific experiments. Nearly duplicate microcosm ecosystems can be obtained for treatments and controls by intermixing of their contents before testing. Self-organization has been extensively studied in micro-

cosms, especially in projects funded by environmental protection agencies to anticipate impacts on ecosystems.[3] Usually, many species are added to some container
while conditions of temperature, inputs, and outflows are set, often with some
environmental condition outdoors in mind. The species go through self-selection
of reinforcing loops, producing a stable metabolism and a complex network within
a few weeks. An example is the self-organizing terrarium in fig. 12.4. Although
miniaturized, nature creates new designs, just as it does over longer time periods
in larger-scale systems such as lakes, rivers, and landscapes.

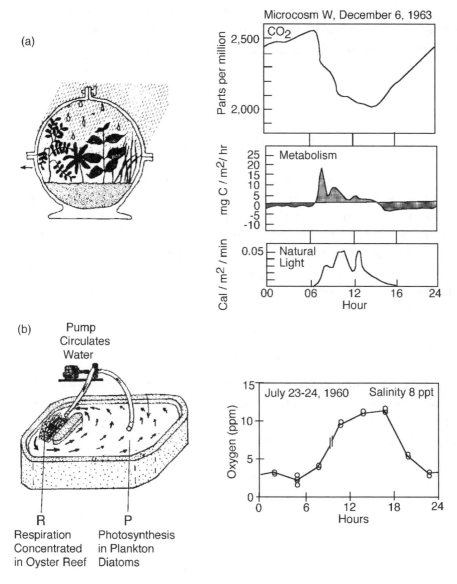

FIGURE 12.4 Graph of the characteristic rise and fall of metabolic quantities with alternating daytime photosynthesis and nighttime consumption. **(a)** Small terrestrial microcosm (terrarium)
(Odum and Lugo 1970); **(b)** reef of oysters and plankton in a circulating tank (Odum et al. 1963).

For example, small closed terrestrial microcosms (fig. 1.3b and 12.4a) balance the net photosynthesis of the day with the total consumption during the night. The measurements of oxygen and carbon dioxide show that such systems reestablish gaseous balance in 48 hours and have a steady undulation of atmospheres in a 24-hour rhythm. This is the case even when the microcosm is subjected to such disruption as gamma irradiation (short x-rays) of 25,000 roentgens, a dosage 30 times that necessary to kill people. The dependence of photosynthesis on respiration and vice versa is so well coupled in some aquatic systems that changes in temperature have little effect (see fig. 2.7; Beyers 1962). The increase in respiratory rates at higher temperature is slowed when the photosynthetic rate of providing products is less. The regulating mechanisms, equations, and electrical analogs of this cycle are given in appendix fig. A8. Because they contain whole ecosystems, ecological microcosms help people gain an overview of humanity and nature.

LANDSCAPE ORGANIZATION AND MANAGEMENT

Since the dawn of history, humans and their supporting ecosystems have been changing the countryside. The surges of human activity and concentration, first in one place and then another, sometimes seem as complex as the waves in a restless sea. Principles of spatial organization are sought by investigators in the fields of city and regional planning and landscape ecology. However, the surface of the earth, the ocean, the atmosphere, and the new ecosystems with people all fit a common plan recognized when patterns are studied with energy hierarchy in mind (chapter 4).

Spatial Hierarchy of People and Landscape

The organization of towns and cities on the landscape is familiar to all, with villages converging to towns and towns to cities and with people, materials, information, and money converging inward and diverging outward (figs. 4.8, 7.14, 9.14, and 10.5). The earth has been organized as a result of the converging of land through geologic processes to the mountain centers (fig. 5.9), of rainwaters into streams of larger and larger size (fig. 5.10), and of snows into glaciers. Figure 12.5 shows some other patterns of landscape organization.

Ecosystems are organized similarly, with small centers feeding to larger centers and materials recycling outward. For example, in a tree the small leaves and small roots are small centers that converge to larger centers, the limbs and trunk roots, and these all converge to the tree trunk. Each animal is a center where foods converge, and the animals converge their pathways to and from central places, such as nests. Many animals converge and diverge to animal cities.

(a) Interaction of Mountain and Stream Hierarchy

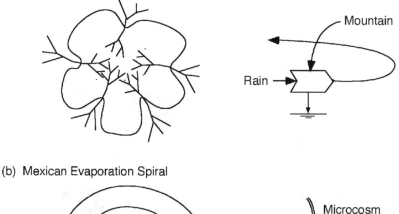

(b) Mexican Evaporation Spiral

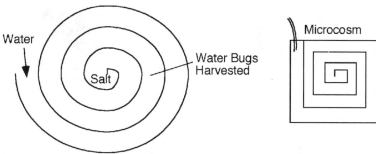

FIGURE 12.5 Hierarchical landscape designs. **(a)** Streams diverging from mountain center; **(b)** salinity-enhancing spirals.

Especially interesting is the way all these separate energy transformation hierarchies of society and environment are becoming connected. For example, the Luquillo Mountains of eastern Puerto Rico are a natural center of the earth from which rainfall forms several stream hierarchies, flowing outward but reconverging into small rivers before reaching the sea (fig. 12.5a). Located only a few miles away is a new emergy center, the city of San Juan (fig. 12.6a).Both the city and the mountains have high annual empower budgets and high transformities. Now, mountain waters are diverted to the city, and people go to the mountains for aesthetic renewal.

Although the hierarchical organization of old landscapes has been known a long time,[4] the free-thinking creativity of the high-energy age has encouraged leaders and governing boards who are responsible for city planning and transportation to experiment with interesting ideas. Unfortunately these experiments have not always been consistent with the energy hierarchy laws. (See chapter 4.) For example, maximizing traffic movement on throughways wrecked the hierarchical organization and functions of cities. Transportation should be managed to maximize the functions of the whole system. The recent "new urbanism" movement to increase concentrations of people by infilling (building houses on all vacant land in the city) may be contrary to the need to retain small hierarchies within the pattern of large hierarchies.

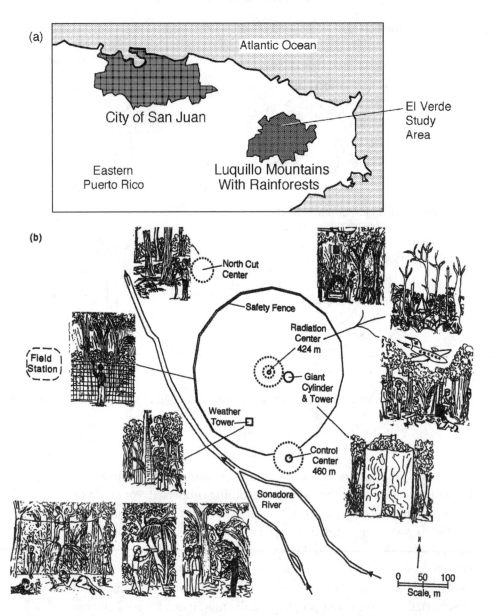

FIGURE 12.6 High-emergy centers in eastern Puerto Rico. **(a)** Urban metropolis and volcanic mountain covered with rainforest; **(b)** rainforest study sites of ecosystem task forces at El Verde in the Luquillo Mountains. (Sites were mapped with circular polar coordinates.) The site of irradiation with 10,000-curie cesium source is in the center. Nearby is the site with giant plastic cylinder for measuring metabolism (fig. 6.15c). The mechanically cut control center is downhill to the north, and the undisturbed control center is uphill to the south. Sketches by Ann Odum show some of the research activities, clockwise from upper left: 8-foot hog fence erected around the radiation area; pruning the cut center as a control area; helicopter brings in the 10,000-curie source in a lead pig; the irradiated center in successional recovery; visiting scientists swarm back after the irradiation; giant cylinder measures forest respiration; Atomic Energy Commission committee on site visit; U.S. Army teams map the trees and boulders; visiting scientists measuring the forest and the canopy with a cable car; meteorological tower and telemetering site.

Empower Density

The empower density increases from the rural periphery to the hierarchical centers (fig. 7.14). With a map of empower density in hand, planners can select locations for new enterprises where empower densities are compatible.

Environmental Systems Management

Environment, Power, and Society advocated that environmental management be on an ecosystems basis, to be preceded by ecosystem-level research, large-scale experiments, and synthesis, with energy indices and simulation models. The example given was the Rainforest Radiation Project, 1962–1970, in the Luquillo Mountain rainforests of eastern Puerto Rico (fig. 12.6)[5] (Odum and Pigeon 1970). A map of the research study area is shown with sketches of scientists at work . With summarizing network diagrams and quantitative data for the flows, it was possible to understand responses of the ecosystem to changes such as pollution, economic demands, addition of species, fire, radioactive isotopes, fertilization, harvests, hurricanes, earthquakes, catastrophic destruction, and changes in water regimes.

Since then, ecosystem projects of the International Biological Program, Hubbard Brook, and the National Science Foundation's Long Term Ecological Research projects, and many others all over the world, have provided the scientific basis for managing most types of environments (biomes in fig. 6.10). In 2000, the U.S. secretary of the interior put "environmental management on an ecosystem basis" as a main objective.

The Effect of Damaged Information

The Rainforest Ecosystem project in Puerto Rico was one of several that compared the sensitivity of ecosystems in different biomes to damage of their inherited information. In addition to metabolism measurements (fig. 6.6) and radioactive tracer studies of biogeochemical cycles, a high-intensity gamma irradiation source was put into the forest for ninety days and the effect of genetic damage studied (fig. 12.6). Organisms died in the 0.3 ha close to the center, but there was rapid selection of normal species and consumption of those damaged. Normal successional recovery was delayed about a year, the time required for reseeding with normal species from surrounding areas. The tropical forest, with its higher diversity, was more resistant and more resilient than the temperate ecosystems that received similar treatments.[6]

ECOSYSTEM SERIES AND LONGITUDINAL SUCCESSION

Because waters flow across the landscape in streams, strands, lakes, and wetlands, they carry exports of substances and species of one area to the next so that ecosystems form a functionally connected series. For example, the organic matter running off from forest areas supports stream ecosystems, with animal food chains using that organic matter. The change in an ecosystem in one place over time is called succession (chapters 4 and 6). When change in ecosystems is stretched out over the landscape with the flow of energy or matter, it can be thought of as longitudinal succession. The outflows of one zone determine the ecosystem in the next. Aquatic ecosystems in longitudinal series are part of the larger-scale hierarchical organization of the hydrological cycle. Several kinds of serial ecosystems are given in fig. 12.7. The scientific study of stream limnology has described many patterns of longitudinal change in conditions, species, and migrations. Failures in environmental management often come from attempts to manage one ecosystem alone, instead of as a part of a series.

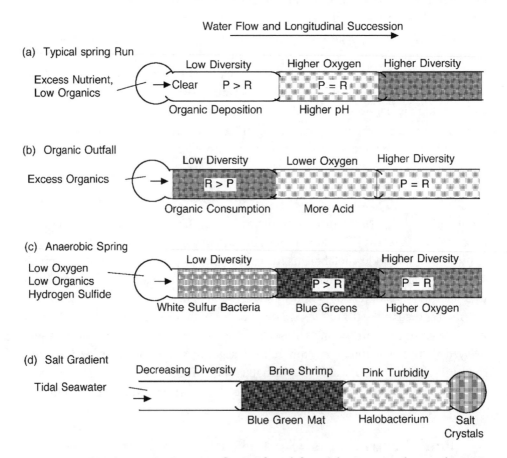

FIGURE 12.7 Ecosystem series in waters flowing from left to right. *P*, gross photosynthesis; *R*, ecosystem respiration. **(a)** Aerobic spring run; **(b)** release of waters containing organic matter; **(c)** anaerobic spring run; **(d)** seawater series with increasing salinity leading to salt deposition.

Photosynthesis/Respiration Ratio in Ecosystem Series

When there is more photosynthesis than respiration (high P/R ratio) there is net production of organic matter (chapter 2). If that organic matter flows to the next area, it stimulates more respiration. Conversely, if an ecosystem with flowing water has more respiration than photosynthesis, it generates nutrients (e.g., phosphorus, available nitrogen), and waters flowing to the next zone stimulate higher photosynthesis. Thus, longitudinal succession is toward a balance of photosynthesis and respiration ($P/R=1$). In other words, a line representing P and R in a longitudinal series moves toward the diagonal line in fig. 2.5. Field studies determine the kind of metabolism by measuring the dissolved oxygen through day–night periods (figs. 2.3).

When pools are seeded with inorganic nutrients, they develop a succession of algae initially, with a high P/R ratio and net production due to blooms of algae such as *Chlorella*. After several days, the algal cells age, and the ecosystem shifts to a state with more respiratory consumption than production.[7] If pools are connected in series and with steady inflow of inorganic nutrients, the first ponds have high P/R ratios, and the ones receiving the outflow waters several days later have an arrested successional stage of respiration greater than photosynthesis. If the inflow is organic waste, the first pond has a low P/R ratio and later ponds high P/R ratios (Oswald et al. 1953).

Overgrowth Zones and Nutrient Removal

The low-diversity overgrowths that occur in ecosystems while they have an excess of unused energy and materials (chapter 3) appear in the longitudinal series, using up the excess so that the next zone in the series receives less resources and develops the higher-diversity, efficient recycle mode. Today, much of south Florida, including the Everglades, has waters with high nutrients that drain agriculture, towns, and highways, causing a longitudinal series, first with low-diversity eutrophic overgrowth ecosystems dominated by species such as water hyacinths, cattails, water lettuce, and milfoils. These species take up the excess nutrients, depositing part into organic sediments. The waters flowing out of these areas are lower in nutrients, and the next ecosystems are oligotrophic (low nutrient) and dominated by sawgrass.

Everglades Series

Draining lands and channelizing Florida's Kissimmee and other rivers upstream from Lake Okeechobee and in the outflow streams from agricultural lands feeding the upper Everglades caused this watershed to receive many more nutrients than in the past, with some nutrients coming from the oxidation of peat. The ecosystems in Lake Okeechobee and the upper Everglades developed eutrophic overgrowths.

Large federal and state appropriations were authorized for Everglades restoration. In addition to restoring water regimes, the main task is to reduce the excess nutrient inflows and allow the overgrowth zones to take out the remaining excess. Restoring a nutrient absorption zone just south of Lake Okeechobee is the best plan (eutrophic slough alternative in fig. 12.8).[8] The figure shows that more than 5 billion emdollars are contributed by the Everglades ecosystems each year.

Spring Runs

Spring waters emerging from the ground, have nutrients but little organic matter and support ecosystems with clear water and a high P/R ratio, as explained with fig. 6.2. Fish food chains are based on periphyton (algal microzoa growths on plants). Downstream, the organic matter in the water increases, as do particles and turbidity, and the food chains include plankton-eating fishes. Recently, more and more of the Florida springs are losing their dissolved oxygen as the land ecosystems of the springs' drainscapes are disturbed.

Hot Springs

Hot springs emerge from the earth with temperatures too high for most life and cool as they flow downstream. A few species of bacteria and blue-green algae (Cyanophyta) are well adapted for photosynthetic production in the headwaters, as the diurnal oxygen curve shows (fig. 12.9a). Downstream, diversity of plants and animals increases. See studies by R. R. Wiegert (1975) and T. D. Brock (1978). The algal mats grow for a time and then break off in pulses, thus fitting the temporal regimen of all ecosystems (chapter 3).

Salt-Forming Series

All over the world, where the climate is dry, there are water flows through series of ecosystems that become more and more salty through evaporation (fig. 12.7d).[9] Nutrients also are concentrated in this process. These ecosystem series have been extensively domesticated to form salt at the end. When seawaters flow through a series of very shallow ecosystems, about 0.1–0.3 m deep, daily sunlight is concentrated in the small amount of water, causing very high temperatures favoring evaporation. In these shallow waters, at night, the oxygen reaches nearly zero soon after sunset. If the nutrients are adequate, a blue-green algal mat (like that in fig. 7.10) develops on the bottom, which helps catch almost all the sunlight. The increasing salt concentration inhibits the consumers, except for brine shrimp and black protozoa. Organic matter accumulates in the water and on the bottom.

In later stages of the series, the algae are replaced by pink blooms of *Halobacterium*, which have rhodopsin pigment. They use the sunlight to accelerate proton production and respiration. In the early morning, when the sun comes up on these high-

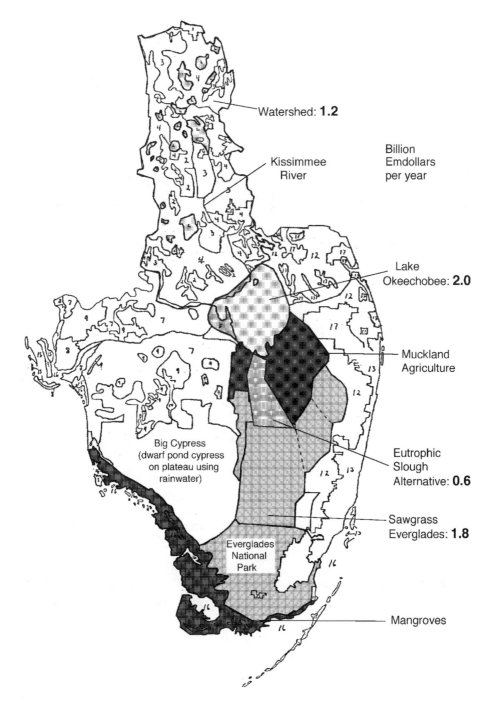

Watershed: **1.2**

Kissimmee
River

Billion
Emdollars
per year

Lake
Okeechobee: **2.0**

Muckland
Agriculture

Big Cypress
(dwarf pond cypress
on plateau using
rainwater)

Eutrophic
Slough
Alternative: **0.6**

Sawgrass
Everglades: **1.8**

Everglades
National
Park

Mangroves

FIGURE 12.8 Watershed of the Kissimmee River, Lake Okeechobee, and the Everglades showing the
eutrophic slough recommended for nutrient removal.

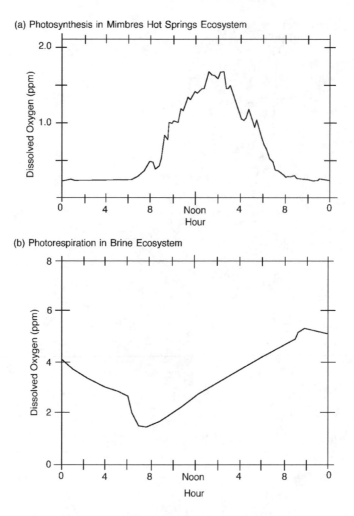

FIGURE 12.9   Diurnal records of oxygen. **(a)** Mimbres Hot Springs ecosystem, New Mexico, July 27, 1965 (Duke 1967); **(b)** brine ecosystem with photoxidation (Nixon 1969).

brine systems, the oxygen goes down at first (fig. 12.9b). Salt crystallizes out in the last stage and is raked up into piles and harvested for some final refining and sale.

WASTE-RECEIVING ECOSYSTEMS

Society has three kinds of recycle (chapter 8): The most valuable material objects of society are *reused,* the still concentrated material wastes of society are returned to the *material processing* industries, and the dilute wastes are *recycled* back to the environment. The high concentrations of chemical substances in wastes actually are rich reserves of energy and substrates capable of supporting special processes for special purposes. When they are released to the environment, the usual result is an interface ecosystem that reorganizes to use the byproducts. Wherever self-organization results in a functional interface ecosystem that uses the wastes and develops useful products, humans can learn from it and then transplant it to similar sites.

The standard waste treatment processes used in environmental engineering, such as trickling filters and activated sludge processes, are self-organized domestic ecosystems. These systems were put into concrete in the last century, researched to learn their properties, and used as "technology" ever since. Ecological engineering uses the same principle, without the pipes and concrete, allowing more flexibility to better fit different conditions.

While society is running on the great excesses of fossil fuels, there are inputs and outputs that never existed before: insecticides, detergents, pulp waters, toxic heavy metal flows, waters loaded with heavy organic loads, and wastes with radioactive isotopes. Where wastes are new, an ecosystem capable of using them may have to evolve using whatever suitable organisms exist somewhere on Earth. Wastes from the economy are now reaching the lakes and coastal seas, changing the microbes that predominate, displacing the ecosystem food chains to which the larger species are adapted. The newspapers are full of "red tides," "brown tides," "green blooms," fish kills, fishery collapses, sick marine mammals, and swimmers' health warnings. Microbial populations self-organize, even evolving genetically in days and months, but much more time is required for self-organization to find or evolve compatible large species.

Figure 12.10 shows the outflows of society passing through waste-receiving ecosystems capable of becoming partners to the main economy by supplying benefit and closing the loop. The sketch describes what is evolving in the whole earth on a large scale and locally, with or without human planning. Where solid wastes are released, strange new land ecosystems develop, including the gas-generating glop in the landfills. With the help of tides, ships, and storms, there is seeding of microbes, plankton, fishes, and even terrestrial animals, from all over the world. These waste interface ecosystems close the material cycles, which converge to the economy and diverge to the environment in the process (figs. 4.8 and 4.12).

Wetlands for Tertiary Waste Treatment

A large project of the Rockefeller Foundation and the National Science Foundation[10] in 1973, starting with cypress domes (fig. 12.10b), showed the feasibility of using Florida and Michigan wetlands to interface with treated municipal wastewaters, recharging groundwaters for return of water to urban use (Ewel and Odum 1984). In these wetlands and those studied elsewhere, species adapted, growth accelerated, and nutrients were absorbed in woody and peat layers. National and international workshops and a documentary National Science Foundation film about these projects cascaded into hundreds of projects around the world. Examples are the packaged wetland gardens used for treatment in Puerto Rico by Greg Morris and in Mexico (Nelson 1998). Kadlec and Knight (1996) summarize and provide guidelines for ecological engineering use of freshwater wetlands.

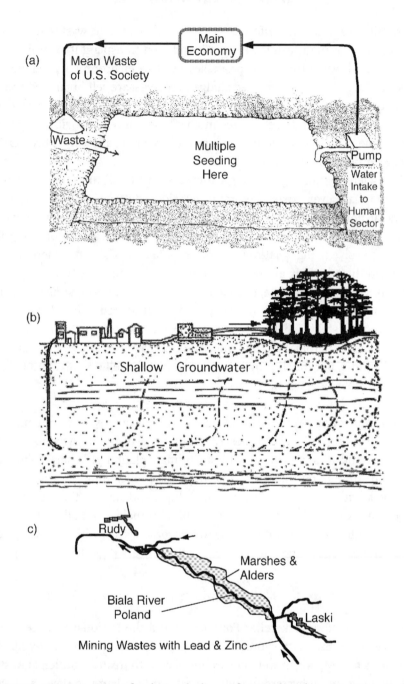

FIGURE 12.10 Overview concept for closing the loops of materials and services connecting society and environment. **(a)** Concept sketch from *Environment, Power, and Society;* **(b)** cypress dome waste recycling project in Florida in 1973 (Ewel and Odum 1984); **(c)** Biala River Wetland in Poland, which has been filtering lead and zinc wastes from mining for 400 years (Odum et al. 2000b).

Organic Waste Outfall in Streams

Environmental engineering textbooks usually include the longitudinal series that is found downstream from the release of wastes with high concentrations of organic matter. First, rapid decomposition uses up the oxygen, which may reach nearly zero. Then the oxygen rises again as respiration decreases, oxygen diffuses in, and algae develop, contributing photosynthesis. The textbook oxygen sag curve is an example of longitudinal succession. The characteristic pattern of organisms found is called a Saprobe series (Kolkwitz and Marsson 1902).

Pond Series for Industrial Waste Treatment

Passing industrial wastes through a series of ponds has become a common method of interfacing concentrated chemical wastes with the environment. For example, a series of ponds receiving refinery wastes have high respiration rates but gradually develop a more balanced pattern of photosynthesis and respiration (Martin 1973; Copeland 1963).

Mesocosm Series for Treatment

John Todd developed systems of waste treatment using mesocosm tanks in series. Each unit in series has an entirely different kind of ecosystem so that substances not used up in one unit pass to an ecosystem with species adapted to use the remainders. Other series are reviewed in *Ecological Microcosms* (Beyers and Odum 1993).

Estuarine Ecosystem for City Nutrient Discharge

As society harvests more and more fishery products from the sea, it does little to reinforce the fishery ecosystems. What it does do is return wastes to the sea. A large-scale reorganization is taking place in which marine ecosystems and their products are becoming compatible with the wastes, and the coastal seas are changing to fit a fossil fuel–driven economy. The new patterns fit the concept in fig. 12.10. In the future, when fuels and urban wastes are less important, the seas may reorganize again.

A conscious effort to induce and observe the process of self-organization of a marine interface ecosystem adapting to secondarily treated municipal wastes was carried out from 1969 to 1970 in North Carolina marshes with support from the Sea Grant Division of the National Science Foundation (Odum 1985, 1989). Secondarily treated sewage of Morehead City, North Carolina, was mixed with estuarine waters and continuously discharged into three ponds, which were well seeded with estuarine organisms (fig. 12.11). A regime of fluctuating algal blooms developed

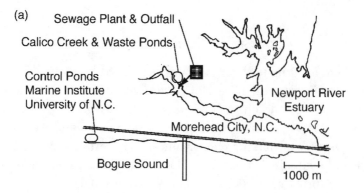

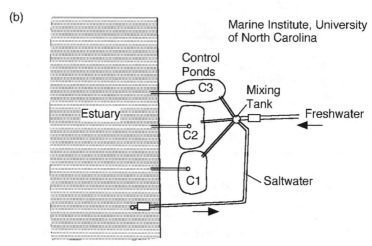

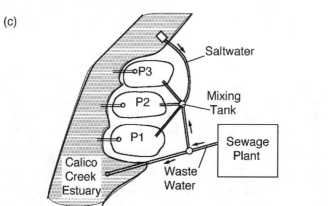

FIGURE 12.11 Experimental estuarine ponds receiving wastewaters at Morehead City, North Carolina. Salinity was regulated between 17 and 22 parts per thousand salt (Odum 1985, 1989). **(a)** Location map; **(b)** experimental design: 3 control ponds received city freshwater mixed with saltwater; **(c)** 3 ponds received treated sewage wastewaters mixed with saltwater; **(d)** net daytime photosynthesis and night respiration data of the experimental ponds from diurnal oxygen curves measured by Smith (1972).

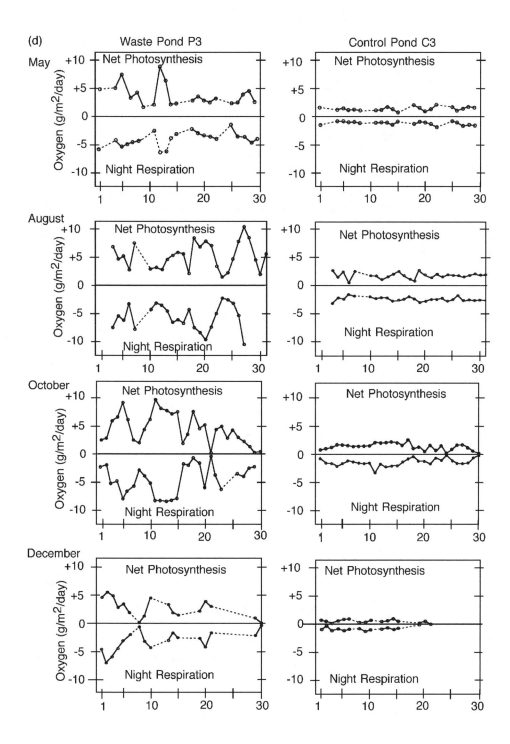

and with it some members of the natural system that were adapted to varying oxygen. An extremely green *Monodus* bloom reformed each winter. The food chain in the fertile waste pond had low diversity, with substantial numbers of blue crabs at the top. In this experiment the wastes were being mineralized by the pond ecosystem, providing free tertiary treatment. Control ponds receiving tap water and estuarine water had an ordinary kind of estuarine community. On a small scale, the pond experiments suggested that helping self-design is useful.

Figure 12.11d records the net photosynthesis and night respiration. The two are coupled and self-regulating (chapter 2); more of one causes more of the other. Waste ponds were much more productive, but the low oxygen at night required a specialized food chain (blue-green algae and animals capable of obtaining oxygen directly from air).

Marine Salt Marshes with City Nutrients

Some of the secondarily treated wastewaters from Morehead City were released into the salt marshes of Calico Creek and flowed in and out of the marshes with the tides. David Marshall (1970) studied the plants and animals of this marsh and compared them with marshes in a similar but unfertilized creek used as a control (fig. 12.12). Results showed that the marsh grass and periwinkle snails were more productive in the treatment marshes. McMahan et al. (1972) found a high diversity of terrestrial insects in the treatment marshes.

Mangroves with City Nutrients

In several places in south Florida, nutrients in municipal wastewaters were absorbed by mangrove swamps, which showed increased growth rates as a result (fig. 12.3) (Sell 1977). Mangroves along the "gold coast" of the Yucatán were effective in keeping nutrient wastes produced by hotels from reaching coral reefs (Nelson 1998).

Heavy Metal Uptake in Wetlands

Hundreds of published papers have shown that wetlands are effective filters for heavy metals, which they store in their peat. For example, our recent book reviews literature, experiments, and field studies of lead uptake, including detailed case histories of lead filtration by wetlands in Florida and Poland (fig. 12.10c) (Odum et al. 2000b). There are many mechanisms of uptake, including sulfide binding, binding to lignin, particle filtering, and oxide precipitation.

Gaia Design of Wetlands

Because the self-organization of wetland ecosystems is so versatile in taking up inorganic and organic substances harmful to ordinary life (Fuhr 1987), it is apparent

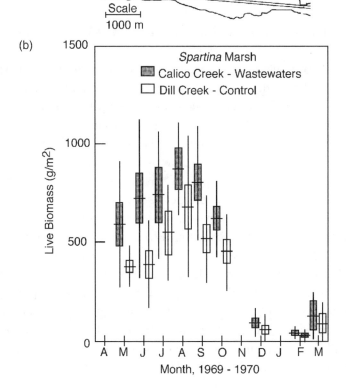

FIGURE 12.12 Net production of salt marsh vegetation (*Spartina alterniflora*) in Morehead City, North Carolina 1969–1970, measured by David E. Marshall (1970). **(a)** Location map; **(b)** comparison of Calico Creek Marsh, receiving secondarily treated municipal sewage waters, with Dill Creek Marsh, used as control.

that the evolution of wetland species and wetland ecosystems fits the Gaia idea that ecosystems designs that support life are reinforced. In other words, *life organizes the biosphere for life* (Lovelock 1979), although many scientists don't like to use such purposeful language to describe system mechanisms. Because many peat deposits eventually become coal, wetlands are the means to return toxic materials to the geologic cycle. Digging up peat deposits containing toxic materials is a bad policy in environmental management.

A discovery was made when the sunlight reflected by wetland vegetation was monitored. Species that are adapted to high nutrients, such as bald cypress, grow faster and absorb much of the visible and infrared light energy, which accelerates their transpiration and productivity (fig. 6.16). In contrast, wetland species such as pond cypress, which live in headwater swamps on tablelands and receive mainly rainwater, reflecting 80% of the near infrared, grow very slowly, thus saving water that runs off to make surrounding areas more productive (McClanahan 1991). Water budgets can be controlled by the wetland species planted. This is an example of self-organization of ecosystems for regional performance.

Recycling Pulp and Paper Mill Wastes to the Trees

In the natural decomposition of wood in swamps, brown lignin is deposited in peat, with some of the lignin decomposing slowly and coloring the runoff waters. In the manufacture of paper from trees, much of the lignin that constitutes about a third of the logs is discharged as black wastewater. Studies in Florida (Keller 1992) show that these highly concentrated wastes can be recycled back into the pine lands, draining through their small scattered swamps joining the natural black waters. After filtration by the swamps, some of this water recharges the groundwaters, helping to restore the large water volumes used by the paper industry.

For example, for 50 years pulp mill wastes from a plant near Perry, Florida, passed down the Fenholloway River (fig. 12.13), which had been declared an "industrial river." Normal ecosystems of the river and the estuarine grass flats were replaced with microbial blooms, eliminating most recreation and fishing. Three alternatives for restoration of the river were considered in 2000: piping the wastes to the estuary, pumping the lignin waters back into the pinelands from which the pines were harvested, and constructing a strand (flat wetland bed of vegetation) from the mill into coastal marshes. Much higher emdollar values were estimated for the latter two choices (table 12.1), ultimately with less cost to the industry.

Humans in Closed Chambers and Life Support in Space

Spaceship Earth is a mostly closed-to-matter ecosystem supporting 6 billion people on sunlight, tide, and earth energy. But what does it take to support a human in a closed-to-matter chamber to go into space? Published estimates range from 1 square meter per person to 3 million square meters (740 acres).

A controversy has raged for years between those who try to design a simple life support system and those with ecological views who suggest that a complex self-organizing system is required. Extensive studies of ecological microcosms on many scales and types have established the stability of high-diversity, self-organizing ecosystems. We explained the issues in *Environment, Power, and Society* in 1971 and reviewed the progress again in 1993 (Beyers and Odum 1993).

Environmental scientists at several international symposia recommended the complex ecosystems (Cooke et al. 1966; Odum 1963). Despite repeated failures of

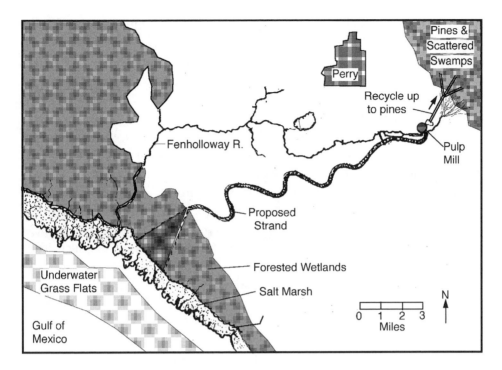

FIGURE 12.13 Map showing alternatives for discharge of pulp mill wastes in north Florida, including the Fenholloway river to the Gulf of Mexico, pumping northeast into pine and swamp lands, and a proposed wetland strand into coastal marshes.

TABLE 12.1 Evaluation of Alternatives for Pulp Paper Wastes in North Florida (million $/year)

ALTERNATIVE	EMPOWER BENEFIT[a] (EM$/YR)	COSTS ($/YEAR)	BENEFIT – COST (EM$ — $)
Fenholloway River, existing	−15.1	−?	<−15.1
Pipe to estuary	−13.4	−3.0	−16.4
Recycle uphill into pine lands	+17.0	−4.8	+12.2
Construct wetland strand	+13.5	−1.7	+11.8

Note: Evaluations made with S. Medina, M. T. Brown, and J. Justan.

[a] Emergy increases minus emergy losses.

simple systems to supply stable life support, the National Aeronautics and Space Administration (NASA) spent millions of dollars for research to develop simple systems but almost nothing for the self-organizing complex ecosystem approach. By the year 2001, some microcosms with selected species were sent into orbit, but still no complex self-organizing ecosystems were tried. The idea of relying on the power of unspecified complexity apparently clashed with a NASA tradition of single species, simple mechanisms, and precise knowledge of whatever is done. This myopia is related to the arrogance of 20th-century urban culture, which tries to control nature rather than have faith in it.[11]

Environment, Power, and Society used the sketch in fig. 12.14a to call for a large closed-to-matter chamber, a mesocosm with people, to be set up as a living model of the earth and a prototype for space. Research proposals to government agencies were not funded.[12] A complex ecosystem was criticized as taking too much space, but apparently space is necessary.

Biosphere 2

Starting around 1985, a diversified team of engineers, scientists, and communicators, led by philosophical engineer John Allen and financier Edward P. Bass, launched an amazing project of site development, symposia, collaboration with Russian space scientists, module testing, and plans to use complex ecosystems to support humans. Finally in 1989, a giant glass mesocosm covering three acres, costing a half billion dollars, was built in the mountains of Arizona. Starting in 1991, it supported eight people for two missions spanning 3 years, based on solar energy and outside electrical power. It proved the self-organizing complex concept. It was remarkably successful at showing what is necessary for a colony in space. It is now operated by Columbia University and should be a national laboratory. It could be linked to NASA with laws requiring that self-organizing complex ecosystems be tested in space.

Because respiration of organic soils exceeded photosynthesis at first, it mimicked our current earth condition, which has more consumption with carbon dioxide release than photosynthetic uptake. Thus, the oxygen was reduced, and some had to be added from a tank truck for people to survive this transient period. (Fortunately, spaceship earth is protected from temporary imbalance by a larger store of oxygen in its atmosphere.) Like smaller microcosms, Biosphere 2 was self-correcting. Figure 12.15b shows the high rates of metabolism nearly balanced in 1994. There were lush vegetative growths and soils beginning to develop tropical characteristics. Because of limitations for importing tropical insects, the biosphere had only some ordinary insects, ants, and cockroaches. Without the usual tropical insects and thus without birds and pollinators, tropical plants predominated that can spread vegetatively. Originally, the mesocosm was multiple seeded with hundreds of species. In the rainforest biome more than 400 plant species were started (fig. 12.15c). By the year 2000, 125 of these survived the self-organization to

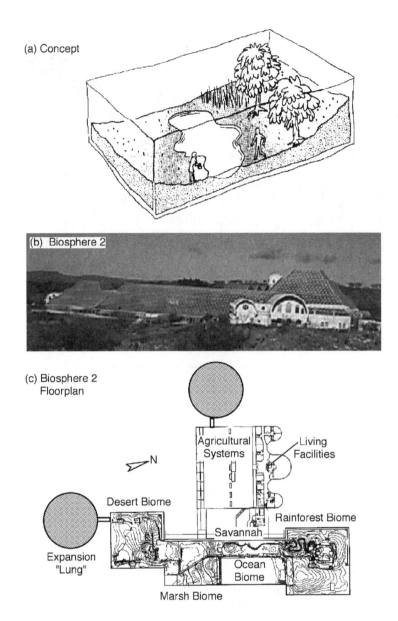

(a) Concept

(b) Biosphere 2

(c) Biosphere 2
Floorplan

Agricultural
Systems

Living
Facilities

N

Desert Biome

Rainforest Biome

Savannah

Expansion
"Lung"

Ocean
Biome

Marsh Biome

FIGURE 12.14 Concept of supporting people in closed high-diversity, self-organizing life support systems in space. **(a)** Sketch from *Environment, Power, and Society;* **(b)** Biosphere 2, Oracle, Arizona; **(c)** floorplan and pressure equalization chamber. Utility ducts and underground access tunnels for maintenance are not shown (Marino et al. 1999).

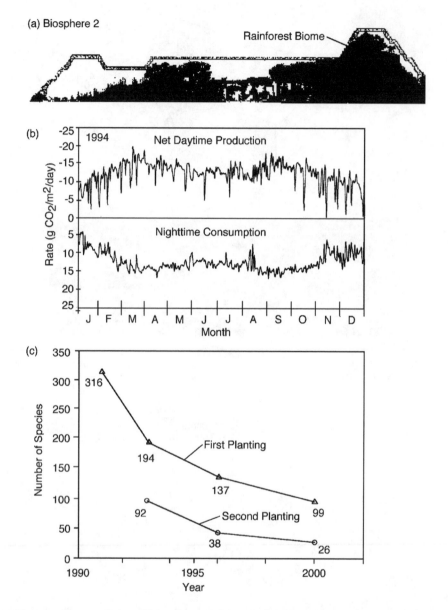

(a) Biosphere 2

Rainforest Biome

(b)

1994   Net Daytime Production

Nighttime Consumption

Rate (g CO_2/m^2/day)

J F M A M J J A S O N D
Month

(c)

350

316

300

250

194

200

First Planting

150

137

92

100

Second Planting   99

50

38

26

0

1990   1995   2000
Year

Number of Species

FIGURE 12.15 Characteristics of Biosphere 2. **(a)** View; **(b)** metabolism measured from diurnal changes in carbon dioxide provided by S. Pitts and D. Kang (Marino et al. 1999). Net daytime production is plotted upward and night respiration plotted downward so that the distance between the lines estimates gross production; **(c)** diversity of plant species in the rainforest biome section (fig. 12.14) from Linda Leigh (1999).

contribute to life support. Much about this strange new world is summarized in a recent book (Marino et al. 1999).

Ecological Engineering

The mission of engineering is sometimes described as *developing designs to solve problems.* Whereas environmental engineering concerns primarily technology and processes before connecting with nature, *ecological engineering matches technology with the self-design of connecting ecosystems.* An ecological engineered interface accepts some byproducts of civilization so that they are environmentally beneficial and returns something of benefit to society (fig. 12.2) (Mitsch and Jorgensen 1987). Like jujitsu, ecological engineering uses small energy to have large effects by going with, rather than against, the self-organizing tendencies of nature. As energy becomes less available for the current high-intensity agriculture, forestry, aquaculture, and biotechnology, productivity may be sustained by using more of the lower-energy, delicate, and elegant techniques of ecological engineering, which let the ecosystems do most of the work. The following are initiatives for ecological engineering:

Encourage Industrial Ecology

Industrial ecology concerns the relationship of industry to environment. Many initiatives from industry have sought to make a better fit with the earth (Lowe et al. 1977). Ecological engineering can help industry serve society and reduce costs with better environmental interfaces. Polluting wastes can be processed to be environmentally beneficial. More of each material cycle can be managed by industry. In place of environmental regulations, tax incentives can make it possible for industries to generate environmental wealth for the public and still compete economically.

Manage with High Diversity

Conditions without excess resources are required to sustain landscapes, clean waters, and provide climax, high-diversity ecosystems with tight recycle (chapter 7). For example, beautiful coral reefs provide biochemical diversity, stability, recreation, and tourist attractions. Sustaining high-diversity tropical forests protects watersheds and their discharges.

Increase Yield by Pulsing

Where a yield is needed, a sharp pulse in conditions may be introduced to which ecosystems respond with pulses of net production for repair. For example, alternating pulses of growth and harvest are normal in aquaculture and forestry. Pulses of freshwater alternating with saltwater maximize oyster production in estuaries of the Gulf of Mexico. Extremes in wastewaters can be turned into benefits.

Channel Energy with Extremes

Energy can be channeled into yield through addition of an extreme condition to which most species are not adapted. For example, if waters are managed to maintain a briny condition with salinity of about 4–5%, the consequence is an abbreviated community of algae, plankton, and a few species of fish with energies channeled into high production yields. A briny ecosystem can supply more protein and food, as in the Laguna Madre of Texas and Mexico (fig. 6.16).

Connect Products to Decomposition

Combining organic wastes and nutrients develops compost and natural silage useful for soil enrichment. Worms for fishing, edible mushrooms, and detritus fishes are examples of yields derived from organic decomposition. In the canals among the sugar cane fields of Guyana, leafy materials from the land support fermentation and decomposition and large fish populations. E. P. Odum (1962) suggested a detritus agriculture.

Control with Limiting Materials

Supply small amounts of limiting materials to control ecosystem self-designs. For example, adding fertilizer nutrients to sterile, high-salinity, shallow tropical lagoons causes benthic mats of blue-green algae to form that support flamingos and the increased evaporation necessary to make salt (fig. 12.7d) (Nixon 1969).

Control Ecosystems with Key Species

Managers of complex factories manipulate the outputs with fingertip exertions because they know how the controls work. Managing environments is possible by manipulating control species.

Larger animals of high transformity (e.g., lions, tigers, elephants, tuna, whales, hawks), control the productivity of many ecosystems through the timing of their behavior. Operating down through the energy hierarchy, they exert control over vast areas. Given enough understanding of ecosystems, humans can apply small energy to control by means of these larger species (sometimes called keystone species). Conversely, disorganized ecosystems can be restored to functional structure through restoration of control species. For example, disorganized coral reefs overgrazed by sea urchins in the Indian Ocean were restored through protection of triggerfish (McClanahan 1990; McClanahan and Obura 1996).

Increase Yield with Cross-Continent Transplants

When species are transplanted from one continent where they were part of a stable system to another, higher yields may result. At least for a time, the diseases and cues from the environment that provide limits are absent. Plantations of eucalyptus trees have grown especially well away from Australia, and plantations of Brazilian rubber have done better in Africa and Asia. Sometimes imports do too well, becoming destructive pests.

The temperate salt marsh grass *Spartina anglica,* covering former mud flats in the western United States, Europe, Asia, and the southern hemisphere, has been attacked in some countries with poison as a dangerous exotic and welcomed in other countries as a new source of coastal production and protection. Our studies in Havelock, New Zealand, showed that it was enriching coastal fisheries and wildlife.[13] Figure 12.3e shows the area of rapid spread over mud flats, with high net production, extensive flowering, self-organization of tidal canals, and huge concentrations of tiny snails (22,000 to 72,000/m^2).

Modify Elemental Ratios to Induce a Product

To stimulate desirable species, modify ratios of chemical substances, such as the nitrogen-to-phosphorus ratio.

Remove Unusual Organic Growth Substances

Charcoal has long been used to remove organic substances in water treatment engineering. Dispersal of charcoal into open waters may be feasible to control blooms of odd species based on unusual organics. Maryanne Robinson (1957) found that zooplankton growth increased when they ingested charcoal in the concentration range of 5–30 ppm (and they looked cute with their black intestines). Clay minerals and ground glass were also beneficial.

Use Microcosms to Anticipate Smaller-Scale Properties of New Systems

Where new conditions are anticipated, set up microcosms or mesocosms with those conditions, seed them with species believed to be available, and study the properties that emerge. To scale up the observed behavior, multiply space dimensions and rates by 10 to infer possibilities on the larger scale with larger species.

Manage Whole Cycles

Assign to industries the management of the whole cycle of materials, from mining through industrial production, use, and back through reuse, reprocessing, and environmental recycle. Whole cycle material management may be aided by tax incentives.[14] Authorization for the establishment of a new industry might require that the complete loop of the manufacturing process be specified, including provisions for the regeneration of wastes and their connection to another part of the economic system where they are usable.

Evaluate Ecological Engineering Projects Using Emergy

The success of ecological engineering measures cannot be evaluated with money alone because it does not include the effects on real wealth coming from the environment. However, evaluating with emergy or emdollars includes both the work of environmental systems and that by humans in the economy. (Emdollars are the economic equivalent of emergy; see chapter 9.)

Provide Incentives for Environmental Reinforcement

When people harvest fish, quail, timber, or other beneficial items from ecosystems for their use, they drain energy, which is a stress. Where a stressed population was formerly competitive, it may be displaced by some other species. In other words, harvesting damages the loop that people find useful unless equivalent emergy is returned to reinforce the energy chain. The second dashed line in fig. 9.13b shows the flow of money needed as incentive. Enough money is needed to return emergy equal to that drawn from the environment.

One way to administer the reinforcement is to have an environmental agency receive the money from environmental users and provide the emergy-evaluated equivalent amount of appropriate services to the environment owner or manager. Sometimes reinforcement can be made with nutrient waters. Wildlife managers use funds from hunters to plant seed-bearing vegetation for quail.

RECREATION AND ECOTOURISM

To understand what happens when people use a wilderness area for recreation and aesthetic benefit, consider the tourist system (fig. 12.16). Note the contributions of nature (ecosystems and scenic lands), the money contributed by tourists (dashed lines), and the purchased inputs of fuels, electric power, goods, and services.

The diagram shows how ecotourism can be a mixed blessing. The tourists interacting with the environment (thick pathways) drain resources from the environment as they carry out experiences. Most of the money they bring goes right out to buy other inputs, with very little money going to local people. Demand by the tourists raises the price of food, land, and other environmental products, which reduces local living standards. Rarely are there arrangements to reinforce the environment in exchange for the load placed on it.

Emergy flows of all the pathways in fig. 12.16 contribute to recreational value. Table 12.2 is an example of emergy-emdollar evaluation of a coastal recreation system near Corpus Christi, Texas.

In many areas, tourists are attracted by historic sites and art, which carry the emergy of work done in the past, the work of preservation, and the emergy used to develop appreciation and recognition. Some treasures seem priceless because great energies were used in the past to develop, save, and create an appreciation of the site or object (see fig. 8.12). Battlefields and their monuments are examples.

EMERGY-EMDOLLAR EVALUATIONS OF THE ENVIRONMENT

Environmental values to society often are underappraised in planning of economic developments. They have no easily recognized dollar value that can be compared with the figures available for the proposed new enterprise. However, emdollars can evaluate all uses in the same terms so that planning boards can recognize real wealth in the public interest.

In order to make environmental choices for the public good, *the general policy is to select alternatives with the most production and use of emdollars (including society and its supporting environment)*. This also means maintaining the natural capital necessary for sustained environmental production. For more on concepts, details, and examples see the author's book *Environmental Accounting*.

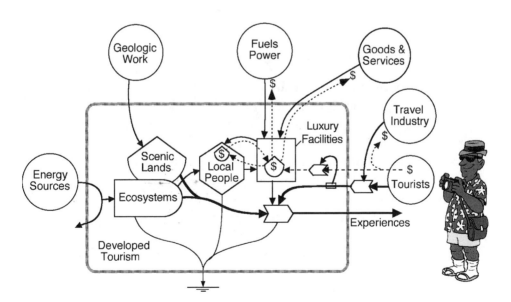

FIGURE 12.16 Energy systems diagram of recreation and tourism showing the contributing pathways. See table 12.2 for an evaluation of a coastal area with a tourist economy near Corpus Christi, Texas.

TABLE 12.2 Coastal Values, Corpus Christi, Texas (year 2000 dollars)

Economic analysis: $4,862/acre/yr[a]	Emergy-emdollar evaluation (em$/acre/yr)
Environment, geologic processes[b]	2,290
Fuels, power[b]	8,640
Recreational expenditures[a]	2,150
State and federal management[a]	16,300
Visitor participation[c]	2,630
Industries[a]	57
Total emergy-emdollars	32,067

[a] Economic evaluation by Anderson (1960), $374/acre/yr in 1960 dollars expressed in year 2000 dollars by multiplying by 13, the ratio of emergy per dollar for 1960 to that for 2000 (Odum 1996). 1 acre = 4,047 m^2. Emdollars of paid services were calculated equal to the cost of services in dollars.

[b] Emergy contributions of environment and energy from Texas Coastal Study by Odum et al. (1987).

[c] Visitor contributions estimated by multiplying time spent times metabolism times the transformity 7 E7 secal/cal and dividing by 2.4 E11 solar ecal/2000 $.

An example of emergy-emdollar evaluation is the recreation system based marine bays near Corpus Christi, Texas (table 12.2), which evaluates the principal pathways in the interface web contributing to tourists' experiences in fig. 12.16.[15] The important pathways from the ecosystems would be ignored in typical economic evaluations. The valuation web, with its many connections and branches, is complex. It is little wonder that evaluating nature leads to argument and controversy.

Replacement Value of Ecosystems and Natural Capital

Monetary cost–benefit ratios often are used to justify the removal of ecosystems. However, in many public hearings and court cases in Florida, emdollar evaluations have been used to show how original ecosystems are much more valuable than estimated with market prices.

In an incident in North Carolina, surveyors cut a swath through a climax forest in a public recreation park. They evaluated their damage to the public to be $64 per acre for the wood they cut. Public protest reflected the deep sense that this was not just. The real stored value was the energy times the time required to replace the forest. To replace a complex, diverse, and beautiful forest takes about 100 years. If the photosynthesis in a forest is 40 kcal/m^2/day, the solar transformity of old trees is at least 20,000 secal/cal, and the emdollar equivalent of the paid human work is about 1.9 E11 solar emcalories/dollar, the value of replacing an acre of this forest is about 620,000 emdollars. A single tree about 100 years old is estimated in this way to be worth 3,000 emdollars.[16]

Life Support Values of Diversity

There is conflict over the importance of preserving the complexity and diversity of birds, flowers, and rarer members of the natural ecosystems. To many, the birds and bees seem to be a luxury compared with the economy and human survival. Too often, people have said that waste is not bad until it is killing people.

Ecosystems package their biochemistry within the species, each one being different and each one occupying a different pathway of processing of materials and energy. The more diversity there is in the biochemistry, the more special abilities there are to generate products and mineralize wastes. As long as there are complex ecosystems, humans are protected. Diverse ecosystems are each human's diversified life support. Ecosystems take the toxins first, and often plant and animal populations are killed. As long as complexity remains, there is some protection for people. As the levels of waste rise beyond the capacity of the life support system to process them, the diversity is destroyed. Then waste toxicity can pass the natural front lines to kill people directly. Sensitive species are an early warning system.

The species–area graphs introduced in fig. 6.11 can be used to estimate the minimum area to support species diversity for a short time. The areas can be multiplied by empower density and divided by emergy/money ratios to obtain emdollar values of diversity. To sustain diversity indefinitely requires larger areas in order to support enough individuals of each species population to maintain genetic adaptation (chapter 8).

HUMAN LIFE SUPPORT

The relationship of humans to their life support can be considered from the individual's point of view and from the global perspective. Emergy use per person was compared in chapter 7, table 7.4.

Metabolic Share and Imbalance

The basic world life support production and consumption is at least 1 g of oxygen per square meter per day (Sundquist 1993). With a world population of 6 billion people and a world surface area of 5.2 E14 m^2, we can divide to obtain about 86,666 m^2 per person (21.4 acres). Each human's portion of the earth's life support system is 8.7 E4 g oxygen per day, or 3.5 E5 kcal of metabolic energy.[17]

With the rise of the Industrial Revolution based on fossil fuel use and the cutting of vegetation, the earth's metabolism has been out of balance, with consumption of oxygen and organic matter greater than production. Large storages of oxygen in air and sea protect us from immediate difficulty in running low on oxygen, but the imbalance has caused a steady rise in carbon dioxide. More carbon dioxide and related gases have increased the atmosphere's greenhouse effect in storing atmospheric heat, increasing the temperature of the seas, which has added more vapor to the air and caused stronger storms. Views differ on how much melting of ice from Greenland and the Antarctic will occur, causing sea level rise. Increased vapor may add more snow and ice on top, while there is more melting at sea level.

In 2001, international discussions sought national actions to reduce carbon dioxide emissions. However, decreasing availability of fossil fuels was causing rises in energy prices. Many predict that this will cause the global economy to climax and world fuel consumption to decrease soon. Less carbon dioxide is released per unit of electric power now than earlier because more of our energy is coming from natural gas and nuclear power plants.

The model in fig. 12.17 shows how carbon dioxide, which is a weak acid in water, reacts with calcium carbonate of soils, limestones, and the ocean to make bicarbonates, a process that buffers the effects of change. If part of the earth's bare lands are revegetated and the consumption of fuels decreases 1% per year, simulation of this model generates a peak in atmospheric carbon dioxide in about 30 years.

We recently evaluated alternative ways of reforestation of degraded tropical lands (Odum et al. 2000a). The most rapid method, with the highest ratio of emdollars to dollar cost, was natural reforestation when high-diversity seed sources are adjacent; otherwise, spread of successional exotics is the quickest way to establish a canopy to aid the later colonization of diversity (Lugo 1997).

Global Impact of Civilization

Earth no longer has unlimited reserves of protecting ecosystems. The role of humans and their civilization is best measured in emergy units so that natural and urban processes are on the same basis. Seventy percent of the global empower is from fossil fuels, mostly from the urban civilization (fig. 9.4). The other 30% of our empower is from global photosynthetic production, which has been reduced or is threatened. As energy costs rise and the rate of fuel use climaxes and decreases, there will be new danger to the environment as too many people try to use these limited reserves to replace the rich fossil fuels. See chapter 13.

Ecological footprint is an attractive way to explain global impacts to people (Wagernagel and Rees 1995). It is the area of the earth required to support something. Emergy units must be used to make such calculations rigorous.[18] Omitting the support area that is required to replace fossil fuels and minerals is a major error that underestimates the footprint of society. The footprint of urban civilization is several times larger than the present earth surface. In other words, it is unsustainable without nonrenewable fuels.

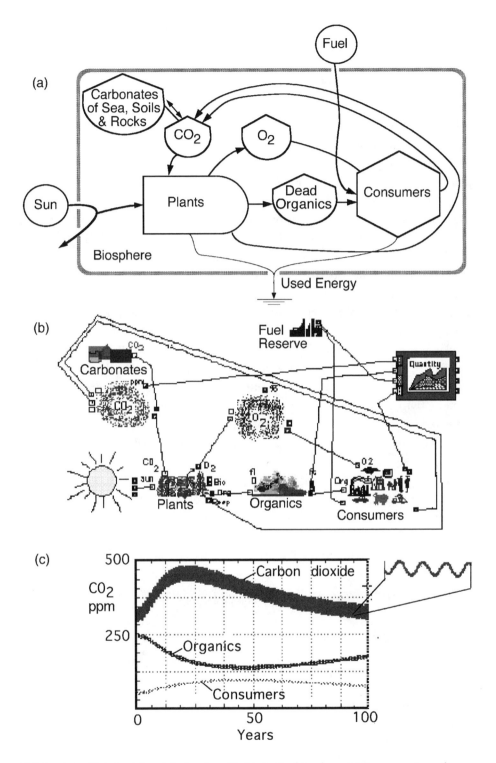

FIGURE 12.17 Global minimodel of carbon dioxide in the biosphere. **(a)** Energy systems diagram; **(b)** computer screen of the simulation program EXTEND showing the blocks (pictorial icons) programmed to represent the systems symbols of the model. Connecting the blocks with a mouse automatically sets up the equations (Odum and Odum 2000); **(c)** simulation with 1% decrease in fuel use per year and reforestation.

Rights to Global Life Support and the Changing Energy Basis

Surviving cultures give individuals their life support, which may come to be regarded as a right (Stone 1972/1988). The human right to share safe life support is basic to many of the legal battles in the defense of the environment. There can be nothing more important to an individual than the opportunity to breathe, drink water, eat nontoxic food, and move about with safety. Long taken for granted, these rights are not free but are paid for daily by the metabolic works of the life support system processing wastes and byproducts. The water and mineral cycles, the complex of complicated organisms that process varied chemicals, and the panorama of ecological subsystems that organize and manage the earth's surface are not the property of individuals but are part of the essential basic right to share the life support system. During times of growth, when population density was small, it was not very harmful for religions to give humans dominion over the earth and property owners rights to ecosystems (chapter 11).

Today actions that foul the life support system threaten the lives of individuals everywhere. As chemicals causing cancer and other debilitating diseases become more concentrated and widely distributed, the threat of stress turns formerly private activities into public menaces. Riparian owners who waste or pollute water take the right of life from people who have never seen them. People are threatened when laws allow nature's genes to be patented and withheld from free public use.

In a time of economic climax and descent, the present rights to develop are likely to be replaced by rights to share environmental support. In the future the rights of landowners may not include the right to remove the landscape's role in sustaining soils, waters, and atmosphere. Conservation easements are becoming popular as a way to write protection of life support into land deeds. In the United States, a constitutional amendment may be required for protecting nature and its life support when it is realized that individual rights to survive and natural rights are the same.

SUMMARY

Many new kinds of ecosystems are developing at the interface of society and environment, examples of adaptive self-design. Knowledge of the principles of adapting ecosystems has developed from field studies and experiments with ecological microcosms. Biosphere 2 provided a living model of the biosphere that showed what is required for humans in space. Self- organized waste-receiving ecosystems are now in worldwide use, an application of ecological engineering. To aid plans for maximizing real wealth, contributions of environment and economy are evaluated quantitatively on a common basis as emergy or emdollars. Many practical ways

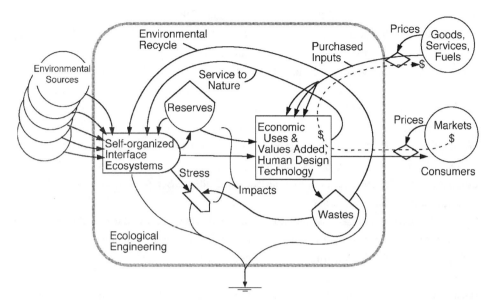

FIGURE 12.18 Systems diagram summarizing ecological engineering and the relationships of the human economy and environment.

are now available to keep emerging systems of nature and society symbiotic as the urban economy passes through its climax and descent. Carbon dioxide reaches a maximum and decreases with reforestation and diminishing fuel use. Figure 12.18 summarizes the relationship of self-adapting ecosystems to the economic and technological designs of society.

Bibliography

Anderson, A. 1960. *Marine Resources of the Corpus Christi Area*. Research Monograph No. 21. Austin: Bureau of Business Research, University of Texas.

Beyers, R. J. 1962. Relationship between temperature and metabolism of experimental ecosystems. *Science*, 136: 980–982.

Beyers, R. J and H. T. Odum. 1993. *Ecological Microcosms*. New York: Springer-Verlag.

Bradshaw, A. D., T. S. McNeilly, and R. P. G. Gregory. 1965. Industrialization, evolution and development of heavy metal tolerance in plants. In *British Ecological Society Symposium No. 5*, 327–343. New York: Wiley.

Brock, T. D. 1978. *Thermophilic Organisms and Life at High Temperatures*. New York: Springer Verlag.

Brooks, R. R., ed. 1998. *Plants That Hyperaccumulate Heavy Metals. Their Role in Phytoremediation, Microbiology, Archaeology, Mineral Exploration and Phytomining*. Cambridge, UK: CAB International, University Press.

Brown, M. T. and S. Ulgiati. 1999. Emergy evaluation of the environment: Quantitative perspectives on ecological footprints. In S. Ulgiati, *Advances in Energy Studies, Proceedings of the International Workshop at Porto Venere, Italy,* 223–240. Rome: Museum of Science.

Cooke, G. D., R. J. Beyers, and E. P. Odum. 1966. The case for the multispecies ecological system with special reference to succession and stability. In J. F. Saunders, ed., *Bioregenerative Systems,* 129–139. NASA SP-165. Washington, DC: National Aeronautics and Space Administration.

Copeland, B. J. 1963. *Oxygen relationships in oil refinery effluent holding ponds.* Ph.D. dissertation, Oklahoma State University, Stillwater.

Christaller, W. 1933 (1966 edition). *Central Places in Southern Germany,* trans. by C. W. Baskin. Englewood Cliffs, NJ: Prentice Hall.

Crow, T. 1976. Effects of gamma radiation on the biomass structure of arboreal stratum in a northern forest. In J. Zavitkovski, ed., *The Enterprise, Wisconsin, Radiation Forest. Radioecological Studies,* 68–78. Washington, DC: Energy Research and Development Administration.

Darwin, C. 1859 (1996 edition). *The Origin of Species.* Oxford, UK: Oxford University Press.

Doxiadis, C. 1968. Man's movement and his city. *Science,* 162: 326–332.

Duke, M. E. L. 1967. *A Production Study of a Thermal Spring.* Ph.D. dissertation, University of Texas, Austin.

Elton, C. S. 1958. *The Ecology of Invasions by Animals and Plants.* London: Methuen.

Ewel, K. C. and H. T. Odum, eds. 1984. *Cypress Swamps.* Gainesville: University of Florida Press.

Fuhr, F. 1987. Non-extractable pesticide residues in soil. In T. R. Roberts, *Pesticide Science Biotechnology,* 381–389. Boston: Blackwell Scientific.

Kadlec, R. H. and R. L. Knight. 1996. *Treatment Wetlands.* Boca Raton, FL: CRC Press.

Keller, P. A. 1992. *Perspectives on interfacing paper mill wastewaters and wetlands.* M.S. thesis, University of Florida, Gainesville.

Kolkwitz, R. and M. Marsson. 1902. Grundsätze für die biologische Beurteilung des Wassers nach seine Flora und Fauna. *Mitt. Kgl. Prufungsanstalt f. Wasserversorgung und Abwasserbeseitigung,* 1.

Leigh, L. S. 1999. *Diversity limits and Biosphere 2.* Ph.D. dissertation, University of Florida, Gainesville.

Lovelock, J. E. 1979. *Gaia: A New Look at Life on Earth.* Oxford, UK: Oxford University Press.

Lowe, E. A., J. L. Warren, and S. R. Mann. 1977. *Discovering Industrial Ecology, an Executive Briefing and Sourcebook.* Columbus, OH: Batelle.

Lugo, A. E. 1997. The apparent paradox of reestablishing species richness in degraded lands with tree monocultures. *Forest Ecology and Management,* 99: 9–20.

Mann, C. 2000. *Emergy evaluation of the Everglades restoration.* Non-thesis paper for master's degree, University of Florida, Gainesville.

Marino, B. D. V., H. T. Odum, and W. J. Mitsch, eds. 1999. *Biosphere 2: Research Past and Present.* Amsterdam: Elsevier.

Marshall, D. E. 1970. *Characteristics of Spartina marsh receiving treated municipal sewage wastes.* Incomplete master's thesis, University of North Carolina, Chapel Hill.

Martin, G. D. 1973. *Optimal control of an oil refinery waste treatment facility: A total ecosystem approach*. M.S. thesis, Oklahoma State University, Norman.

McClanahan, T. R. 1990. *Hierarchical control of coral reef ecosystems*. Ph.D. dissertation, University of Florida, Gainesville.

McClanahan, T. R. 1991. Infrared reflectance and water budgets of vegetation in reclamation after phosphate mining. In *Evaluation of Alternatives for Restoration of Soil and Vegetation on Phosphate Clay Settling Ponds*, 159–172. Publication No. 03-076-094. Bartow: Florida Institute of Phosphate Research.

McClanahan, T. R. and D. O. Obura. 1996. Coral reefs and near shore fisheries. In T. R. McClanahan and T. P. Young, eds., *East African Ecosystems and Their Conservation*, 67–100. New York: Oxford University Press.

McMahan, E., R. L. Knight, and A. R. Camp. 1972. A comparison of microarthropod populations in sewage-exposed and sewage free *Spartina* salt marshes. *Environmental Entomology*, 1(2): 244–252.

Mitsch, W. J. and S. E. Jorgensen, eds. 1987. *Ecological Engineering: An Introduction to Ecotechnology*. New York: Wiley.

Murphy, E. F. 1967. *Governing Nature*. Chicago: Quadrangle.

Myers, R. 1975. *The relationship of site conditions to the invading capability of* Melaleuca quinquenervia *in southwest Florida*. M.S. thesis, University of Florida, Gainesville.

——. 1983. Site susceptibility to invasion by the exotic tree *Melaleuca quinquenervia* in southern Florida. *Journal of Applied Ecology*, 20: 645–658.

Nelson, M. 1998. *Ecological engineering of saline wastewater treatments for tropical coastal developments*. Ph.D. dissertation, University of Florida, Gainesville.

Nixon, S. 1969. *Characteristics of some hypersaline ecosystems*. Ph.D. dissertation, University of North Carolina, Chapel Hill.

O'Brien, R. D. 1967. *Insecticides*. New York: Academic Press.

Odum, E. P. 1962. Relationship between structure and function in the ecosystem. *Japanese Ecology*, 12: 108–118.

Odum, H. T. 1963. Limits of remote ecosystems containing man. *American Biology Teacher*, 25: 429–443.

——. 1985. *Self Organization of Ecosystems in Marine Ponds Receiving Treated Sewage*. Chapel Hill: University of North Carolina Sea Grant SG-85-04.

——. 1989. Experimental study of self organization in estuarine ponds. In S. E. Jorgensen and W. J. Mitsch, eds., *Ecological Engineering, an Introduction to Ecotechnology*, 291–340. New York: Wiley.

——. 1996. *Environmental Accounting, Emergy and Decision Making*. New York: Wiley.

Odum, H. T., S. J. Doherty, F. N. Scatena, and P. A. Kharecha. 2000a. Emergy evaluation of reforestation alternatives in Puerto Rico. *Forest Science*, 45(4): 521–530.

Odum, H. T., G. A. Knox, and D. E. Campbell. 1983. *Organization of a New Ecosystem, Exotic Spartina Salt Marsh in New Zealand*. Technical Report #38, Center for Wetlands, University of Florida, Gainesville. A complete manuscript was prepared in 1999 by G. A. Knox, H. T. Odum, and D. Campbell, "The ecology of a salt marsh at Havelock, New Zealand, dominated by the invasive marsh grass *Spartina anglica*."

Odum, H. T. and A. Lugo. 1970. Metabolism of forest floor microcosms. In H. T. Odum and R. F. Pigeon, *A Tropical Rain Forest*, section I, 35–54. Oak Ridge, TN: Division of Technical Information, U.S. Atomic Energy Commission.

Odum, H. T. and E. C. Odum. 2000. *Modeling for All Scales, An Introduction to Simulation.* San Diego, CA: Academic Press.

Odum, H. T., E. C. Odum, and M. Blissett. 1987. *Ecology and Economy: "Emergy" Analysis and Public Policy in Texas.* Publication #78. Austin: Lyndon B. Johnson School of Public Affairs, The University of Texas.

Odum, H. T. and R. F. Pigeon, eds. 1970. *A Tropical Rain Forest.* Oak Ridge, TN: Division of Technical Information, U.S. Atomic Energy Commission.

Odum, H. T., W. L. Siler, R. L. Beyers, and N. Armstrong. 1963. Experiments with engineering of marine ecosystems. *Publications of the Institute of Marine Sciences, University of Texas,* 9: 373–453.

Odum, H. T., W. Wojcik, L. Pritchard Jr., S. Ton, J. J. Delfino, M. Wojcik, J. D. Patel, S. Leszczynski, S. J. Doherty, and J. Stasik. 2000b. *Heavy Metals in the Environment, Using Wetlands for Their Removal.* Boca Raton, FL: Lewis.

Oswald, W. J., C. G. Gotaas, H. B. Golueke, and W. R. Kellen. 1957. Algae in waste treatment. *Sewage and Industrial Wastes,* 29(4): 436–457.

Oswald, W. J., H. B. Gotaas, H. F. Ludwig, and V. Lynch. 1953. Algal symbiosis in oxidation ponds. *Sewage and Industrial Wastes,* 25: 684–691.

Owen, D. F. 1966. *Animal Ecology in Tropical Africa.* San Francisco: W.H. Freeman.

Robinson, M. 1957. The effects of suspended materials on the reproductive rate of *Daphnia magna. Publications of the Institute of Marine Science, University of Texas,* 4(2): 265–277.

Sell, M. G. 1977. *Modeling the response of mangrove ecosystems to herbicide spraying, hurricanes, nutrient enrichment and economic development.* Ph.D. dissertation, University of Florida, Gainesville.

Smith, M. 1972. *Productivity of marine ponds receiving treated sewage wastes.* M.A. thesis, University of North Carolina, Chapel Hill.

Stone, C. D. 1972 (1988 edition). *Do Trees Have Standing Toward Legal Rights for Natural Objects?* Stanford, CA: Tioga.

Sundquist, E. T. 1993. The global carbon dioxide budget. *Science,* 259: 934–941.

Wadsworth, F. H. 1997. *Forest Production in Tropical America.* Washington, DC: USDA Forest Service Agricultural Handbook #710.

Wagernagel, M. and W. E. Rees. 1995. *Our Ecological Footprint: Reducing Human Impact on Earth.* Philadelphia: New Society Publishers.

Wiegert, R. 1975. Simulation modeling of the algal–fly components of a thermal ecosystem. In B. C. Patten, ed., *Systems Analysis and Simulation in Ecology,* 157–181. New York: Academic Press.

Zipf, G. E. 1919. *Human Behavior and the Principle of Least Effort.* Reading, MA: Addison-Wesley.

Notes

1. As explained in *Environment, Power, and Society*, a conscious effort to study ecological engineering and natural self-organization was begun with ponds at the Institute of Marine Science of the University of Texas at Port Aransas, 1956–1963 (Odum et al. 1963), with terrestrial microcosms at the Puerto Rico Nuclear Center, University of Puerto Rico, Rio Piedras, 1963–1966 (Odum and Lugo 1970), and with the ponds of the University of North Carolina Institute at Morehead City, 1966–1970 (Odum 1985, 1989).

2. Charles Elton's *The Ecology of Invasions* (1958) documents the many famous cases of species from other areas having large success and becoming principal members of new areas.

3. The summarizing book *Ecological Microcosms* describes the ecosystems that developed, the processes measured, and principles involved in experiments with living models of most kinds of ecosystems (Beyers and Odum 1993).

4. For example, the hexagonal shapes of areas supporting villages, villages converging to towns, and so on (Christaller 1933/1966), the distribution of many small cities and few large ones (Zipf 1919), and the increasing building structure and human circulation toward city centers (Doxiadis 1968).

5. Studies of the Atomic Energy Commission, the Puerto Rico Nuclear Center, University of Puerto Rico, and U.S. Forest Service, which later became a Long Term Ecological Research project of the National Science Foundation, led by Robert Waide and Ariel Lugo. The experience suggests the following steps for ecosystem management task forces:

Description and identifications. Well-trained subprofessional systematists under the technical supervision of senior systematists at museums and elsewhere, but under the field direction of a project supervisor, focus on a study area, map and tag the plants, make reference collections of the animals, identify bacteria, and publish identification manuals to aid other studies.

Network diagram. Systems ecologists coordinate those with knowledge of the natural history and autecology to help prepare initial network diagrams of the principal living and nonliving storages and flows, including dominant species. These models should include one with interactions with economic development on the next larger scale.

Chemical and energy flows. Field measurements and literature data are used to estimate biomass chemical constituents, chemical cycles, and metabolic rates. These data are used to make models quantitative and allow preliminary computer simulation.

Whole system experiments. Large-scale field experiments are arranged to determine the effect of changes that may be expected from the anticipated trends in economy and environment. Ecosystem experiments designed to determine how the whole network adapts are different in concept from the "null hypothesis, one-factor experiments" done by those studying processes one at a time on a smaller scale.

Evaluation and management. If the models adequately simulate observed changes, they can be rerun to consider recommendations for changes involving the areas. The quantitatively evaluated models can be used to calculate contributions and impacts in emergy and emdollars to determine choices that contribute most to the public good.

6. The cesium source was sent to the U.S. Forest Service laboratory in Rhinelander, Wisconsin, when it was removed from the Luquillo rainforest in Puerto Rico (Crow 1976).

7. Many experimental algal ecosystems were studied, stimulated by possibilities of efficient food production from waste nutrients (Oswald et al. 1957).

8. Emergy and emdollar evaluations were made of this system by Charles Mann (2000) with M. T. Brown and independently by the author. These values are much larger than the economic costs of the newly authorized plans to restore the Everglades. The proposed eutrophic slough (trapezoidal area below the lake in fig. 12.8) yields more value than the plan to leave that zone in agriculture and build new treatment wetlands and underground circulation for nutrient uptake.

9. Scott Nixon (1969) studied microcosms in arrested succession at constant salinities and succession along a brine gradient in a series of microcosms. Photo-oxidation in brines was discovered and simulated from the metabolism (fig. 12.9) independent of later physiological studies.

10. After a planning grant in 1972 a national workshop was held, and the Center for Wetlands of the University of Florida was founded in 1973. In Florida, cypress dome swamps were studied with and without wastewaters from a package treatment plant, and with and without burning. After 6 years, wastes were connected to city sewage lines, and the swamps were allowed to adapt again. Results were included in a book (Ewel and Odum 1984) and many papers. A sister project in Houghton Lake, Michigan, stored wastewater in winter and sprayed it on marshes in other seasons.

11. A review of the efforts to support people in closed systems up to 1994 is given in chapter 19 of *Ecological Microcosms* (Beyers and Odum 1993). Also see the book on research in Biosphere 2 (Marino et al. 1999).

12. It is a normal attitude of scientists to seek an understanding of mechanisms in the phenomena that they study and to be skeptical of anything they don't yet understand. This causes many scientists and engineers to reject self-organization and even ecological engineering that uses self-design. These attitudes were picked up by journalists seeking controversy with the ignorance and small-scale prejudices of their age. They misled the public to believe that the project was questionable science and a failure.

13. Results of a joint project of the International Divisions of National Science Foundations of U.S. and New Zealand (Odum et al. 1983). *Spartina anglica* (a fertile variety of a sterile hybrid) rapidly covered muds in the intertidal zone below the native *Leptocarpus* marsh. Production of organic matter increased greatly, and tide traps showed juvenile fishes using the marshes as a nursery. Habitat for shore birds that use mud flats decreased.

14. Dispersing concentrations so that ecosystems can be stimulated, rather than stressed, generates much more emergy-emdollars of public benefit than economic benefit. Public and private enterprises need some economic incentive, estimated from the emdollar evaluations, so that they can complete recycle to the environment and still stay in business.

15. An economic evaluation made of a coastal recreation area near Corpus Christi, Texas, by A. Anderson in 1960 was reported in *Environment, Power, and Society*. The area of 228,480 acres received 13,914 visitors per day. Annual economic values found were recreation, tourism, and fishing ($166/acre) and oil, gas, shells, cooling waters, and savings on shipping

($203/acre). Management by state and federal agencies was estimated at $1 million annually. The earlier analysis was modified as an emergy-emdollar evaluation in table 12.2 using data on environmental energy sources and fuel use per acre from our 1987 evaluation of the Texas coastal counties (Odum et al. 1987). Time of visitors was multiplied by their metabolism and transformity. Emergy values in solar emcalories were divided by 2.4 E11 solar emcalories/$ in 2000 to obtain year 2000 emdollars.

16. Accumulated emergy in an acre of forest:

(100 yr)(4,047 m²/acre)(4 E4 cal/m²/day)(365 days/yr)(20,000 solar emcal/calorie)/(1.9 E11 solar emcal/$) = 621,960 em$.

17. Metabolic share of the earth, including the oceans:

(1 g oxygen/m²/day)(21.4 acres/person)(4,047 m²/acre) = 8.7 E4 g oxygen per person per day.

18. It is incorrect to give equal weight to energy of different kinds or concentrations. Brown and Ulgiati (1999) evaluated footprint as the emergy of the area required to absorb the products of all or part of human society.

THE GROWTH of civilization on the nonrenewable reserves of the earth is surging to a climax of information miracles, stormy economics, turbulent populations, concentrated wealth, and bewildering complexity. Although the future is always masked by the oscillations of smaller scale, the empower of society may be at climax in transition to times of receding energy. This last chapter uses principles of energy hierarchy and pulsing to anticipate the future, suggest adaptive policies, and seek a prosperous way down.[1]

MODELS AND WARNINGS OF THE GLOBAL FUTURE

Figure 13.1a summarizes how the economy of society is supported by renewable energies of environmental production and reserves of fuels and minerals accumulated over geologic time. Many authors who compare the remaining nonrenewable resource reserves with the steep growth of population, consumption, and the economy warn with frantic voices that the present civilization is unsustainable.[2]

Computer Simulation Scenarios

Since *Environment, Power, and Society* was published, hundreds of models and computer simulations relating the global future to resources have been published, some simple and some complex. They all generate a similar curve of growth (fig. 13.1b), like that anticipated in *Environment, Power, and Society*,[3] cresting as the fuels diminish and then descending. Many scientists, including this author, take the top-down overview of Earth and resources and make aggregated, simple minimodels of resources and assets.[4] As explained by the energy hierarchy concepts of this book, they believe the larger-scale trends are determined at the larger scale into which the smaller-scale mechanisms have to fit. They obtain smoothed curves of growth, climax, and descent like the example in fig. 13.1b. See also figs. 3.12, 9.11, and 10.2e and f.

(a) Class of Models

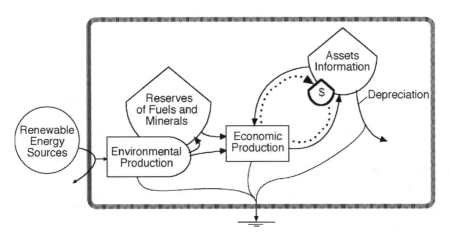

(b) Typical Simulation

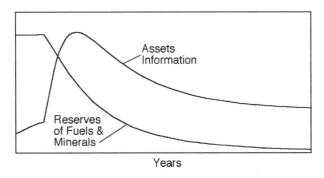

Years

FIGURE 13.1 Growth curves of typical simulation minimodels on nonrenewable and renewable energy. **(a)** Energy systems diagram; **(b)** pulse of typical simulations.

Others, who think from the parts up, put in more of the complexity that they see around them.[5] They believe that the causality of the larger scale is generated by the smaller-scale mechanisms. Their simulation curves have more short-range ups and downs. These may be appropriate as long as they are realistic in driving the models with renewable and nonrenewable energy resources. If these simulations are smoothed, the overall shapes of the curves are similar to the one in fig. 13.1b. The timing of the peak varies because of different estimates of nonrenewable reserves, different assumptions about the efficiencies of energy conversion to assets, and different assumptions about depreciation rates.[6]

The fact that most of the resource-based models peak sooner if more fuel reserves are assumed sometimes is regarded as a paradox. More available energy causes a steeper growth frenzy, which uses up resources faster. If this principle is correct, then finding more nonrenewable resources will not delay descent.

Voices of Concern

Ever since the fuel-driven Industrial Revolution began, and especially since energy concern became intellectual mainstream in the 1970s, more and more voices have warned of the eventual downturn caused by resource depletion. Some come from resource scientists who calculate and project reserves; some come from historians and anthropologists who record the demise of past civilizations. Because these warnings have come so often and for so long, present society concerned with the short run ignores "doomsayers," whom they believe to have been wrong before. It's the situation of the boy who cried wolf too soon and too often.

Warnings started with Malthus in 1798, predicting disaster from exponential population growth. Other influential examples include Nussbaum (1983), Diamond (1998), Catton (1982), Ehrlich (1968, 1974), Fey and Lam (2000), Hardin (1993), Henderson (1991, 1996), Georgescu-Roegen (1977), Kaplan (1996), and Tainter (1988). Duncan (1989) called boom and bust "the Olduvai Theory." Fey and Lam (2000) called for a Manhattan Project–style national effort to prepare. Recently, there have been many warnings of catastrophic descent on the Internet at Jay Hanson's Web site, DIEOFF, and discussion group (energyresources@onelist.com) and Bruce Thomson's RunningOnEmpty@onelist.com.

Energy Crisis Inquiry

Today, government and the collective media have little institutional memory of the extensive discussions of energy before and after the 1973 energy crisis. The new generation can learn from Adams (1988), Ayres (1978), Cook (1976), Cottrell (1955), Fluck and Baird (1980), Freeman (1974), Hall et al. (1986), Herendeen and Bullard (1975), Hubbert (1949), Krenz (1976), Lovins (1977), Odum and Odum (1982, 2000, 2001), Peet (1992), Pillet and Odum (1987), Pimentel and Pimentel (1979), Slesser (1978), Smil (1991, 1998), Spreng (1988), Stanhill (1984), Steinhart and Steinhart (1974), and the *Annual Reviews of Energy* issued each year.

Emergy Reserves and Alternative Sources

Consensus seems to be developing among resource geologists as to the reserves of global fuels available and the time when the cost of processing will allow prices to rise and the economy to turn down. Campbell (1997) provided a record and extrapolation of fuel production and use (fig. 13.2), with a peak in about 2009. It shows how growth of the global economy was arrested in the late 1970s with high inflation caused by the oil embargo's high fuel prices. Fleay in *The Decline of the Age of Oil* (1995), Campbell in *The Coming Oil Crisis* (1997), Youngquist in *Geodestinies* (1997), and Magoon in *Are We Running Out of Oil* (2000) reviewed the history and availability of oil and other resources required for civilizations,

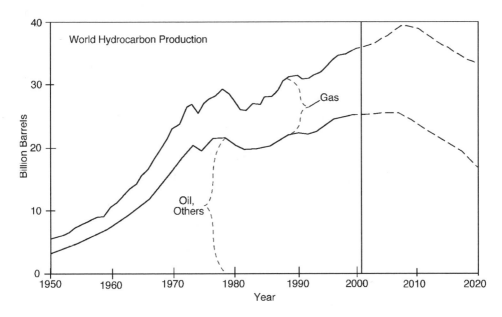

FIGURE 13.2   Past and future trend in fuel use by the United States (Campbell 1997).

also concluding that the present trends in population and economic growth are not sustainable.

The majority of people on Earth have developed an almost religious faith that technology can provide alternative energy sources. They don't understand that technology requires available energy. Even leaders in science and engineering think available energy of low transformity such as sunlight can be substituted for high-transformity fuels (Brown 2000). They have not learned the law of energy transformation hierarchy (chapter 4). Chapter 7 already explained that there are no large energy sources known with enough net emergy contribution to replace the fossil fuel bonanza as it recedes.

Gradual Descent or Crash

Although many simulation scenarios show some kind of gradual descent to a lower-energy world (fig. 13.1b), most of the voices of concern anticipate or warn of a crash, with apocalyptic collapse of society, decimation of population, loss of knowledge, and destruction of the environment. Some hold the view that crashing and starting over is more humane. The weakened civilizations may be abruptly removed by conquest. The conquest of Rome by less developed "barbarians" was analogous to the clean removal of a weakened oyster reef by epidemic microbial diseases.

Our modern culture is so oriented to the goodness of growth that it regards any kind of descent as the antithesis, as bad, evil, catastrophic, unacceptable, or unthinkable. Fortunately, shared human beliefs and social resolutions can change suddenly when the public attention receives the small shocks from smaller-scale pulses. The danger is that people panic as growth stops and act out a self-fulfilling prophecy.

The rate of descent in our "renew plus nonrenew" simulation models depends on the depreciation rate of the structural storage of assets. The higher the assets are in the energy transformation series, the larger the storage, and the less the percentage depreciation per year (chapter 4, fig. 4.9). Depreciation rates are compared in fig. 13.3a. A decrease of 8% per year is quick enough to be called a crash. Many people think of economic assets as depreciating 5% per year (turnover time 20 years). Infrastructure of Texas highways is 2% per year (turnover time 50 years; Lyu

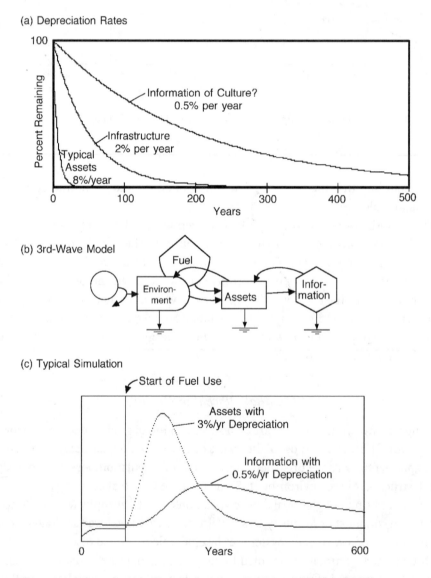

(a) Depreciation Rates

Information of Culture?
0.5% per year

Infrastructure
2% per year

Typical
Assets
8%/year

(b) 3rd-Wave Model

Fuel

Environ-
ment

Assets

Infor-
mation

(c) Typical Simulation

Start of Fuel Use

Assets with
3%/yr Depreciation

Information with
0.5%/yr Depreciation

FIGURE 13.3 Depreciation and its effect on the pulse of civilization. **(a)** Graph of depreciation rates; **(b)** main blocks of model simulating cascade to information; **(c)** typical simulation.

1986). But the essence of our modern society is in its widely shared information (religion, Internet, language, education, high-tech knowledge), which has taken two or three centuries to develop. Figure 13.3a shows the decline if information decrease was 0.5% per year. Whereas many futurists find inertia when they extrapolate human populations, the point was made with fig. 10.2 that populations can decline as fast as the economy when there is loss of energy support.

The self-accelerating information system emerging at the beginning of the 21st century is like a large consumer at the end of the energy transformation chain (fig. 13.3b). Its development is sometimes called a third wave (Toffler 1980). A simulation minimodel using renewable and nonrenewable sources generated the curves in fig. 13.3, with information developing a maximum after the peak of material assets.

Glimpses of Fuel-Limited Futures

Three episodes in the history of energy use in the last century provide insights. First, after the crash of the stock market in 1929, the world had a decade of no growth. Energy use declined. In the United States, the pause was enough to create antigrowth attitudes, frugal living, respect for real wealth, and initiatives for social welfare. But the Great Depression was not caused by the lack of energy resources; instead, disruption of money circulation, displacement of people, and other causes (chapter 9) were responsible. In Europe, faith in governments was lost, and desperate people embraced dictatorships. Like weeds, they eliminated diversity and accelerated military invasions for resources to support fanatic missions of conquest (overgrowth). The outcome was World War II. However, in the no-growth times ahead, there will not be much energy available to support overgrowth societies.

Second, shocks caused by short-range oscillations of economic pulsing and resource use have already given us glimpses of the future of an energy-based decline. The oil-supplying countries had enough monopoly of developed oil wells in 1973 to raise international oil prices. Because the price of oil jumped 60 times in a short time, the real wealth received in trade by many countries dropped 60 times, causing at least 25% inflation and economic declines (called stagflation). Societies became energy conscious, with programs of energy conservation. But the oil monopoly could not be sustained long because the high prices accelerated use of the undeveloped remaining oil elsewhere (e.g., in Alaska), development of pipelines to use natural gas, return to coal use in electric power plants, and dependence on nuclear power plants. In 2001, increasing demand and decreasing energies allowed the Organization of Energy Producing Countries (OPEC) to control prices, once again stopping growth.

Third, in 1991, when the Iraq dictatorship invaded Kuwait, the world was threatened with loss of access to most of the oil reserves on which all developed countries depend (25% of the emergy of the United States but almost all of the emergy of many developed economies of Europe and Asia). The United States led a Persian Gulf War to which most developed countries contributed (Odum 1995). At least for a time, the war prevented a dictatorship from taking over the main remaining fuel reserves, most of which are in adjacent countries of the Near East. The U.S. military capacity that won the war had been built up by a previous administration using borrowed money and deficit financing, which contributed to a subsequent recession. In other words, the emergy of social welfare was diverted to the war.

Whether fuel-dependent nations of the world can always keep fuel reserves on the international free market is an open question. During the 1973 energy crisis, the U.S. Navy fleet in the Mediterranean was unable to go to sea for lack of fuel. When available energy is scarce, only local and limited wars are possible (chapter 10).

Whereas national security policies should hold a nation's native resources in reserve and use imported Near East fuels first, many nations in recent years have done the reverse, stupidly using their last native reserves to be "energy independent" now, ensuring their energy destitution, military weakness, and national prostration later.

Empower Wave of Information Splendor

In fig. 13.1, the global production and accumulation of assets are lumped together, whereas the real world is an energy hierarchy. When large reserves of fuel energy are injected, energy passes through a series of transformations spreading through the energy hierarchy as a wave upscale and downscale, as explained in chapter 4.[7] Part of the concentrated minerals and fossil fuels that have entered the economy of industrial cities was transformed into products that dispersed outward (downscale in the hierarchy) to intensify agriculture, fisheries, aquaculture, forestry, and the ecosystems. Even more of the new fuel emergy spread upscale through autocatalytic energy transformations to higher and higher transformity (fig. 13.4). As explained in chapter 4, energy that is transformed to high transformity converges spatially to centers and may achieve its highest level in a sharp, unsustainable pulse.

In common language, the Industrial Revolution captured the economy starting with wood, charcoal, coal, and water power, followed by oil, nuclear fuel, and natural gas. People and processes moved into the cities, transformed the chemistry of society, intensified rural production, and created a brilliance of information processing centers by the end of the 20th century. Figure 13.4 shows the wave along a graph with increasing transformity to the right. The information explosion of the 21st century is based on the high emergy of the earlier Industrial Revolution. While the public dreams of science fiction futures, the roots that support the information wave are already declining.

Policies for Climax

After the centuries of growth, many people sense that civilization is leveling off and succession is reaching a climax. Public policy is already engaged in trying

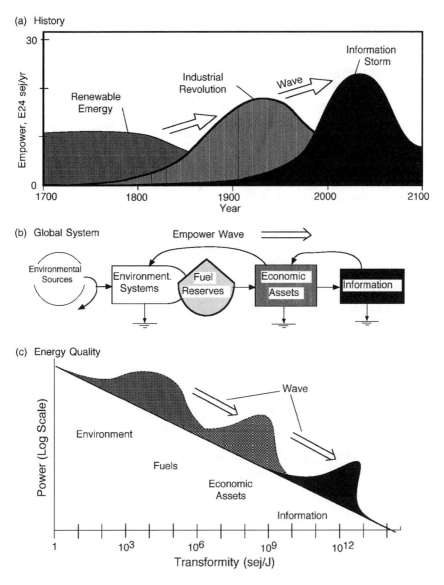

FIGURE 13.4 Wave of emergy passing from resources to the Industrial Revolution to the apex of information. **(a)** Empower over time; **(b)** stages passed by the wave along the system; **(c)** wave of transformity increase.

to *sustain* society at the present levels, although this is not possible on known re-sources. Table 13.1 summarizes growth policies that have been operating for two centuries. But societies that continue growth policies when growth is not possible will waste emergy and be displaced.

POWER CONTROL AT THE MINE AND WELLHEAD

The greatest need for stabilizing climax is to gain control of the fossil fuel supplies for the public interest. There must be enough international agreement to prevent

TABLE 13.1 Policies Appropriate During Growth

Maximize empower through low-diversity overgrowth.

Reproduce.

Encourage immigration.

Compete, displace.

Borrow and loan.

Permit large differences in income and wealth.

Minimize regulation.

Emphasize increasing wages by labor unions.

Maintain defense forces, expect war with other expanding centers.

Increase money supply.

Build temporary-quality construction rapidly.

Allow growth to absorb unemployment.

Encourage stocks, bonds, and other unearned income.

Couple environment to fuel-using economic production.

Accelerate centers of development with unequal exchange.

Set priorities for development of fuel, transportation, and water resources.

runaway growth of any country that seems temporarily good for it but becomes cancerous for all in the long run. We need to explain the energy policies needed to the citizens of the world. Religious programs need to adopt energy ethics (chapter 11). If the economists advising our planning endeavors will incorporate the basic energy drives in energy-econometric models, as indicated in chapter 7, then we may substitute more certainty for the heretofore elusive ability to predict economic details using an incomplete model. The rising levels of education and worldwide television communication may make all this possible.

To keep an economy healthy after growth requires new policies. Recommendations that are based on energy principles from previous chapters are suggested in table 13.2. These increase the efficiency of energy processing, increase diversity and division of labor, eliminate waste and luxury, reinforce environment, and stabilize international exchanges. The United States formed its institutions while overgrowing a frontier. To remain vital, it will have to change the beliefs, laws, and constitutions that were originally reinforced by growth.

WAYS OF DESCENT

In modern history there is little precedent for a diminishing economy. The simulations of most models of society descend gradually (because the modelers leave out

TABLE 13.2   Policies for Sustaining Climax

Maximize empower through high-diversity, efficient cooperation.[a]
Change industries concerned with new construction to maintenance.[a]
Reduce borrowing to that concerned with replacement (not growth).[a]
Hold money supply constant.
Don't expect much unearned income from interest and dividends.[a]
Adopt a zero population growth rate.
Replace low-quality growth structure with structure of lower depreciation.
Provide incentives to eliminate luxury use of fuels, cars, and electric power.[a]
Regulate foreign exchanges for emergy equity between nations.[a]
Provide public works programs for the unemployed.
Provide part-time jobs for the retired as long as health permits.[a]
Place a ceiling on individual income.[a]
Share information without profit.[a]
Develop a national campaign to respect people by their service, not income.

[a]Policies to continue during descent.

crash mechanisms). We know from energy principles, models, and ecosystems that a gradual descent is possible (fig. 13.3b), but so is a crash. If society does not succeed in changing attitudes and institutions for a harmonious descent, the alternative is to prepare information packages for the contingency of restart after crashing. Television dramas often show pathological disintegration and violence that writers imagine will follow social disorganization.

The fossil record is full of systems that rose and fell to extinction. Biological specialization in organic body development tends to be one-way, with new developments coming from the undifferentiated, unspecialized cells that are tucked away in the powerful main structures of action such as muscle and nerve. In analogy with these facts, many authors after Hegel have suggested the dangers of extinction of the main civilization that might follow any effort to decrease its activity. Can complex civilization de-differentiate? Are uncommitted youth society's means for programming change?

Gradual Descent

There is plenty of precedent in ecological systems for programmed, organized descent when resources decrease. Many ecosystems decrease without disaster each season. They reduce populations, store their critical genetic information in spores, seeds, hibernating animals, and temporarily transport some out by migration. When available energy returns, there is regrowth.

Prosperous Way Down

Policies for the prosperous way down are suggested in table 13.3 If population is reduced as quickly as available energy, the standard of living (empower per person) need not diminish. Extensive discussion of these policies is in the book *A Prosperous Way Down* (Odum and Odum 2001).

Crash and Restart

Crashes are also normal in ecosystems. Sometimes they are caused by the population pulses of control species and sometimes when the systems of the next larger scale spread catastrophic impact. Having evolved by adapting to ecosystem dynamics, the diverse pool of species supplies specialists for rapid restoration as needed. Recovery depends on the pool of species information.

After World War II, the American overseas policy known as the Marshall Plan succeeded in rapidly restarting the economies of Japan, Germany, Italy, and other countries. Available energy was cheap, and the level of education was high. With a pool of information, restart was rapid.

But what happens if descent is pathologically disorganized? Global society can learn from the epidemic diseases and social disorder in Africa and other places where society has lost its structure, information, and adequate emergy basis. The organization of global society in descent is not safe if one continent is festering with military pathology, overpopulation, rampant disease, and imperial exploitation of its resources. By helping Africa recover now, the world could learn how to treat sick societies and be prepared with tested policies for a healthy descent on all continents.

Transmitting Information

All the scenarios for saving the essence of civilization entail transfer of essential information to the future (fig. 13.5). The condensation of information from a blooming society to manageable smaller quantities is somewhat analogous to the brain's condensation of excess information of short-term memories into selected essentials to store in long-term memory. Information centers are likely to reorganize around the hydroelectric power of mountains and other renewable energy resources. There may be enough electric power to sustain the global information network and its role in the organization of society (fig. 10.12d). But centers of useless power dissipation, such as the gaudy night lights of the casinos of Las Vegas, may have to give up their electric power.

Sustaining information is essential to a unified global future, and electric power is essential for sharing information through television and the Internet. Consequently, hydroelectric dams may be given priority over salmon. Higher empower

TABLE 13.3 Policies for Prosperous Descent

Maximize empower through environmental production and efficient use.
Endorse lifestyles that limit reproduction.
Control population to keep empower per person from decreasing.
Downsize by reducing salaries rather than discharging employees.
Place an upper limit on individual incomes.
Redefine progress as adaptation to earth restoration.
Restore natural capital and associated environmental production.
Restore environmental reserves, forests, fisheries.
Use ecological engineering self-design for environmental–economic interfaces.
Use agricultural varieties that need less input.
Limit the power of private cars.
Plan for more population moving from cities to agricultural towns.
Decentralize organizational hierarchy.
Select hierarchically organized roads and railroads for maintenance.
Direct electric power for useful information processing and sharing.
Select and consolidate information for libraries.
Reinforce respect for polycultural pluralism.
Reduce money circulation to sustain emergy/money ratio.
Replace plastic discard packaging with reuse–recycle containers.
Plan for annual reduction in budgets.
Select for maintenance structures with low depreciation rates.
Follow policies indicated by footnote in table 13.2.
Share free information for unified cooperation.
Balance emergy trade equity to replace free exploitation.
Set a priority for ecological net production over consumption.
Use capital investment for downsizing.
Redefine medical ethics that interfere with genetic selection.
Reuse or recycle according to transformity.

was found in hydroelectric geopotential in the Umpqua River in Oregon compared with its original salmon run of 400,000 fish per year (Odum 2000).

Although energies are still in excess, adequate preparations can be made for preserving and holding the needed knowledge and cultural memory in libraries and universities. Then, plans can be made for a more agrarian system, benefited by the knowledge we now have about them. We can plan for smaller cities, fewer cars, greater ratios of agricultural workers to town consumers, and fewer problems with pollution.

The times that will follow descent may be too far ahead to interest us now. However, it is important for present morale, during transition, to believe that

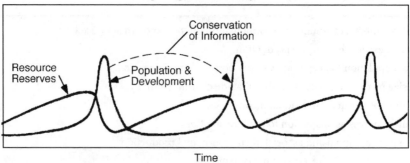

FIGURE 13.5 Transmission of essential information between pulses of progress.

the mission of information progress of our era can continue far into the future, albeit more slowly.

PERIOD OF LOW-EMERGY REGENERATION

After descent may come a long period of subsisting on renewable energies with a neoagrarian economy until enough emergy is stored by earth systems to support more pulses of civilization and progress. On a global basis in 1997, the empower in renewable energy driving society was about one-third of the total budget (fig. 9.4). In the future, with less nonrenewable emergy, developed countries must adapt to nonre-newable empower that is only 10% or 20% of their present use. There will be enough emergy to support the best of the current civilization minus the wastes, luxuries, and excessive population. If essential relevant knowledge of our current civilization is carried forward, the low-energy agrarian society of the times ahead will be based on much more knowledge than the primitive agrarian cultures of earlier centuries.

Later, as in the past, small bursts of innovation and progress can occur here and there where regions have rebuilt emergy stores in soils, virgin forests, peat deposits, and so on. The geologic cycles of the earth will still be concentrating minerals and fu-els and moving them upward to the earth's surface, where accumulations can support flashes of cultural innovation from time to time. Veizer used ingenious ways to esti-mate these rates.[8] However, the very large concentrations of oil, natural gas, coal, and uranium ores that generated the recent era may not occur again for a very long time.

SUMMARY

The human society of the planet is reaching the climax of its succession. Reversals of attitudes, policies, and laws are to be expected in the transition from the era of

FIGURE 13.6 Monuments of extinction on Easter Island. Photo by Martin Gray (http://sacredsites.com)

growth to a time of descent. By developing explanations and plans now for making descent prosperous, we can be ready when the shocks of change galvanize the attention of society. Some can have faith in the future that comes from understanding energy principles. Others will find faith in religions that adapt the necessary commandments for once again fitting culture to the earth.[9] The people of Easter Island disappeared, leaving only their monuments (fig. 13.6; see Catton 1982) as an example to the world of what happens when culture cannot downsize to fit its environmental production.

BIBLIOGRAPHY

Adams, R. N. 1988. *The Eighth Day, Social Evolution as the Self Organization of Energy.* Austin: University of Texas Press.

Ayres, R. U. 1978. *Resources, Environment & Economics, Applications of the Materials/Energy Balance Principle.* New York: Wiley.

Barney, G. O. 1980. *Global 2000 Report to the President.* Washington, DC: U.S. Government Printing Office.

Brown, L. R., ed. 2000. *State of the World.* Washington, DC: Worldwatch Institute.

Brown, M. T. 1980. *Energy basis for hierarchies in urban and regional landscapes.* Ph.D. dissertation, University of Florida, Gainesville.

Campbell, C. J. 1997. *The Coming Oil Crisis.* Brentwood, Essex, UK: Multi-Science.

Catton, W. R. 1982. *Overshoot.* Champaign: University of Illinois Press.

Cook, E. 1976. *Man, Energy, and Society.* San Francisco: W.H. Freeman.

Cottrell, F. 1955. *Energy and Society.* New York: McGraw-Hill.

Diamond, J. 1998. *Guns, Germs, and Steel.* New York: W.W. Norton.

Duncan, R. C. 1989. The life expectancy of industrial civilization: The decline to global equilibrium. *Population and Environment,* 14(4): 325–257.

Ehrlich, P. R. 1968. *The Population Bomb.* New York: Ballantine.

——. *The End of Affluence.* New York: Ballantine.

Fey, W. and A. Lam. 2000. *The Bridge to Humanity's Future.* Available at EcocosmDynamics. org.

Fleay, B. 1995. *The Decline of the Age of Oil, Petrol Politics: Australia's Road Ahead.* Annandale, NSW: Pluto.

Fluck, R. C. and D. C. Baird. 1980. *Agricultural Energetics.* Westport, CT: AVI.

Forrester, J. 1971. *World Dynamics.* Cambridge, MA: Wright Allen.

Freeman, D. 1974. *Energy, The New Era.* New York: Vintage.

Georgescu-Roegen, N. 1977. The steady state and economic salvation: A thermodynamic analysis. *BioScience,* 27: 266.

Hall, C. A. S., C. J. Cleveland, and R. Kaufmann. 1986. *Energy and Resource Quality, The Ecology of the Economic Process.* New York: Wiley.

Hardin, G. 1993. *Living Within Limits, Ecology, Economics, and Population Taboos.* New York: Oxford University Press.

Henderson, H. 1991. *Paradigms in Progress: Life Beyond Economics.* Indianapolis: Knowledge Systems.

Henderson, H. 1996. *Building a Win–Win World: Life Beyond Global Economic Warfare.* San Francisco: Berrett-Koehler.

Herendeen, R. A. and C. W. Bullard. 1975. Energy costs of goods and services: 1963 and 1967. *Energy Policy,* 3: 268.

Hubbert, M. K. 1949. Energy from fossil fuels. *Science,* 109: 103–109.

Kang, D. 1998. *Pulsing and Self Organization.* Ph.D. dissertation, University of Florida, Gainesville.

Kaplan, R. 1996. *The Ends of the Earth.* New York: Random House.

Krenz, J. H. 1976. *Energy Conversion and Utilization.* Boston: Allyn & Bacon.

Lovins, A. B. 1977. *Soft Energy Paths: Towards a Durable Peace.* New York: Harper & Row.

Lyu, W. 1986. Emergy analysis of highway transportation in Texas. In *Student Reports for Policy Research Project Course,* 130–157. Austin: L.B.J. School of Public Affairs, University of Texas.

Magoon, L. B. 2000. *Are We Running Out of Oil.* Menlo Park, CA: Geological Open File 00-320.

McClanahan, T. R. 1990. *Hierarchical control of coral reef ecosystems.* Ph.D. dissertation, University of Florida, Gainesville.

Meadows, D. H., D. L. Meadows, and J. Randers. 1992. *Beyond the Limits*. Post Mills, VT: Chelsea Green.

Meadows, D. H., D. L. Meadows, J. Randers, and W. W. Behrens III. 1972. *The Limits to Growth*. New York: Universe Books.

Nussbaum, B. 1983. *The World After Oil, The Shifting Axiom of Power and Wealth*. New York: Simon & Schuster.

Odum, H. T. 1983 (1994 edition). *Systems Ecology*, chapter 23. New York: Wiley. Reprinted as *Ecological and General Systems*. Boulder: University Press of Colorado.

——. 1987. Models for national, international, and global systems policy. In L. C. Braat and W. F. J. van Lierop, eds., *Economic–Ecological Modeling*, 203–251. Amsterdam: North Holland.

——. 1995. Emergy policies for a new world order. In H. Abele, ed., *Energy and Environment: A Question of Survival*, 177–200. Acta Forum Engelberg 1993. Engelberg, Switzerland: Verlag Stiftsdruckerel.

——. 1996. *Environmental Accounting, Emergy and Decision Making*. New York: Wiley.

——. 2000. Emergy evaluation of salmon pen culture. In *International Institute of Fisheries Economics & Trade 2000 Proceedings*. CD-ROM publication. Corvallis, OR: Department of Agriculture and Resource Economics.

Odum, H. T. and E. C. Odum. 1982. *Energy Basis for Man and Nature*, 2nd ed. New York: McGraw-Hill.

Odum, H. T. and E. C. Odum. 2000. *Modeling for All Scales*. San Diego, CA: Academic Press.

Odum, H. T. and E. C. Odum. 2001. *A Prosperous Way Down*. Boulder: University Press of Colorado.

Peet, J. 1992. *Energy and the Ecological Economics of Sustainability*. Washington, DC: Island Press.

Pillet, G. and H. T. Odum. 1987. *Energie, Ecologie, Economie*. Geneva, Switzerland: Georg Editeur.

Pimentel, D. and M. Pimentel. 1979. *Food, Energy, and Society*. New York: Halstead, Wiley.

Slesser, M. 1978. *Energy in the Economy*. New York: St. Martin's Press.

Smil, V. 1991. *General Energetics, Energy in the Biosphere and Civilization*. New York: Wiley.

Smil, V. 1998. *Energies, an Illustrated Guide to the Biosphere and Civilization*. Cambridge, MA: MIT Press.

Spreng, D. T. 1988. *Net-Energy Analysis and the Energy Requirements of Systems*. New York: Praeger.

Stanhill, G., ed. 1984. *Energy and Agriculture*. Berlin: Springer-Verlag.

Steinhart, J. S. and C. E. Steinhart. 1974. *Energy Sources: Use and Role in Human Affairs*. North Scituate, MA: Duxbury.

Tainter, J. 1988. *The Collapse of Complex Societies*. New York: Cambridge University Press.

Toffler, A. 1980. *The Third Wave*. New York: William Norcrop.

Veizer, J., P. Laznicka, and S. L. Jansen. 1989. Mineralization through geologic time: Recycling perspective. *American Journal of Science*, 289: 484–524.

Watt, K. 1992. *Taming the Future*. Davis, CA: Contextured Web Press.

Youngquist, W. 1997. *Geodestinies, the Inevitable Control of Earth Resources over Nations and Individuals*. Portland, OR: National Book Co.

Notes

1. Also see our recent book *A Prosperous Way Down* (Odum and Odum 2001) for other policies, discussion, and details.

2. In chapter 3 of *A Prosperous Way Down* (Odum and Odum 2001) the authors quote 111 writers about the future.

3. From *Environment, Power, and Society* (1971:304):

The future can be divided into alternatives according to energy supplies. There is the *future of power expanding,* the *future of power constant,* and the *future of power receding.* . . . We are not now sure which future will be next; perhaps they will follow each other in a step-by-step sequence. . . . National and international planning task forces should be assigned to each.

4. Some of these models and simulations are found in the author's previous papers and books (Odum 1983, 1987, 1996; Odum and Odum 1982, 2000, 2001) and those of many other authors. The curves in fig. 13.1b were generated by a minimodel PRICEPOLY with the following differential equations:

Reserves $= F$: $dF/dt = -X * K_2 * F * R * Q$;
Assets $= Q$: $dQ/dt = (X * K_3 * R * F * Q) + (K_4 * R * Q) + Pw$;
Renewable energy $= R$; $R = J/[1 + (X * K_1 * F * Q) + (K_0 * Q)]$, where J is energy inflow, Pw is natural production, and X becomes logic 1 when use of the reserves begins.

5. Complex models developed by combining mechanisms: Meadows et al. (1972, 1992), Barney (1980), Forrester (1971), Watt (1992).

6. Unpublished model 3RDWAVE contains two production functions converting renewable energy and available materials into the agrarian–environmental block, one using feedbacks from assets and information; four production functions using agrarian–environmental stores into the assets block, one linear, two using fossil fuels, and two using feedback interactions from information; and (3) two production functions converting assets into the information block, one linear and one autocatalytic.

7. The effect of stored energy on an energy transformation hierarchy was simulated by Brown (1980), McClanahan (1990), and Kang (1998), among others.

8. Veizer et al. (1989) estimated rates from survival curves of geologic fuel-generating deposits going back in time as if they were populations of surviving animals.

9. *Environment, Power, and Society* ended with the following text: "A more powerful morality may emerge through the dedication of the millions ... who have faith in the new networks and endeavor zealously for them. Prophet where art thou?" By *prophet* the author meant someone who could speak the right language to excite the public in energy systems policies. In response to the 1971 book, several religious-oriented writers and students came to us in Gainesville. I regret that they were turned off by our analog computers.

THE BEHAVIOR of each of the energy system modules was described with definitions of symbols in fig. 2.10. For each module there are equations that describe the operation of the module in its simplest form, that is, when it has no special complications because of its sources and links to other units. The equations implied by the systems symbols will remind various scientific and engineering readers of familiar equations in their fields and help connect their knowledge to the symbols. More complete discussion of each symbol is given elsewhere in relation to energy and systems concepts (Odum 1983).

SOURCE (FIG. 2.10A)

Figure A1 has two main kinds of sources, one that delivers a force and one that delivers a flow. The following are some of the appropriate equations:

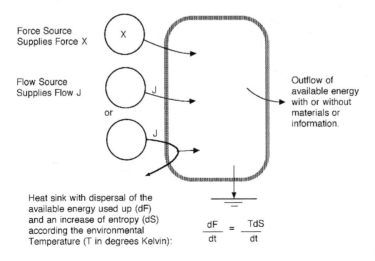

$$\frac{dF}{dt} = \frac{TdS}{dt}$$

FIGURE A.1 Frequently used sources and sinks.

$X = k$ Force constant

$J = k$ Flow constant

$X = kt$ Force as ramp

$J = kt$ Flow as ramp

$X = \sin kt$ Sine function

STORAGE (FIG. 2.10B)

Figure A2 shows the storage process and ways it is represented in several systems languages when input is a constant flow J. It can be called a Von Bertalanffy module, after the author who made its equation the center of general systems theo-

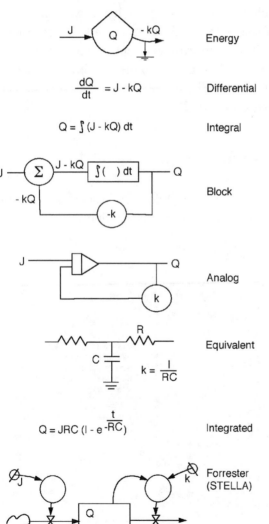

Energy

$$\frac{dQ}{dt} = J - kQ$$

Differential

$$Q = \int (J - kQ)\, dt$$

Integral

Block

Analog

Equivalent $\quad k = \dfrac{1}{RC}$

$$Q = JRC\,(1 - e^{-\frac{t}{RC}})$$

Integrated

Forrester (STELLA)

FIGURE A.2 Simple storage (tank symbol, the Von Bertalanffy module) expressed in several ways (fig. 2.10b).

ry (Von Bertalanffy 1968). The outflow path combines second law dispersal and outflow of usable energy. With a constant source there is exponential asymptotic growth, called a first-order process in engineering.

Figure A3 shows the rate at which potential energy is generated and stored (power) and its efficiency of energy transformation e as a function of the backforce loading X_2 described in figs. 3.1 and 3.2b. The available energy dispersed to the environment, with its entropy increase, is the difference between useful power input minus the useful power transformed and restored. The derivation starts with an expression for the power of output storing potential energy, with one term for the effect of the driving force and one for the effect of the backforce from the storage. Coefficients are simplified by consideration of the special case when forces are equal and the system is stalled and by the coupling principle (Onsager's principle) that the effect of one force on a second flow equals the effect of the second force

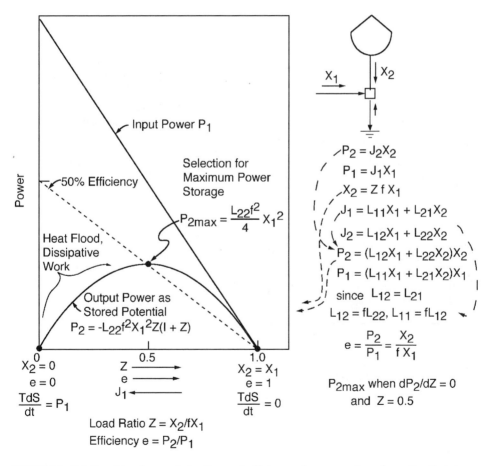

FIGURE A.3 Relationship of power to loading and efficiency of an energy transformation that stores potential energy when the input force X_1 is constant (Odum 1983; Odum and Pinkerton 1955). See fig. 3.2.

on the flow of the first. The loading ratio Z is substituted to obtain power output as a function of loading, and the input power is held constant. An expression for maximum power was found by setting the derivative to zero. When loaded for maximum power storage at steady state, the ratio of output to input force and efficiency is 50% minus the energy losses due to second law dispersal, represented by f in the equations.

However, most transformations in the real world, if not all, also involve a hierarchical concentration that requires additional available energy to be used. Thus, the equations in fig. A3 represent an upper limit to efficiency that is possible when energy transformations are selected for maximum power. A more complete explanation of the derivation was given in our original article (Odum and Pinkerton 1955).

Using the equations in fig. A3, fig. A4 relates power and efficiency of potential energy storage when the backforce loading X_2 is constant but the input force X_1 varies, as in devices that receive solar energy.

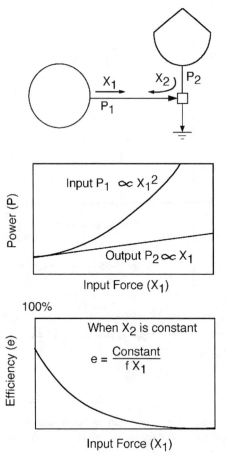

FIGURE A.4 Relationship of power and efficiency of an energy transformation that stores potential energy when the output load (backforce X_2) is constant but the input force is varied. See equations in fig. A3 and examples in figs. 3.3c and 3.5.

Heat Sink (Fig. 2.10c)

The heat sink module (fig. 2.10c) represents the dispersal of available energy accompanied by an increase in entropy S. Heat sink equations relate the dispersing heat Q to the loss of available potential energy F and to the rate of entropy change S, where T is the sink temperature on the Kelvin scale.

$$dF/dt = dQ/dt = TdS/dt$$

Miscellaneous Box (Fig. 2.10d)

No equations are implied until details are diagrammed with other symbols in the box.

Interaction (Fig. 2.10e)

The interaction module in its simplest form has a production output that is the product of the local input forces at the site. Figure A5 explains the output of an interaction module as a function of a varying input force X_1 when one of the input forces is supplied by a local storage X_2 that becomes limiting because it is supplied by a constant, limiting flow J.

Cycling Receptor (Fig. 2.10f)

Figure A6 shows the production output of the cycling receptor module, in which energy transformation is limited by the quantity of material recycling within the

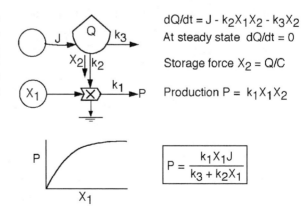

FIGURE A.5 Performance characteristics of a productive interaction (work gate) symbol (fig. 2.10e) when one of the input flows J affecting the local available energy Q is constant and becomes a limit when the other force X_1 increases.

$dQ/dt = J - k_2 X_1 X_2 - k_3 X_2$
At steady state $dQ/dt = 0$

Storage force $X_2 = Q/C$

Production $P = k_1 X_1 X_2$

$$P = \frac{k_1 X_1 J}{k_3 + k_2 X_1}$$

(a)

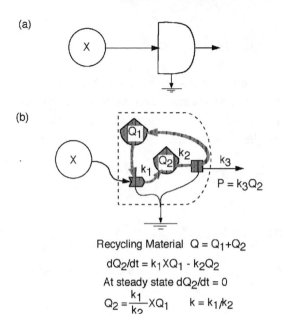

(b)

FIGURE A.6 Properties of the cycling receptor module (fig. 2.10f) with energy transformations that use an internal cycle of materials to interact with the input energy source. **(a)** Summarizing symbol; **(b)** component pathways and relationships; **(c)** output response with varying input intensity; **(d)** passive electrical analog circuit using a photovoltaic cell.

$P = k_3 Q_2$

Recycling Material $Q = Q_1 + Q_2$

$$dQ_2/dt = k_1 X Q_1 - k_2 Q_2$$

At steady state $dQ_2/dt = 0$

$$Q_2 = \frac{k_1}{k_2} X Q_1 \qquad k = k_1/k_2$$

(c)

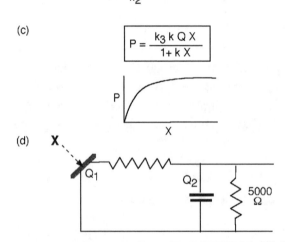

$$P = \frac{k_3 k Q X}{1 + k X}$$

(d)

module. The relationship sometimes is called the Michaelis–Menten equation, after the first application to a biochemical processes in which the enzyme was recycling (Michaelis and Menten 1913). This module can receive pure energy flows (without materials). In photovoltaic cells electrons recycle as part of the energy transformations of incoming light.

PRODUCER (FIG. 2.10G) AND CONSUMER (FIG. 2.10I)

Figure A7 shows the response of autocatalytic configurations of self-maintaining units such as producers and consumers. In fig. A7a, with a constant *force* source (energy unlimited), the growth is exponential. In fig. A7b, with a constant *flow*

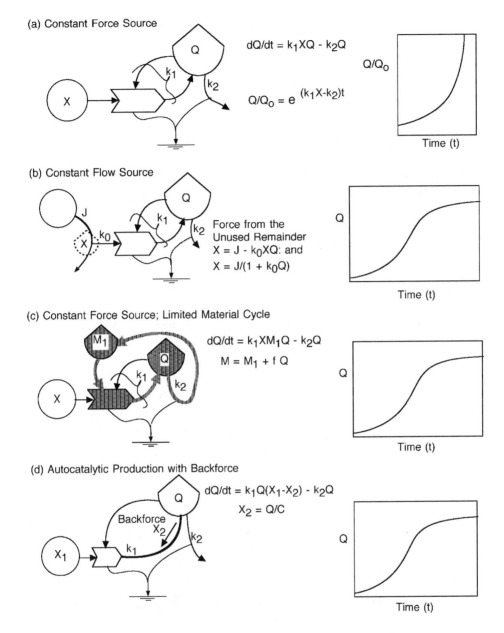

FIGURE A.7 Autocatalytic configuration often implied within a producer symbol (fig. 2.10g) and the consumer symbol (fig. 2.6i). **(a)** Exponential growth when the source is constant (constant force with unlimited energy); **(b)** S-shaped or logistic growth when the source is limited to a constant flow (J = constant) at the source; **(c)** logistic growth when there is constant input force, but the quantity of necessary recycling material is limited; **(d)** logistic growth when there is a backforce and a constant input force.

source, the growth is S-shaped. In fig. A7c, S-shaped growth results in a logistic equation when there is a recycling material that becomes limiting even though the input force is maintained constant. In fig. A7d, S-shaped growth results when there is backforce from the storage.

GREEN PLANT PRODUCER (FIG. 2.10H)

The combination of modules from a cycling receptor and one or more self-maintenance modules forms a system that combines the separate equations.

SWITCH (FIG. 2.10J)

This module refers to any action that has on or off positions controlled by some code of input combinations that are above or below thresholds according to the various combinations of digital logic, which must be specified for full description (e.g., "AND," "OR," "NAND," "NOR," "XOR").

EXCHANGE (FIG. 2.10K)

This module refers to countercurrent control where J_1 is a flow of commodities, goods, or services, J_2 is money flow, and p is price. J_1 can be in any units.

$$J_2 = pJ_1.$$

Price may be determined from outside markets. On the other hand, in a large system the price may be the variable determined by the ratio J_2/J_1.

CONSTANT GAIN AMPLIFIER (FIG. 2.10L)

This module has the same function used in electrical engineering for operational amplifiers, except as used here there is no automatic change in algebraic sign between input from the left X and transformed output P.

$$P = GX, \text{ where } G \text{ is the gain.}$$

The module applies only when energy flow for the amplification (gain) J can be drawn from an unlimited source through the upper pathway:

$$J = kGX.$$

Active Impedance (Fig. 2.10m)

This module in its simplest form has the response characteristics of an electrical inductance (see texts on electrical engineering). Backforce is proportional to acceleration.

One-Way Valve (Fig. 2.10n)

This module has the response characteristics of an electrical diode with flow proportional to force, with backforces expressed but no backflow.

Adding Junction (Fig. 2.10o)

Two flows of the same kind join, and the output flow is the sum of the inputs. If the two flows are of different kind, then they interact, and the module in fig. 2.10e applies.

Split (Fig. 2.10p)

One flow divides into two flows of the same kind. The sum of the outputs equals the input. If the outputs are of different kind, then the outputs are co-products, and the box symbol in fig. 2.10d or the interaction in fig. 2.10e applies.

Equations for Ecosystem Minimodels

The computer simulations of growth in storage as a function of time in figs. A1, A5, A6, and A7 are given in Odum and Odum (2000), which includes the programs on a CD-ROM.

The minimodel in appendix fig. A8 simulates the coupled processes of production and respiration in four languages, as discussed in chapter 2.

The minimodel PIONINFO in fig. A9 simulates the changes in diversity accompanying cyclic succession, climax, pulsed consumption, and restart (Odum 1999) discussed in chapter 6 (fig. 6.14).

The minimodel PULSE in fig. A10 (Odum and Odum 2000) simulates the repeating oscillation of growth, pulsed consumption, and recycle discussed in chapter 3 (fig. 3.13).

The minimodel CLIMAX in fig. A11 simulates succession, discussed in chapter 6 (fig. 6.12).

(a)

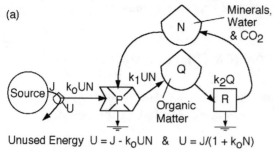

Minerals,
Water
& CO_2

Unused Energy $U = J - k_0UN$ & $U = J/(1 + k_0N)$

Source $\xrightarrow{J}$ k_0UN ... k_1UN ... Q ... k_2Q ... R ... Organic Matter ... P ... U

(b)

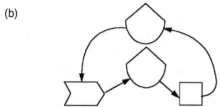

FIGURE A.8 Diagrams of the production (P) and respiration (R) system in 4 languages: the symbiotic relationship of photosynthetic production and total system respiration in the biosphere (or other closed-to-matter system). **(a)** Energy network diagram; **(b)** mineral cycle diagram; **(c)** differential equations diagrammed for operational analog simulation; **(d)** equivalent circuit (passive electrical analog language) (see fig. 2.4).

(c)

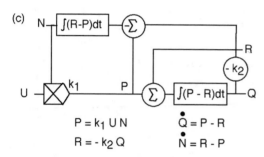

N—$\int(R-P)dt$—$(-\Sigma)$

R

$-k_2$

U —$\boxtimes$ k_1 — P —Σ—$\int(P-R)dt$—Q

$P = k_1\, U\, N$ $\dot{Q} = P - R$

$R = -k_2\, Q$ $\dot{N} = R - P$

(d)   Silicon or Selenium Cell

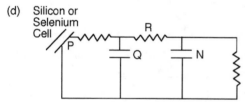

R

P   Q   N

(opposite page)

FIGURE A.9 Details and equations for the simulation model PIONINFO (Odum 1999) discussed in fig. 6.14. The model simulates repeating succession, with pioneers appearing first, replaced with main biomass structure, leading to climax diversity, a surge of consumption, and regrowth again.

FIGURE A.10 Details and equations for the simulation model called PULSE (Odum and Odum 2000) discussed in fig. 3.13. The model simulates the repeated cycle of product accumulation alternating with frenzied surges of consumers.

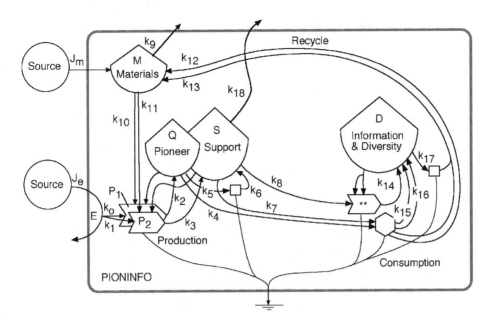

$$E = J_e - k_0*E*M*Q - k_1*E*M*S \qquad \text{Therefore } E = J_e / (1 + k_0*M*Q + k_1*M*S)$$

$$P = P_1 + P_2 \text{ where } P_1 = k_2*E*M*Q \text{ and } P_2 = k_3*E*M*S$$

$$C = k_4*Q + k_7*S + k_8*S*D*D \qquad P_{net} = P - C$$

$$dQ = P_1 - k_4*Q - k_5*Q$$

$$dS = P_2 + k_6*Q - k_7*S - k_8*S*D*D - k_{18}*S$$

$$dM = J_m + k_{12}*Q + k_{13}*S - k_9*M - k_{10}*E*M*Q - k_{11}*E*M*S$$

$$dD = k_{14}*S*D*D + k_{15}*S + k_{16}*Q - k_{17}*D$$

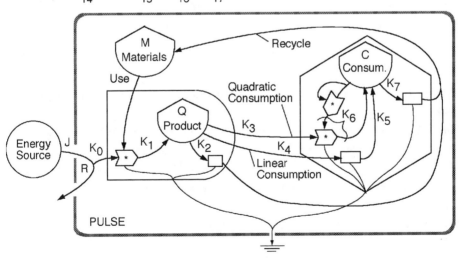

Available energy: $R = J - K*R*M$ and $R = J/(1 + K*M)$
Available Materials: $M = M_t - f_1*Q - f_2*C$
Total Material $= M_t$; f_1 and f_2 are fractions
Consumers C: $DC = K_5*Q + K_6*Q*C*C - K_7*C$
Products Q: $DQ = K_1*R*M - K_2*Q - K_4*Q - K_3*Q*C*C$

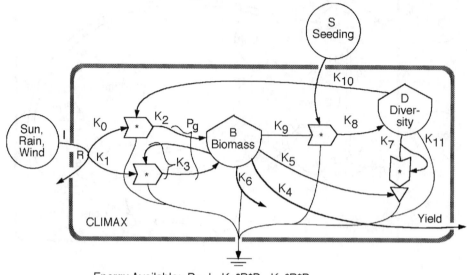

Energy Available: $R = I - K_0{*}R{*}D - K_1{*}R{*}B$
And $R = I/(1 + K_0{*}D + K_1{*}B)$

Gross Production: $P_g = K_3{*}R{*}B + K_2{*}R{*}D$

Net Production Rate =
$dB/dt = P_g - K_9{*}B{*}S - K4{*}B - K_5{*}D{*}D - K_6{*}B$

Diversity D: $dD/dt = K_8{*}B{*}S - K_7{*}D{*}D - K_{11}{*}D - K_{10}{*}R{*}D$

FIGURE A.11 Details and equations for the simulation model CLIMAX (Odum and Odum 2000) discussed in fig. 6.12. The model simulates growth of a system that first generates net production and biomass and uses its structure and seeding of species to develop diversity that reinforces gross production.

References

Michaelis, L. and M. L. Menten. 1913. Die Kinetik der Invertinwirkung. *Biochemische Zeitschrift*, 49: 333–369.

Odum, H. T. 1983. *Systems Ecology, an Introduction*. New York: Wiley. 1994. Reprinted as *Ecological and General Systems*. Boulder: University Press of Colorado.

Odum, H. T. 1999. Limits of information and biodiversity. In H. Loeffler and E. W. Streissler, eds., *Sozialpolitik und Okologieprobleme der Zerkunft*, 229–269. Vienna: Austrian Academy of Sciences.

Odum, H. T. and E. C. Odum. 2000. *Modeling for All Scales, an Introduction to Simulation*. San Diego, CA: Academic Press.

Odum, H. T. and R. C. Pinkerton. 1955. Time's speed regulator, the optimum efficiency for maximum power. *American Scientist*, 43: 31–343.

Von Bertalanffy, L. 1968. *General Systems Theory*. New York: Brazilier.

INDEX